IET ENERGY ENGINEERING 120

Fault Diagnosis for Robust Inverter Power Drives

Other volumes in this series:

Volume 1 **Power Circuit Breaker Theory and Design** C.H. Flurscheim (Editor)
Volume 4 **Industrial Microwave Heating** A.C. Metaxas and R.J. Meredith
Volume 7 **Insulators for High Voltages** J.S.T. Looms
Volume 8 **Variable Frequency AC Motor Drive Systems** D. Finney
Volume 10 **SF_6 Switchgear** H.M. Ryan and G.R. Jones
Volume 11 **Conduction and Induction Heating** E.J. Davies
Volume 13 **Statistical Techniques for High Voltage Engineering** W. Hauschild and W. Mosch
Volume 14 **Uninterruptible Power Supplies** J. Platts and J.D. St Aubyn (Editors)
Volume 15 **Digital Protection for Power Systems** A.T. Johns and S.K. Salman
Volume 16 **Electricity Economics and Planning** T.W. Berrie
Volume 18 **Vacuum Switchgear** A. Greenwood
Volume 19 **Electrical Safety: A guide to causes and prevention of hazards** J. Maxwell Adams
Volume 21 **Electricity Distribution Network Design, 2nd Edition** E. Lakervi and E.J. Holmes
Volume 22 **Artificial Intelligence Techniques in Power Systems** K. Warwick, A.O. Ekwue and R. Aggarwal (Editors)
Volume 24 **Power System Commissioning and Maintenance Practice** K. Harker
Volume 25 **Engineers' Handbook of Industrial Microwave Heating** R.J. Meredith
Volume 26 **Small Electric Motors** H. Moczala *et al.*
Volume 27 **AC-DC Power System Analysis** J. Arrillaga and B.C. Smith
Volume 29 **High Voltage Direct Current Transmission, 2nd Edition** J. Arrillaga
Volume 30 **Flexible AC Transmission Systems (FACTS)** Y.-H. Song (Editor)
Volume 31 **Embedded Generation** N. Jenkins *et al.*
Volume 32 **High Voltage Engineering and Testing, 2nd Edition** H.M. Ryan (Editor)
Volume 33 **Overvoltage Protection of Low-Voltage Systems, Revised Edition** P. Hasse
Volume 36 **Voltage Quality in Electrical Power Systems** J. Schlabbach *et al.*
Volume 37 **Electrical Steels for Rotating Machines** P. Beckley
Volume 38 **The Electric Car: Development and future of battery, hybrid and fuel-cell cars** M. Westbrook
Volume 39 **Power Systems Electromagnetic Transients Simulation** J. Arrillaga and N. Watson
Volume 40 **Advances in High Voltage Engineering** M. Haddad and D. Warne
Volume 41 **Electrical Operation of Electrostatic Precipitators** K. Parker
Volume 43 **Thermal Power Plant Simulation and Control** D. Flynn
Volume 44 **Economic Evaluation of Projects in the Electricity Supply Industry** H. Khatib
Volume 45 **Propulsion Systems for Hybrid Vehicles** J. Miller
Volume 46 **Distribution Switchgear** S. Stewart
Volume 47 **Protection of Electricity Distribution Networks, 2nd Edition** J. Gers and E. Holmes
Volume 48 **Wood Pole Overhead Lines** B. Wareing
Volume 49 **Electric Fuses, 3rd Edition** A. Wright and G. Newbery
Volume 50 **Wind Power Integration: Connection and system operational aspects** B. Fox *et al.*
Volume 51 **Short Circuit Currents** J. Schlabbach
Volume 52 **Nuclear Power** J. Wood
Volume 53 **Condition Assessment of High Voltage Insulation in Power System Equipment** R.E. James and Q. Su
Volume 55 **Local Energy: Distributed generation of heat and power** J. Wood
Volume 56 **Condition Monitoring of Rotating Electrical Machines** P. Tavner, L. Ran, J. Penman and H. Sedding
Volume 57 **The Control Techniques Drives and Controls Handbook, 2nd Edition** B. Drury
Volume 58 **Lightning Protection** V. Cooray (Editor)
Volume 59 **Ultracapacitor Applications** J.M. Miller
Volume 62 **Lightning Electromagnetics** V. Cooray
Volume 63 **Energy Storage for Power Systems, 2nd Edition** A. Ter-Gazarian
Volume 65 **Protection of Electricity Distribution Networks, 3rd Edition** J. Gers
Volume 66 **High Voltage Engineering Testing, 3rd Edition** H. Ryan (Editor)
Volume 67 **Multicore Simulation of Power System Transients** F.M. Uriate
Volume 68 **Distribution System Analysis and Automation** J. Gers
Volume 69 **The Lightening Flash, 2nd Edition** V. Cooray (Editor)
Volume 70 **Economic Evaluation of Projects in the Electricity Supply Industry, 3rd Edition** H. Khatib
Volume 72 **Control Circuits in Power Electronics: Practical issues in design and implementation** M. Castilla (Editor)
Volume 73 **Wide Area Monitoring, Protection and Control Systems: The enabler for smarter grids** A. Vaccaro and A. Zobaa (Editors)
Volume 74 **Power Electronic Converters and Systems: Frontiers and applications** A.M. Trzynadlowski (Editor)
Volume 75 **Power Distribution Automation** B. Das (Editor)
Volume 76 **Power System Stability: Modelling, analysis and control** B. Om P. Malik
Volume 78 **Numerical Analysis of Power System Transients and Dynamics** A. Ametani (Editor)
Volume 79 **Vehicle-to-Grid: Linking electric vehicles to the smart grid** J. Lu and J. Hossain (Editors)
Volume 81 **Cyber-Physical-Social Systems and Constructs in Electric Power Engineering** S. Suryanarayanan, R. Roche and T.M. Hansen (Editors)
Volume 82 **Periodic Control of Power Electronic Converters** F. Blaabjerg, K.Zhou, D. Wang and Y. Yang
Volume 86 **Advances in Power System Modelling, Control and Stability Analysis** F. Milano (Editor)
Volume 87 **Cogeneration: Technologies, optimisation and implementation** C.A. Frangopoulos (Editor)
Volume 88 **Smarter Energy: From smart metering to the smart grid** H. Sun, N. Hatziargyriou, H.V. Poor, L. Carpanini and M.A. Sánchez Fornié (Editors)
Volume 89 **Hydrogen Production, Separation and Purification for Energy** A. Basile, F. Dalena, J. Tong and T.N. Veziroğlu (Editors)
Volume 90 **Clean Energy Microgrids** S. Obara and J. Morel (Editors)
Volume 91 **Fuzzy Logic Control in Energy Systems with Design Applications in MATLAB®/Simulink®** İ.H. Altaş
Volume 92 **Power Quality in Future Electrical Power Systems** A.F. Zobaa and S.H.E.A. Aleem (Editors)
Volume 93 **Cogeneration and District Energy Systems: Modelling, analysis and optimization** M.A. Rosen and S. Koohi-Fayegh
Volume 94 **Introduction to the Smart Grid: Concepts, technologies and evolution** S.K. Salman
Volume 95 **Communication, Control and Security Challenges for the Smart Grid** S.M. Muyeen and S. Rahman (Editors)
Volume 97 **Synchronized Phasor Measurements for Smart Grids** M.J.B. Reddy and D.K. Mohanta (Editors)
Volume 98 **Large Scale Grid Integration of Renewable Energy Sources** A. Moreno-Munoz (Editor)
Volume 100 **Modeling and Dynamic Behaviour of Hydropower Plants** N. Kishor and J. Fraile-Ardanuy (Editors)
Volume 101 **Methane and Hydrogen for Energy Storage** R. Carriveau and D.S.-K. Ting
Volume 104 **Power Transformer Condition Monitoring and Diagnosis** A. Abu-Siada (Editor)
Volume 106 **Surface Passivation of Industrial Crystalline Silicon Solar Cells** J. John (Editor)
Volume 107 **Bifacial Photovoltaics: Technology, applications and economics** J. Libal and R. Kopecek (Editors)
Volume 108 **Fault Diagnosis of Induction Motors** J. Faiz, V. Ghorbanian and G. Joksimović
Volume 110 **High Voltage Power Network Construction** K. Harker
Volume 111 **Energy Storage at Different Voltage Levels: Technology, integration, and market aspects** A.F. Zobaa, P.F. Ribeiro, S.H.A. Aleem and S.N. Afifi (Editors)
Volume 112 **Wireless Power Transfer: Theory, technology and application** N. Shinohara
Volume 115 **DC Distribution Systems and Microgrids** T. Dragičević, F. Blaabjerg and P. Wheeler
Volume 117 **Structural Control and Fault Detection of Wind Turbine Systems** H.R. Karimi
Volume 119 **Thermal Power Plant Control and Instrumentation: The control of boilers and HRSGs, 2nd Edition** D. Lindsley, J. Grist and D. Parker
Volume 123 **Power Systems Electromagnetic Transients Simulation, 2nd Edition** N. Watson and J. Arrillaga
Volume 124 **Power Market Transformation** B. Murray
Volume 126 **Diagnosis and Fault Tolerance of Electrical Machines, Power Electronics and Drives** A.J.M. Cardoso
Volume 128 **Characterization of Wide Bandgap Power Semiconductor Devices** F. Wang, Z. Zhang and E.A. Jones
Volume 130 **Wind and Solar Based Energy Systems for Communities** R. Carriveau and D.S.-K. Ting (Editors)
Volume 131 **Metaheuristic Optimization in Power Engineering** J. Radosavljević
Volume 132 **Power Line Communication Systems for Smart Grids** I.R.S. Casella and A. Anpalagan
Volume 905 **Power System Protection, 4 volumes**

Fault Diagnosis for Robust Inverter Power Drives

Edited by
Antonio Ginart

The Institution of Engineering and Technology

Published by The Institution of Engineering and Technology, London, United Kingdom

The Institution of Engineering and Technology is registered as a Charity in England & Wales (no. 211014) and Scotland (no. SC038698).

First published 2018

The Institution of Engineering and Technology
Michael Faraday House
Six Hills Way, Stevenage
Herts, SG1 2AY, United Kingdom

www.theiet.org

British Library Cataloguing in Publication Data
A catalogue record for this product is available from the British Library

ISBN 978-1-78561-410-1 (hardback)
ISBN 978-1-78561-411-8 (PDF)

Typeset in India by MPS Limited

To all who live life with the insatiable sense of curiosity and discovery and even more to those who feed this sense in others. They are the creators of mankind.

Contents

Preface

It has been fascinating to see the evolution of the concept of reliability in power drives and power electronics in the last decades. The new vision that considers power semiconductors or capacitors not fundamentally different than a bearing or a tire, even though it looks self-evident to many of us, still needs to win recognition across a larger audience in this field. So often, I find myself explaining that "a 1,000 Amp IGBT, in reality, has very little to do with how much current the IGBT can reliably operate over a specific period of years." The current that they can handle over time has more to do with the cooling system, ambient temperature, and humidity, among other factors. Consequently, these factors have significant implications in the specification of the device that need to be tied to its life expectancy. Nowadays, IGBTs' specifications should include at minimum the IGBT current over time at specific ambient temperature, so I usually emphasize as an example that "an IGBT with 1,000 Amp for 20 years at 50°C ambient temperature is the meaningful specification when designing reliable power electronic systems." Without this minimal specification, the "1,000 Amp" current intensity alone does not have any important meaning if your goal is to design for reliability. This book deals with the concept that the specification of a component needs a life expectancy associated with other important factors, to allow the appropriate component selection when designing robust and reliable systems. An analysis of the fundamental physics of aging in power drives components and the markers that diagnose the process towards failure is presented to illustrate to the reader state-of-the-art approaches to increase reliability.

There are many factors playing an important role in designing a robust power drive. The main factor deals with how long the power drive will last at the conditions in which it will be operating. I hope the effort in this book helps to bring clarity and tools for early diagnostics to the designers in the field.

I want to take this opportunity to recognize Dr. Harley, who passed away during the execution of this book, and to all his legacy: to his students in electrical engineering and his contributions to the engineering field. I had the opportunity and honor to have a great and talented group of professionals as collaborators in this book, among them my undergrad advisor Dr. Jose Manuel Aller, my Ph.D. advisor Dr. Thomas Habetler, and my PHM mentor (who coincidentally was my wife's Ph.D. advisor), Georgia Tech Professor Emeritus, George Vachtsevanos.

Antonio Ginart

September 2018

Chapter 1

Fundamentals for reliability and early diagnosis for inverter power drives

José Manuel Aller[1,2], Antonio Ginart[3], and George Vachtsevanos[4]

1.1 Introduction

Several decades ago, the prevalent concept of the community was that there were fundamentally different approaches to diagnosing and maintaining mechanical versus electrical/electronic devices. For mechanical devices, the approaches were more concentrated on wear and tear, based on life-based model estimation such as the Coffin–Manson equation; the second group, especially for electronics, was concentrated only on probabilistic and random phenomena. In other words, electronic approaches were mainly confined to statistics, assigning a probabilistic value of failure to each of the components to determine the overall reliability of the equipment. With some few exceptions, while the statistical approach is extremely valuable, we can do more in understanding reliability because the process of electromagnetic energy conversion of a device requires matter. More specifically, matter is the enabler of the process, channeling and regulating the conversion of one form of energy to another, usually from electric to magnetic or vice versa. The fact that the converters require matter as "the enabler" exposes the fundamental principle: that aging governs electrical elements in the same fashion as do the mechanical parts. Consequently, solid-state materials such as conductors, insulators, or semiconductors transfer energy from molecule to molecule, atom to atom, degrading during the process, which is modified by the level of temperature, electromagnetic fields, humidity, and other factors. Therefore, small cracks that appear due to impurities or "hot collisions" progress over time and are manifested as aging in the material, which usually shows as a loss of elasticity or an increase in the losses associated with energy transfer. This process creates a degradation "marker" that can be identified in many cases at the early stages of the degradation process. The understanding, identification, and progression of these degradation markers are the focal point of this chapter.

[1]Electrical Engineering Department, Universidad Politécnica Salesiana – Sede Cuenca, Ecuador
[2]Electrical Engineering Department, Simon Bolivar University, Venezuela
[3]Smart Wires, USA
[4]School of Electrical and Computer Engineering, Georgia Institute of Technology, USA

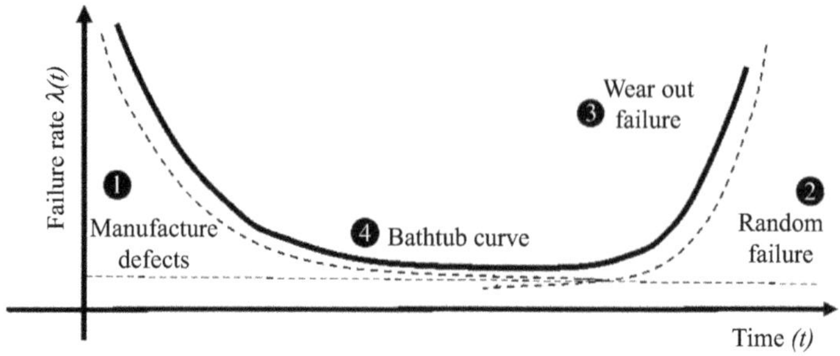

Figure 1.1 Bath curve

The material-based degradation associated with time-before-failure is usually represented by the bathtub curve representation, Figure 1.1. The curve is mainly composed of three processes [1]:

- manufacture defects (early failure, infant mortality)
- random failure
- wear-out failure

1.1.1 Manufacture defects (early failure)

The manufacture of devices and components keeps evolving, but nevertheless there are always very small variations in the process and material, with different levels of impurity that will be reflected in the final products. Sometimes these devices with large variations escape quality control for different reasons, leading to early failure during operation. The early failures are usually defined when they happen during the first year of the life cycle. Early failures are governed by the Weibull distribution [2]:

$$R(t) = e^{-(t/\eta)^m} \tag{1.1}$$

where m is the shape parameter and η is the scale parameter. With the shape parameter m smaller than one, $m < 1$, this is the dominant distribution that defines the initial shape of the bathtub.

1.1.2 Random failure

Random failures occur when average defects embedded during the manufacturing process are exposed to the standard stresses that they were designed to support; consequently, these should not happen. Nevertheless, there are always combinations of random stresses such as extra pick voltages, other noises, and interaction with other systems and devices that can affect a weaker part of the device, producing a random failure. Consequently, in the Weibull distribution from the early failure, the shape parameter m takes a value close to one, $m = 1$, that corresponds to a failure rate function that very closely approximates an exponential distribution, which defines a random failure period.

1.1.3 Wear-out failure

This is the final stage when standard stresses overcome the average manufacturing defects: the material's own limitations and design compromise under the stress that they withstand during the life cycle. Different parts of the structure start to collapse under the standard stresses. The description of this period is ruled by the Weibull distribution with a shape parameter m that is greater than one, $m > 1$.

1.2 Statistical life estimation and failure rate: the bathtub curve

The bathtub curve in Figure 1.1 is defined by three processes. The flat and usually longer period of the life cycle is the random failure period. During this period, the failure rate is kept constant; this is ideal to define time FIT (failure in time), as well as MTBF (mean time before failure). The definition established for the flat period and the other two (manufacture defects and wear-out failure) is based on the following concepts:

1. Reliability $R(t)$ and unreliability $F(t)$ functions;
2. Probability density function and MTBF;
3. Failure rate function;
4. Exponential distribution; and
5. Weibull distribution.

1.2.1 Reliability $R(t)$ and unreliability $F(t)$ functions

Reliability $R(t)$ and unreliability $F(t)$ functions are two complementary functions:

$$R(t) + F(t) = 1 \tag{1.2}$$

$F(t)$ is defined as

$$F(t) = \frac{c(t)}{n} \tag{1.3}$$

where n is the total number of tested components; $c(t)$ is the number of failures to develop in time t.

Consequently, the reliability function is defined as

$$R(t) = 1 - \frac{c(t)}{n} = \frac{n - c(t)}{n} \tag{1.4}$$

1.2.2 Probability density function and medium time before failure

The probability of failure of any device or component during the period t while the device has been in use is defined as probability density function $f(t)$:

$$f(t) = \frac{dF(t)}{dt} = -\frac{dR(t)}{dt} \tag{1.5}$$

Consequently, the reliability or unreliability functions can be computed as

$$F(t) = \int_0^t f(\tau)d\tau \tag{1.6}$$

$$R(t) = 1 - \int_0^t f(\tau)d\tau = \int_t^\infty f(\tau)d\tau \tag{1.7}$$

1.2.3 Failure rate function

The failure rate function, $\lambda(t)$, is defined as the rate of failures that happen over time (t):

$$\lambda(t) = \frac{f(t)}{R(t)} \tag{1.8}$$

In practice, it is very difficult to calculate instantaneous failure; consequently, this value is expressed in a period of time large enough, usually a year or 1,000 h. Because the time required for failure is rather large, it is common to use the concept of FIT in semiconductor. In this calculation, a test is done over a large number of devices, and the number of failures is counted. Specifically, it can be computed as

Number of failures/ (actual number of devices tested × actual operation time)

$$1\,FIT = 1 \times 10^{-9}\,1/\text{h} = 1\,\text{ppm}/1{,}000\,\text{h}$$

The FIT provides the failure rate. To obtain the average life of the product, the MTBF is defined. Usually, the MTBF is used for reparable devices, while the medium time to failure (MTTF) is used for non-reparable devices:

$$MTTF = MTBF = \int_0^\infty R(\tau)d\tau = \int_0^\infty \tau f(\tau)d\tau \tag{1.9}$$

1.2.4 Exponential distribution

The random type of failure that occurs during the flat portion of the bathtub curve is described well by the exponential distribution. For this reason, it is the fundamental distribution that describes reliability life after the initial period of decreasing failure. For exponential distribution, the probability density function $f(t)$ can be expressed as

$$f(t) = \lambda e^{-\lambda t}, \quad \text{for } t > 0 \tag{1.10}$$

Consequently,

$$R(t) = e^{-\lambda t} \quad \text{and} \quad MTTF = \int_0^\infty R(\tau)d\tau = \int_0^\infty \tau\,\lambda e^{-\lambda\tau}d\tau = \frac{1}{\lambda}$$

Figure 1.2 shows the functions $f(t)$, $R(t)$, and $\lambda(t)$ for exponential distributions.

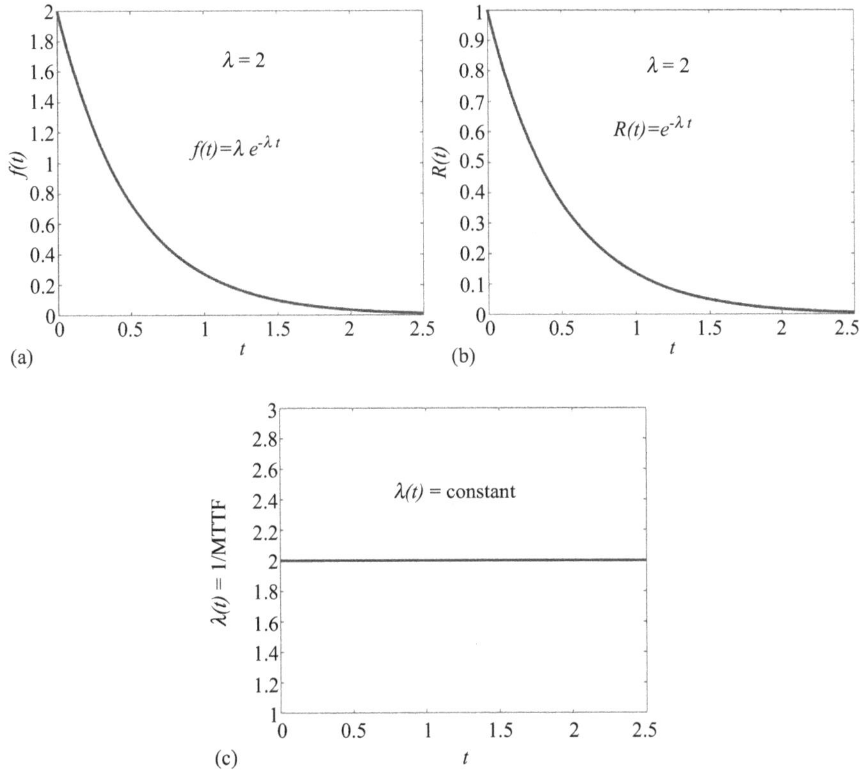

Figure 1.2 Exponential distribution: (a) probability density function f(t), (b) reliability function R(t), and (c) failure rate function λ(t)

1.2.5 Weibull distribution

The initial and final part of the bathtub is represented accurately by the Weibull distribution, proposed by Swedish mathematician Waloddi Weibull, as extension of the exponential distribution that fits failure metal fatigue. In this case, the probability density function $f(t)$ and the reliability function $R(t)$ are expressed by the equations below that include the shape parameter m and the scale parameter η, as is expressed in (1.1)

$$f(t) = \frac{m}{\eta^m} t^{m-1} e^{-(t/\eta)^m} \quad \text{for } t > 0;\ m > 0;\ \eta > 0$$

$$R(t) = e^{-(t/\eta)^m}$$

Failure rate $\lambda(t)$ and its time derivative can be expressed as

$$\lambda(t) = \frac{m}{\eta^m} t^{m-1}$$

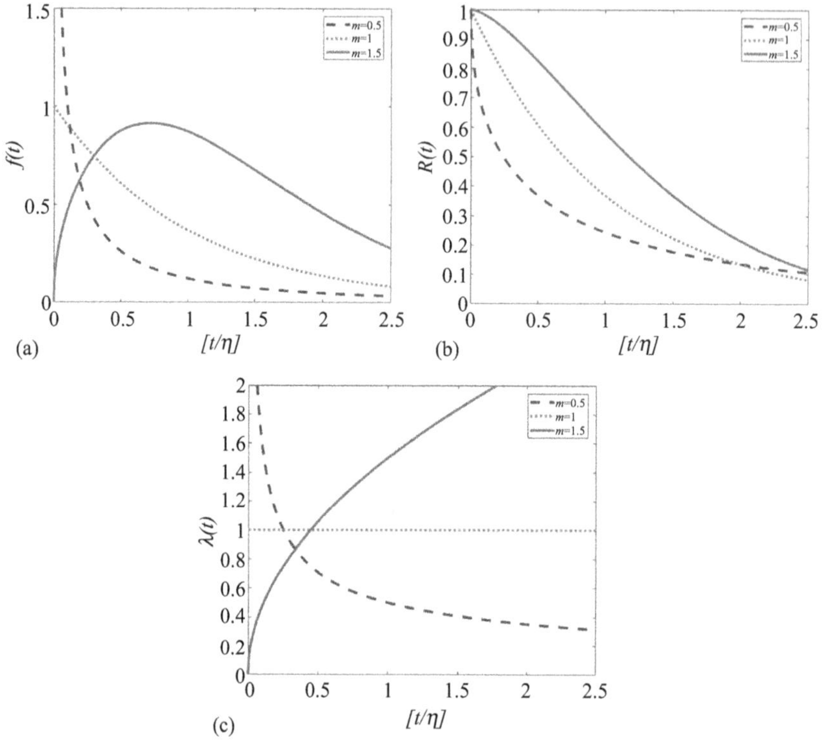

Figure 1.3 Weibull distribution: (a) probability density function f(t), (b) reliability function R(t), and (c) failure rate function λ(t)

and

$$\frac{d\lambda(t)}{dt} = \frac{m(m-1)}{\eta^m} t^{m-2}$$

Adjusting the parameter m in the Weibull distribution can describe the different failure rate: for decreasing failure, m less than one is used; when the failure rate remains constant ($m = 1$) or shows exponential distribution; and when the failure rate increases over time ($m > 1$).

Figure 1.3 shows the corresponding function for $f(t)$, $R(t)$, and failure rate $\lambda(t)$ for the Weibull distribution.

1.3 Degradation, failure mechanisms, and life model estimation

1.3.1 Solid-stare materials

The components that constitute a standard power drive are made by solid-state materials classified as insulator, conductors, and semiconductors.

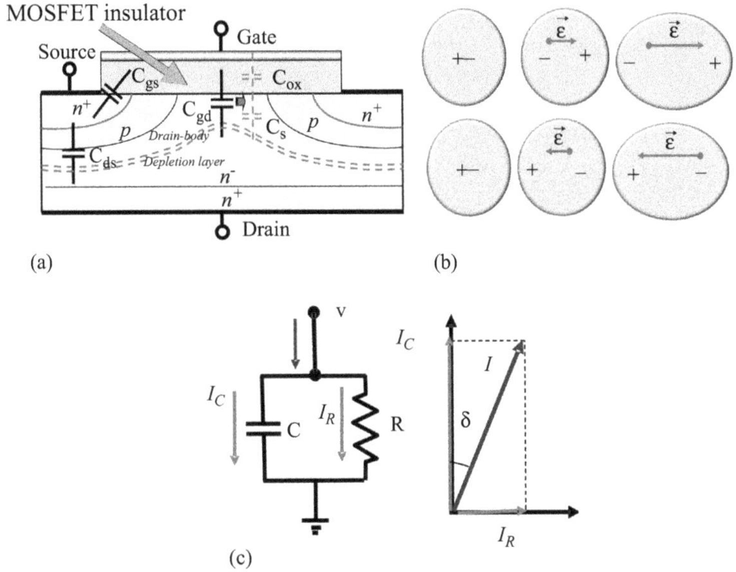

Figure 1.4 Degradation process in power semiconductors insulation: (a) MOSFET insulator, (b) insulator molecules under electric field stress, and (c) leakage current model

1.3.1.1 Insulator

There are insulators in the manufacture of any power semiconductor as well as other electrical or electronic elements. The degradation process is summarized in Figure 1.4. Figure 1.4(a) shows the metal oxide that is formed as insulator between the gate and the semiconductor body of the metal oxide semiconductor field effect transistor (MOSFET). Figure 1.4(b) displays how the molecules of an insulator expand under the stress of electric field; finally, Figure 1.4(c) shows one of the most common methods for evaluating aging in dielectrics, which is related to measurement of the leakage current mostly due to variation in the equivalent capacitances and increases in losses models as resistances. For capacitor insulator, this plays a predominant role in defining life expectancy.

1.3.1.2 Conductors

The conductors in power drives are usually based in copper, which constitutes the main component for cables and busbar. In most cases, copper cables and busbar are very reliable part of the driver system, except for specific interconnection, where particular attention should be provided in the applied torque or the specific washer required for the connections. When the reliability of the interconnection is a critical, temperature monitoring is needed to diagnosis early occurrences of a loosened connection. Connectors are other point of failure, in power drives, a connector with some type self-lock mechanisms are advised: vibrations and the nature of the pulse width modulation (PWM) can create an intermittent fault that can lead to the creation of flashes

(a) (b)

Figure 1.5 Problems interconnecting conductive parts in semiconductors: (a) two MOSFET connection in a power module and (b) fatigue and crack propagation on welding of two materials

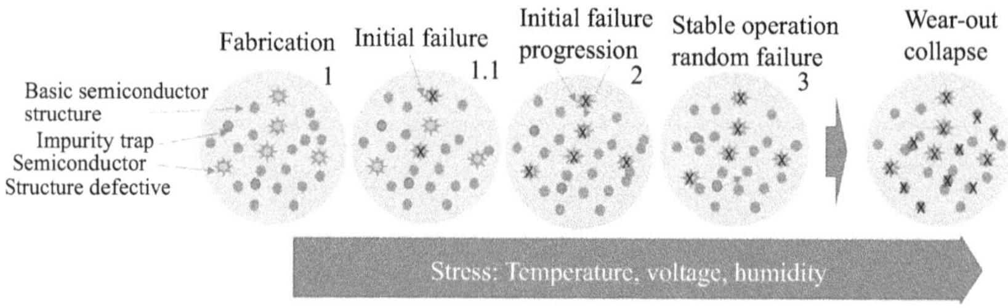

Figure 1.6 Process of aging of a semiconductor under standard stressed

or hot points [3]. In the case of modules of power semiconductors, aluminum plays a significant role as an interconnection between the semiconductor's dice and the outdoor connector. In the past, only aluminum was used as the primary interconnection material because copper is highly contaminous of semiconductors. Recently, most of the manufactures of power modules has begun to use copper as the interconnection material under the claim that the problems of copper migrating into the semiconductor have been solved. The most common failure mechanisms in the dice interconnection are electromigration and bond wire fatigue. Figure 1.5(a) shows the interconnection of a two-MOSFET and its respective diodes in parallel in a 100 A power module, and Figure 1.5(b) shows four steps for the fatigue and crack-propagation process on the welding of two materials.

1.3.1.3 Semiconductors

Semiconductors are the primary material that enables power electronic, because under electrical command, the can must behave as an excellent conductor (switch on) or insulator (switch off). Figure 1.6 shows how the aging of the semiconductor fit the bathtub curve behavior.

1.3.2 Failure modes and physics-based life model calculation

1.3.2.1 Physics-based life models

Physical life models are developed to quantify how a specific stress such as temperature, voltages, or humidity affects the life of devices or equipment. The equations that describe the physical model are based on the reaction rate theory initially studied by Arrhenius [4–6].

The reaction rate formulation depends on temperature and describes a chemical reaction phenomenon that is triggered when a threshold level called energy activation is met. The life model rate of reaction relation expresses the time taken by the substances to break the bonds and come back to their "original components." In other words, it is the estimated time to decay and how fast the process occurs (acceleration) according to a particular stress, which, in the case of Arrhenius, is temperature. Later, Eyring studied more complex reactions, concluding that for a wider range of temperature, the chemical model is more complex and depends on other stress factors besides temperature [7–9].

Both Arrhenius' and Eyring's models are expressed in the form of exponential equations. For simplicity, other researchers and reliability practitioners have used, instead of the exponential equations, empirical equations based on a power law.

1.3.2.2 The Arrhenius model

This model represents a degradation in the structure/material as a function of the temperature. It has been applied in electrical insulation, semiconductor materials, battery cells, lubricants, plastics, and incandescent lamp filaments among other things [8]:

$$L(T) = A\, e^{(E/kT)} \tag{1.11}$$

where $L(T)$ is the life under thermal condition; A is an experimental constant; E is the activation energy; T is temperature in Kelvin; k is the Boltzmann's constant 8.6171×10^{-5} eV/°C.

1.3.2.3 Power model

The power model uses an empirical relation based on power equation forms. It is one of the most widely applied models for estimation of capacitors' life, fatigue, and bearings. It is commonly used by transformer and capacitor manufacturers, with temperature or voltage as main stress factor, or a combination of the two:

$$L(T) = \begin{cases} \frac{\varphi}{(V-V_0)^{\eta}} & \text{if } V - V_0 > 0 \\ \infty & \text{otherwise} \end{cases} \tag{1.12}$$

1.3.2.4 Eyring model

Eyring's model is commonly used as an extension of Arrhenius' model because it describes more complex behavior that extends to a wider range of stress variables, such as mechanical stress, humidity, and voltage [10]. In addition, it can combine two stress factors for the estimation of life, such as temperature and voltage in capacitors, windings, and semiconductors. The application of the original Eyring's complex reaction formula to life estimation is

$$L(T) = a\left(\frac{kT}{h}\right)\, e^{-(Ea/kT)} S\alpha \tag{1.13}$$

where a and α are constants; h is the Planck constant $6.62607004 \times 10^{-34} \mathrm{m}^2$ kg/s; S are the stress factors other than temperature; and T is the absolute temperature in K.

For relatively small temperature range T, this equation can be simplified to the Arrhenius equation with an additional term multiplying a function $S(\alpha)$, dependent on another stresses. The model derived this way is called modified Eyring's equation:

$$L(T) = A\, e^{-(Ea/kT)} S(\alpha) \tag{1.14}$$

1.3.2.5 Thermal cycling

Thermal cycling is one of the main environmental acceleration factors that leads to aging in materials. Device degradation occurs because thermal cycling deteriorates the thermal paths, restricting the internal heat generated by the device. When a device composed of multiple materials is exposed to the stress of thermal cycling, it deteriorates until a fracture or void space is produced. In general, dissimilar materials differ in coefficients of thermal expansion that make the device more susceptible to cracks or fractures due to the forces originated by thermal expansion and contraction. Thermal cycling eventually weakens metallic contacts, triggering wire bond fatigue.

The Coffin–Manson model [11], as shown in the following, can be used to estimate the number of thermal cycles before failure for a specific device:

$$N_f = C_0(\Delta T - \Delta T_0)^{-q} \tag{1.15}$$

where N_f is the number of cycles to failure; C_0 is a material-dependent constant; ΔT is the entire temperature cycle; ΔT_0 is the cycle in the plastic region; and q is a material-dependent constant.

1.4 Inverters failure and power drives

Power inverters are the powertrain of the power drives, constituting a widely used power electronic configuration with the basic function of generating an alternating current (AC) current from a direct current (DC) source. The origin of the word "inverter" comes from the inverse of the rectification process that produces DC from AC. With the cost reduction in semiconductors, power drives have been expanding to new applications in different sectors, exceeding their traditional industrial sector, which is mainly represented by the induction motor controls, and the residential and commercial sector represented by uninterruptible power supply (UPS). These new avenues include the automobile sector, represented by both hybrid and pure electric vehicles. Furthermore, this concept of "electrification" has expanded to other types of vehicles in which DC power distribution and hybrid propulsion systems are gaining ground. Examples of these include the Boeing 777 Dreamliner, the F-35 Joint Strike Fighter, and the US Navy's Zumwalt-class destroyer. Additionally, power drives have gained penetration in the power generation and distribution sector, largely due to the expansion of the solar and wind energy generation sectors. The increasing penetration of power drives makes their reliability, robustness, and early diagnosis the central point of attention in the planning, design, and financing of the sectors in which power drives are present.

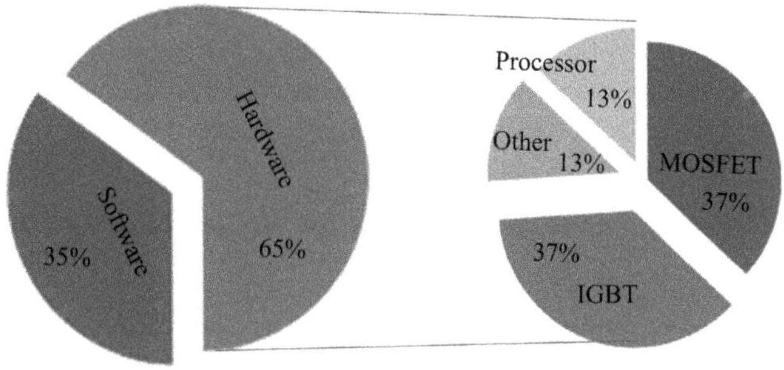

Figure 1.7 Importance of a robust semiconductor for reliability

Historical records of 15 kW string inverters, collected for over 5 years [12], demonstrate that two-thirds of the faulty inverters were caused by hardware-related issues; the remaining one-third was due to software problems. On the software side, most of the faults were produced by communication interruptions and problems during necessary software updates; others were related to intermittent problems classified as "non-fault found." On the hardware side, more than 70% of the problems were associated with power semiconductor failures, and the remaining 30% were divided between 15% microcontroller and 15% other components that mostly include the DC-link capacitors.

From Figure 1.7, the importance of a robust semiconductor to achieve high reliability is very clear.

The critical components in a power drive are summarized as follows:

- Power switching devices: insulated-gate bipolar transistor (IGBT), gate turn-off thyristor (GTO), power MOSFETs;
- Electrolytic / film capacitor bank, DC link;
- Power diodes;
- Controller unit;
- Sensor and isolation device: current sensor (Hall effect), pulse transformer, optocoupler;
- Power switching driver; and
- DC bus or AC power line short circuit.

1.5 Circuit with ideal switches: power switches fundamentals

Power inverters are based in the resilience and marvels of power electronic semiconductors working as a switch. For this reason, in order to understand reliability and diagnosis in inverters, it is of paramount importance to understand their basic principle of operation as a switch.

Power inverters, like all power electronics apparatuses, are based on power semiconductors that work as "fast switches," coordinated and controlled by an algorithm. The on or off state in power semiconductors becomes the core in enabling power electronics for energy conversion based on a modulation that allows different shapes to achieve the desired power signal output. The ideal switch can deliver and manage energy with theoretically zero power consumption. For this reason, the fundamental structure of any power converter is based on two factors:

1. The desired signal to be generated (message) and
2. The type of modulation that usually is PM and serves as the "carrier of the message" enabled by the switch.

The process of switching needs particular attention when using PWM because in nonideal switches, it produces non-intended excitation of "parasitic elements" built into the power devices [3,9,13].

The standard power drives operating frequency can be grouped in three main bands:

- Low frequency range < 200 Hz (main electrical signal for the motor that achieves the main purpose of the power drive);
- Between 10 and 200 kHz in the PWM signal (modulates the main or fundamental signal of the power drive); and
- >1 MHz (the consequence of system's resonance during switching for the interaction among the extra parasitic element of the real power switches and the circuit).

The evolution of power devices has flourished since the late 1950s with silicon-controlled rectifiers (SCRs), subsequently followed by the GTOs, Power MOSFETs, IGBTs, and more recently the silicon carbide transistors (SICs) and Gallium nitrides (GANs) power devices. Figure 1.8 shows how the simple operation of a switch in a predetermined frequency can regulate very precisely the power transfer from the source to the load.

The power transfer to the load in arbitrary period T can be computed as

$$P = \frac{1}{T}\int_{0}^{T} \frac{v(t)^2}{R} dt = \frac{1}{T}\int_{0}^{t_1} \frac{v(t)^2}{R} dt = \frac{t_1}{T}\frac{V^2}{R} \tag{1.16}$$

Note that the value of t_1 is completely arbitrary and lower than T, creating an opportunity to have full regulation of the average voltage applied to the load. In this case, the

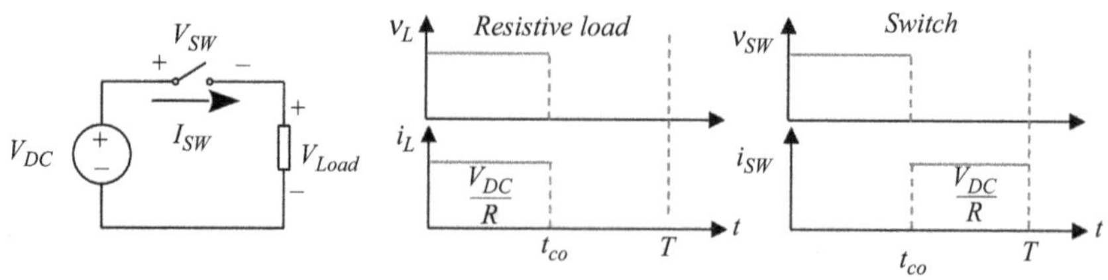

Figure 1.8 Ideal switches under resistive load

load is a resistance, and the voltage can vary from 0% to 100%. During this process, the average power of the ideal switch is

$$P = \frac{1}{T}\int_{0}^{T} v(t)\, i(t)dt = 0 \tag{1.17}$$

For the ideal switch, voltages and current never happen simultaneously; always one of the two is zero. Consequently, it is possible to transfer regulated power without consumption by the switch.

1.6 PWM, the enabler of power electronics

PWM is the preferred modulation in power electronics that allows to compute the average value during a defined interval of time, in an analog form that minimizes the computational time [14,15]. The synthesized train of pulses by the PWM modulation has the same average value as the target signal being synthesized. The PWM has an arbitrary frequency that is always significantly greater than the signal synthesized. There are countless ways to compute PWM in order to obtain a DC or average value of the desired signal to be generated. The three most common ways to generate PWM signals are shown below. These are the most commonly used PWM approaches in microcontrollers developed for power driver controllers.

Figure 1.9 depicts three ways to deliver PWM. In all cases, the average values are the same, but the impact on the system can vary. In majority of the cases, the microcontroller takes the decision for the next step:

$$V_{ave} = \frac{1}{T}\int_{0}^{T} v(t)dt = \frac{1}{T}\int_{t_1}^{T} Vdt = \frac{1}{T}\int_{0}^{t_2} Vdt = \frac{1}{T}\int_{t_3}^{t_4} Vdt = \frac{\Delta T}{T}V \tag{1.18}$$

where $\Delta T = T - t_1 = t_2 = t_4 - t_3$.

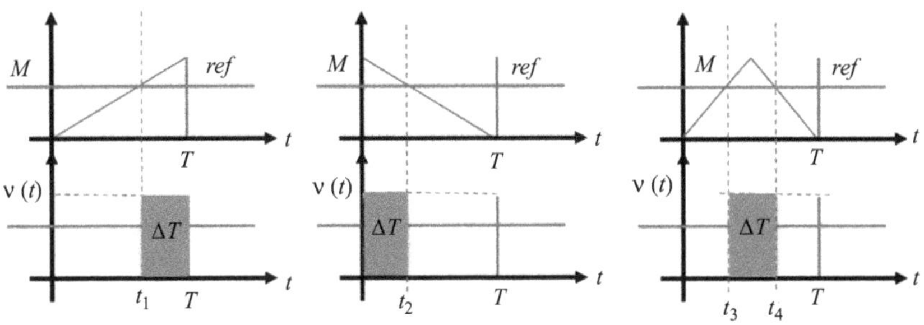

Figure 1.9 Analog way to generate PWM values (average)

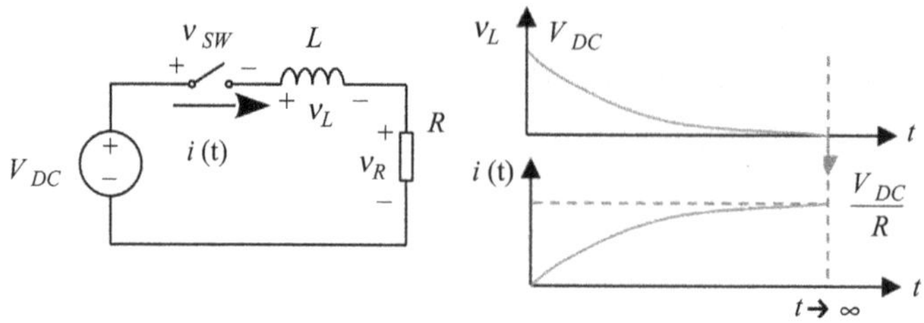

Figure 1.10 RL circuit

When a "dead beat controller" is used [16], the decision is taken based on past information ($t - 1$), and consequently, some type of prediction is required to improve control of the system.

Another possibility consists of combining the three schemes depending on how wide the PWM will be, but generally, the control is one step behind. Not all PWMs are defined by a constant period; some exhibit a variable period such as a hysteresis control, in which the duration of the period is tied to the absolute value of the signal. Other PWM techniques, such as the delta modulation [17], produce a period that depends on the derivative of the signal to be modulated.

1.7 Switching under RL circuit load

Figure 1.10 describes the behavior of the switch interacting with a load composed of a resistance and inductance. This type of load is a typical simple representation of inductors, cables, simplified winding motors, etc. Note that in Figure 1.10 the switch closes at $t = 0$, and the current at this time is zero. The differential equation that describes this phenomenon is

$$V = Ri(t) + L\frac{di(t)}{dt}, \qquad i(0) = 0 \tag{1.19}$$

The solution of (1.19) is

$$i(t) = \frac{V}{R} - \frac{V}{R}e^{-(R/L)t} \tag{1.20}$$

In the circuit shown in Figure 1.10, once the switch is closed, trying to reopen it could be complicated because the magnetic energy stored in the inductor cannot disappear. For this reason, opening the switch will create a large overvoltage that will produce an arc that transfers energy to the air and could destroy the switch. If the switch is built with a semiconductor, then the energy stored will destroy the semiconductor-based switching device. To avoid this problem, a freewheeling diode is incorporated into the circuits as is shown in Figure 1.11, to maintain the current in the inductor until the energy trap reaches zero.

In this case, v_D is close to zero when the switch open at $t = t_1$:

$$0 = Ri + L\frac{di}{dt} \tag{1.21}$$

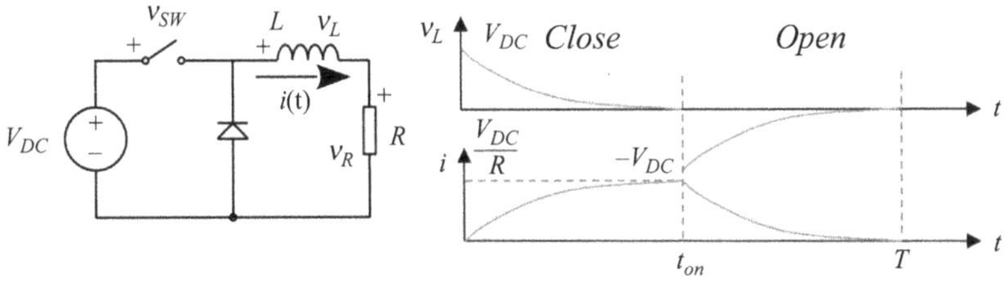

Figure 1.11 RL circuit with a freewheeling diode

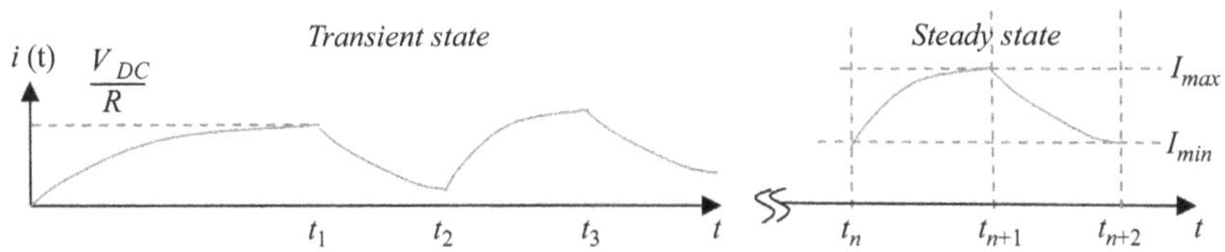

Figure 1.12 RL circuit with a freewheeling—steady state

The solution for the current $i(t)$ is the natural response with the initial current $i(t_1) = \frac{V}{R}e^{-(R/L)t_1}$:

$$i(t) = \frac{V}{R}e^{-(R/L)(t-t_1)} \tag{1.22}$$

In case the switch closes again before the current goes to zero, the same previously derived equation rules, but now we need to consider the initial condition, as is shown in Figure 1.12. Finally, a steady-state condition is reached when I_{max} and I_{min} are the same in each cycle.

1.8 RLC circuit[1]

In a real system, the power semiconductors interact with their own intrinsic components, as well as with the other systems' switches. These intrinsic elements are usually called parasitic and consist of a combination of capacitors, inductors, and resistances. These parameters are prone to change significantly during aging, and for this specific reason, they are very important indicators for early diagnostics [13]. The total combination of standard and parasitic elements during switching generates a nonlinear high-order system that can be simplified and understood as a second-order combination of a series and parallel circuits. The series and shunt representation is shown in

[1] RLC circuit is an electrical circuit consisting of a resistor (R), an inductor (L), and a capacitor (C), connected in series or in parallel. The name of the circuit is derived from the letters that are used to denote the constituent components of this circuit, where the sequence of the components may vary from RLC.

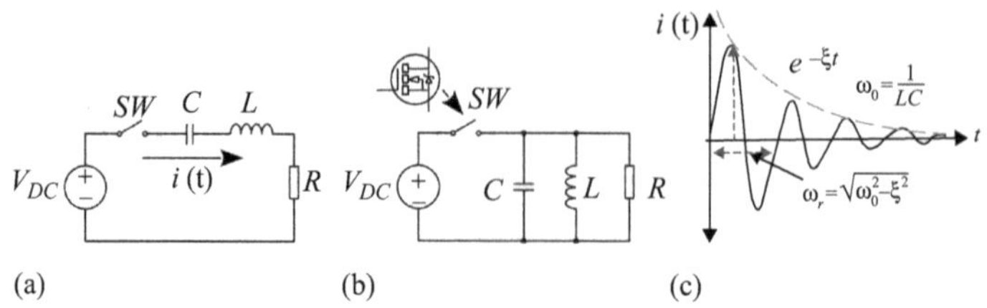

Figure 1.13 Underdamped response for a second order system: (a) series RLC circuit, (b) shunt RLC circuit, and (c) RLC response

Figure 1.13(a) and 1.13(b), respectively, and the current for the underdamped system is shown in Figure 1.13(c).

1.8.1 Series RLC model

The RLC series system is illustrated in Figure 1.13a and can be expressed by a second-order differential equation:

$$\frac{d^2i}{dt^2} + \frac{R}{L}\frac{di}{dt} + \frac{1}{LC}i = 0, \;\; i(0) = 0; \qquad \frac{di(0)}{dt} = \frac{V - V_c(0)}{L} \tag{1.23}$$

where

$$\frac{R}{L} = 2\xi \quad \text{and} \quad \frac{1}{LC} = \omega_0^2 \tag{1.24}$$

The current model in (1.23) could exhibit an oscillatory behavior as shown earlier in Figure 1.13(c). The solution of differential equation (1.23) leads to the expression for the current,

$$i(t) = e^{-\xi t}\left[A_1 e^{\sqrt{\xi^2-\omega_0^2}t} + A_2 e^{-\sqrt{\xi^2-\omega_0^2}t}\right] \tag{1.25}$$

When the damping factor ξ value is less than the resonant frequency ω_0, the solution becomes complex, creating the ringing oscillation. The resulting expression for $\xi > \omega_0$ is

$$i(t) = e^{-\xi t}\left[A_1 e^{j\omega_r t} + A_2 e^{-j\omega_r t}\right] \tag{1.26}$$

where $\omega_r = \sqrt{\omega_0^2 - \xi^2}$.

1.8.2 Shunt RLC model

The RLC system in parallel configuration can be expressed as a second-order differential equation as

$$\frac{d^2v}{dt^2} + 2\xi\frac{dv}{dt} + \omega_0^2 v = 0 \tag{1.27}$$

where ξ and ω_0 defined in (1.24) are the damping factor and resonant frequency, respectively.

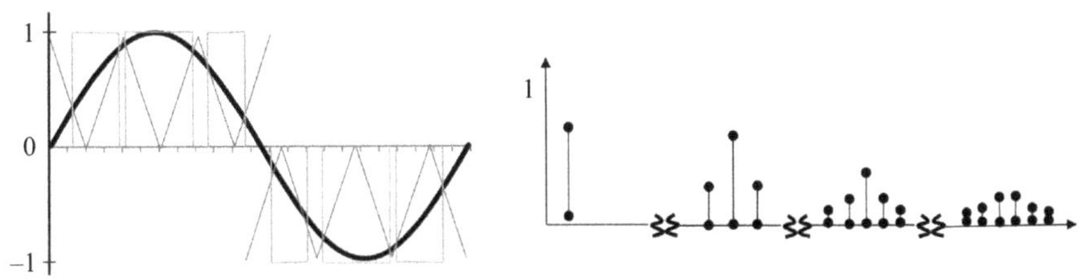

Figure 1.14 PWM synthesis and the resulting signal Fourier spectrum

A general solution to the differential equation has similar form to the solution for current presented in (1.26),

$$v(t) = e^{-\xi t}\left[B_1 e^{\sqrt{\xi^2-\omega_0^2}t} + B_2 e^{-\sqrt{\xi^2-\omega_0^2}t}\right] \tag{1.28}$$

1.9 PWM in inverters

The PWM frequency range is typically between 1 and 200 kHz for generating sine wave types that vary from a few hertz to 50–60 Hz depending the country, and up to 400 Hz in aeronautics system. The pulse width from PWM is an integral part of the inverter system and is created by comparing the sinusoidal signal and a triangular carrier. The periodic train of pulses enables the generation of segmented pulses that have the same average value as the desired signal. Consequently, the voltage Fourier spectrum contains the desired signal produced and a high-frequency signal with harmonics due to the PWM. The explanation of this detail is important to understand because several techniques use this concept of an auto-generated train of pulses to develop early diagnostics of power inverters by analyzing the variation of the frequency response over time due to its own PWM. Figure 1.14 shows the technique applied to generate a PWM signal, comparing a sinusoidal reference with a high-frequency triangular carrier and the Fourier spectrum obtained.

The harmonic content has two important modulation index that modified the magnitude (m_a) and spectrum of frequency (m_f) of the generated PWM signal:

$$m_a = \frac{V_{max\ reference}}{V_{max\ triangular}} \tag{1.29}$$

$$m_f = \frac{f_{reference}}{f_{triangular}} \tag{1.30}$$

1.10 Inverter basic operation

The basic operation is describe as alternate operation between transistors 1, 4 on and transistors 2, 3 on, as is shown in Figure 1.15(a). For inductive loads, the current have

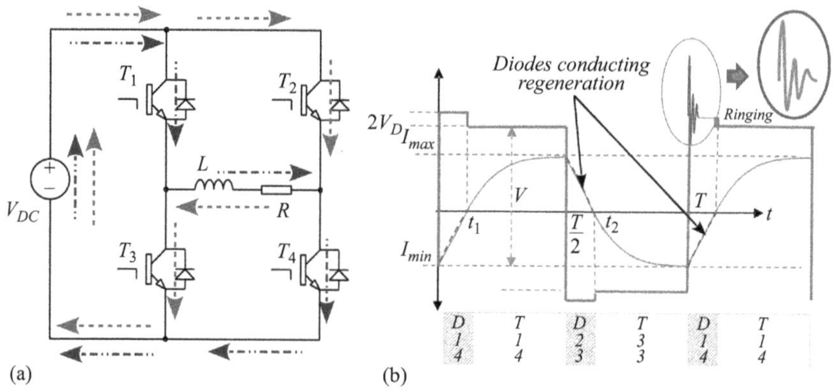

Figure 1.15 Basic inverter topology: (a) inverter circuit and (b) inverter wave form and ringing

to force a pad through antiparallel diodes when the transistors are off, as is shown in Figure 5.15(b). During this process, there is a regenerating current flow from the load to the source.

Figure 5.15(b) shows how the diodes and transistors switch carry the load current. In addition, this figure presents how, during the transition, the overvoltage and second-order oscillation or ringing occurs. The overvoltage and the ringing have been very useful in describing aging in the power system devices. Equations describing the inverter operation shown in Figure 1.15 are

$$2v_D + V = Ri + L\frac{di}{dt}, \quad \text{for } 0 < t < t_1, \text{ when } i(t) = 0 \tag{1.31}$$

$$V = Ri + L\frac{di}{dt}, \quad \text{for } t_1 < t < \frac{T}{2} \tag{1.32}$$

$$-(2v_D + V) = Ri + L\frac{di}{dt}, \quad \text{for } \frac{T}{2} < t < t_2, \text{ when } i(t) = 0 \tag{1.33}$$

$$-V =, \quad \text{for } t_2 < t < T \tag{1.34}$$

Because $2v_D \ll V$, (1.31)–(1.34) can be grouped in just two equations:

$$V = Ri + L\frac{di}{dt}, \quad \text{for } 0 < t < \frac{T}{2} \tag{1.35}$$

$$-V = Ri + L\frac{di}{dt}, \quad \text{for } \frac{T}{2} < t < T \tag{1.36}$$

Solving the differential equations (1.35) and (1.36),

$$i(t) = \begin{cases} I_{min}e^{-(R/L)t} + \frac{V}{R}\left(1 - e^{-(R/L)t}\right) & \text{for } 0 < t < \frac{T}{2} \\ I_{max}e^{-(R/L)(t-(T/2))} + \frac{V}{R}\left(1 - e^{-(R/L)(t-(T/2))}\right) & \text{for } \frac{T}{2} < t < T \end{cases} \tag{1.37}$$

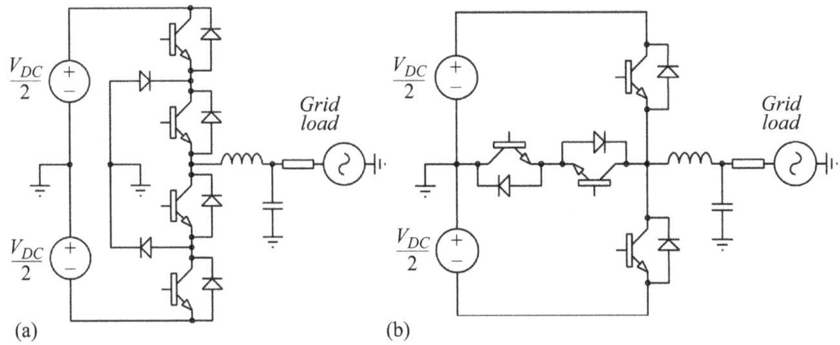

Figure 1.16 Diagram of three level inverter topologies: (a) NPC inverter and (b) MNPC inverter

Because $I_{max} = -I_{min}$, I_{max} can be obtained from (1.37), evaluating the expression for time $t = (T/2)$,

$$I_{max} = \frac{V}{R}\left(\frac{1 - e^{-(R/L)(T/2)}}{1 + e^{-(R/L)(T/2)}}\right) \tag{1.38}$$

1.11 Three-phase and multilevel inverters

The most widely used inverter is the three-phase inverter, but over the past 30 years, multilevel inverters have been part of different research topics focused on several areas of interest including new topologies, new controls, and new modulation schemes. One of the main multilevel inverter topologies is the cascade multicell inverter, which is based on a series connection of H-bridge modules with separated DC sources. This topology was first introduced in 1975 by [18] and became widely used until the mid-1990s [19]. A recent improvement was found in [20], where a nine-level cascade inverter topology is presented. The neutral-point diode-clamped inverter (NPC) topology was developed as an alternative for the cascade multicell concept, and it was described for the first time in [21]. Since then, the three-level (3L) diode clamped inverter has become favorably accepted for the design of high-power medium- and low-voltage applications [22]. New topologies based on the 3L NPC have been proposed and are discussed in [15,23]. At the same time, a new concept of multilevel clamping is introduced mixed voltage neutral point clamped (MNPC), producing an alternative topology in which a classical 3L NPC can be selectively clamped by a dedicated switching unit in order to increase the number of voltage levels. Diagram of NPC and MNPC 3L inverters are shown in Figure 1.16. Nowadays, the MNPC is called type T and NPC type I.

In addition to the diode clamping concept, the flying capacitor topology has also been introduced in different research projects [24]. In such circuits, capacitors are used to clamp the device voltage to one capacitor voltage from a capacitor voltage divider [19]. However, in some cases, flying capacitor topologies have been combined

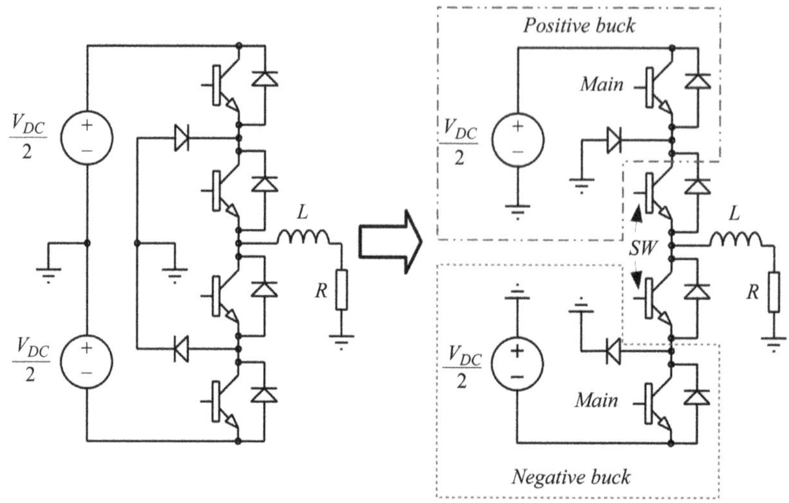

Figure 1.17 Basic operation of the NPC inverter

with the NPC to produce additional voltage levels using fewer elements than in a traditional topology. It is described in [25], where a hybrid five-level topology is proposed. The most common multilevel inverters used in solar applications are the NPC and the modified NPC [26], generally known as MNPC [27] or NPC2. Theoretically, the NPC topology requires components that withstand half voltage, but the MNPC arrangement could deliver a more efficient inverter topology. With the improvement of power semiconductors, the NPC topology has been increasingly implemented in commercial solar inverters.

1.12 Operation principle of multilevel inverters

In both the MNPC and the NPC, the operation principle can be explained as two buck converters, one each for the positive and negative parts of the sinusoidal. The two buck converters disconnect alternately to avoid a short circuit between each other. Using the NPC structure, Figure 1.17 shows how a 3L inverter can be described as two buck converters.

1.13 Dominant topology

The dominant topology in the market is determined by the state of the power semiconductor. Progress in theology and materials make a topology more desirable that others. As example, the comparison among topologies is made assuming IGBT trench stop technology or similar. 600 V technology is assumed for all switches in the NPC topology. For the MNCP topologies, the main switching device is assumed to be 1,200 V. Table 1.1 compares the average conduction and switching losses between

Table 1.1 Comparison between 600 and 1,200 V IGBT's

	600 V—NPC	**1200 V—MNPC**
Conduction (VDS) (%)	100	110
SW losses E_t (%)	100	170

VDS, voltage drain source.

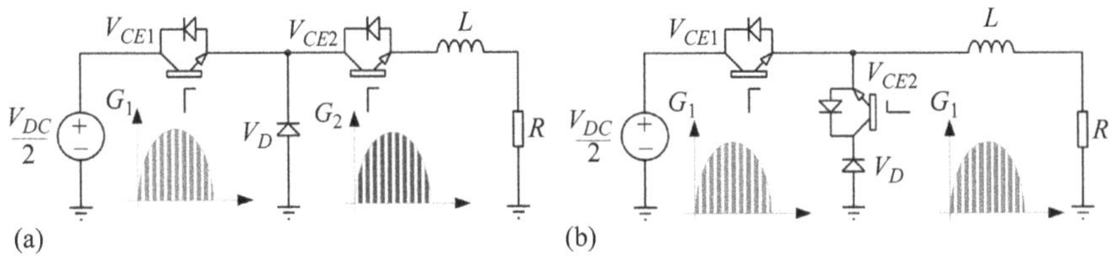

Figure 1.18 Graphical comparison between NPC and MNPC topologies: (a) NPC topology and (b) MNPC topology

Table 1.2 Conduction and switching semiconductors in NPC and MNPC topologies

	PWM ON	**PWM OFF**
NPC	$V_{CE1}+V_{CE2}$	V_D+V_{CE2}
MNPC	V_{CE1} [a]	170%

[a] 1,200 V IGBT

600 and 1,200 V IGBTs for 40–60 A from two leading IGBT manufacturers, Infineon and Fairchild. At other times or for other level of current and voltage, the results could be different.

The graphical comparison of conduction and switching losses for the different switches in MNPC and NPC configuration are shown in Figure 1.18. In the NPC topology, for a $\cos\phi$ close to 1 during PWM **ON** period, V_{CE1} and V_{CE2} present a voltage drop. In PWM **OFF** period, only V_D and V_{CE2} present a voltage drop. In the case of MNPC in PWM **ON** period, only V_{CE1} presents a drop voltage, and in PWM **OFF** period, only V_D and V_{CE2}. In conclusion, it is clear that in the NPC topology, V_{CE2} is on all the time, while in the MNPC, it is only on in the OFF period. For NPC and MNPC, a half-wave sinusoidal passes through the main switches. Table 1.2 shows the conduction and switching devices for NPC and MNPC topologies. Despite the power losses, comparison between NPC and MNPC topologies as seen in Figure 1.19, where it shows that with the present technology of IGBTs, switching losses are more than 40% with respect to conduction losses, MNPC (Type II) is nonetheless a more efficient topology than the regular NPC (Type I). Figure 1.20 shows the three-phase three-lever inverter using MNPC topology.

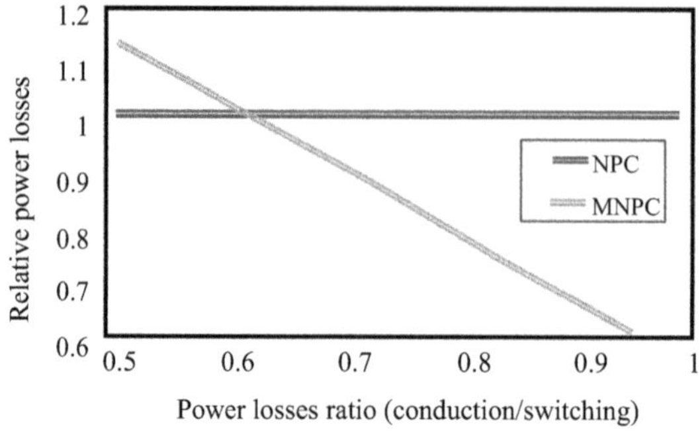

Figure 1.19 Power losses comparison between NPC and MNPC topologies

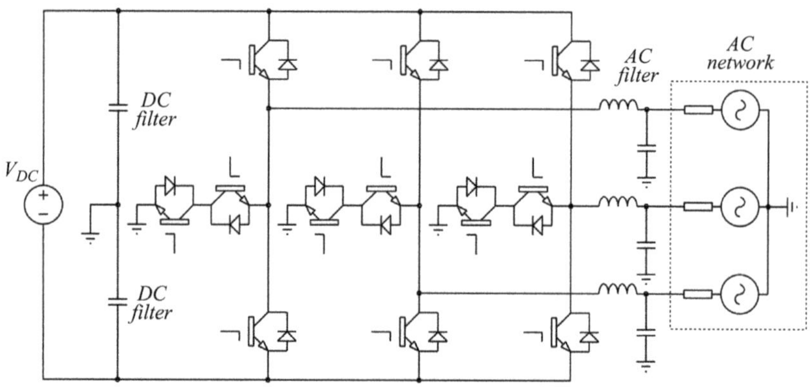

Figure 1.20 Three-phase three-level inverter

1.14 Resonant converters

The classification of converters in power electronics based on the losses associated with the switching are hard and soft switching [28]. Soft switching is achieved with circuits that produce current or voltages at zero crossing and make the switching while taking advantage of this situation to minimize the switching losses. This type is known as resonant converters [29].

The resonant converter can be classified as follows:

- Load resonant converters [30];
- Resonant switch converter [31]:
  - Zero current switching,
  - Zero voltage switching (ZVS), and
- Resonant DC link converter [32]; and
- High frequency DC link converter [33].

This book focuses on hard switching because it is the most widely prevalent technology in the market.

1.15 Real switches: power losses in hard switching

Real switches cannot operate in zero time nor have zero voltage when they are closed, and in reality, they have a very small leakage current when there are open. Consequently, the theoretical goal for the ideal switch of zero power consumption is never reached. As discussed in the previous section, temperature is the main aging factor for solid-stated material. Temperature depends basically on two factors: power losses and the thermal path or system to get rid of the heat. The focus of this section is to explain how, in a power semiconductor, its real structure influences the power dissipated during switching, and how to make a design calculation based on real parameters provided by the manufacturer document commonly called the data sheet.

There are two losses types: conduction losses and switching losses.

1.15.1 Conduction losses

The voltage drop in the semiconductor defines the conduction losses in the inverter. The average power can be computed as

$$P = \frac{1}{T}\int_{0}^{T} v(t)\, i(t)dt \tag{1.39}$$

The relation between current and voltages can be understood for the two major groups of power components based on two technologies: bipolar and electric field (field-effect transistors [FETs]) devices [34]. Technologies based on bipolar devices were the almost exclusively dominant group for a long time; they are represented by power diodes, thyristors, GTOs, and IGBTs. These devices are mostly fabricated in silicon, but recently, some applications based with gallium nitride [35] and silicon carbide [36] have made their way to the market. On the other side, FET technologies have traditionally used the silicon-based power MOSFET that now has significant competition from other new FET devices such as SIC MOSFET and E-GANs.

The relation voltage and current for the bipolar-based devices have a complex system. At low current, they exhibit a behavior ruled by the semiconductor minority carrier's movement characteristic, where higher temperature reduces the resistivity of the material. At higher currents, a different characteristic governs the behavior of these components, and they operate as a conductor material, increasing their resistance as temperature rises. These two behaviors at low and high currents create a transition point (i_T) for the current, at which the device behaves as a semiconductor or a conductor, respectively (see Figure 1.21(a)) allows a relatively easy and reliable paralleling of bipolar-based power devices. Computing the conduction losses for a full power drive with multiple power components in a precise way is a computationally intensive task that requires the use of the voltage–current lookup tables based on temperatures for every power semiconductor. This task demands a long execution time,

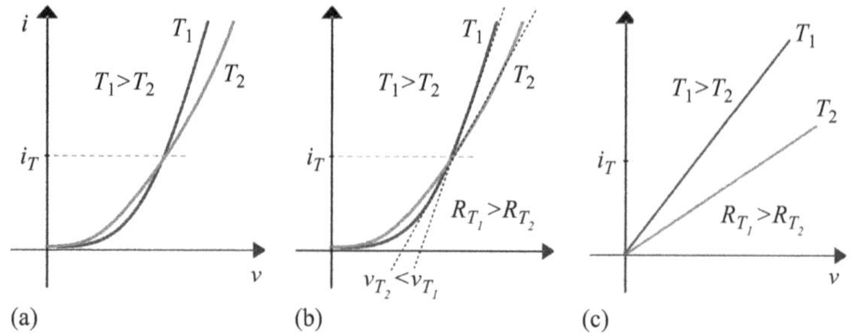

Figure 1.21 Voltage–current transistors relation for temperature change: (a) bipolar transistors, (b) piecewise linear representation, and (c) MOSFET V–I relation

even when it is performed with a powerful computer. To overcome the computational demand, a piecewise linear model has been developed [37], which divides the relation voltage current into two parts. One constant part is called (V_{t0}), which decreases in value with temperature, corresponding to the semiconductor behavior; another part is linear, whose slope (the resistance) increases with temperature as presented in Figure 1.21(b) and corresponds to the conductor behavior.

The FET family is simpler to model because its voltage–current characteristic is mainly resistive, increasing its value at higher temperatures. Due to this characteristic behavior as a conductor material, the MOSFET (HexFETs) device was the first power semiconductor to be easily arranged in parallel because of its self-correcting characteristic that increases the resistance when its temperature rises, which is caused by more current passing through, and produces a self-balancing system as is shown in Figure 1.21(c).

More specifically, parallel FET devices exhibit self-correcting characteristics because the FET device that conducts more current raises its temperature, and consequently, its resistance also increases, pushing the extra current to the other parallel devices. In addition, due to the resistive behavior of FET technology devices, more devices in parallel exhibit less power losses compared to nonparallel FET devices.

In IGBT's, the power dissipated can be computed, based on linearization, as

$$P = \frac{1}{T}\int_{0}^{t_{cond}} [V_{t0} - R_{ds}(T)\, i]\, i\, dt = V_{t0} I_{ave} + R_{ds}(T)\, I_{rms}^2 \tag{1.40}$$

where R_{ds} represents the resistance drain-source during the conducting state and t_{cond} is the time in this state. V_{t0} is assumed to be constant (Figure 1.22).

For MOSFET in conduction to behave as resistor, this implies

$$P = R_{ds}(T)\, I_{rms}^2 \tag{1.41}$$

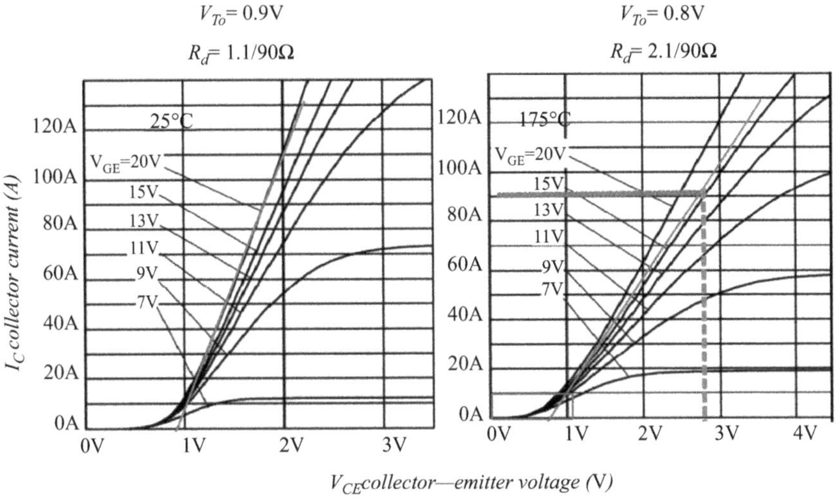

Figure 1.22 IGBT data sheet example with linearization over R_d

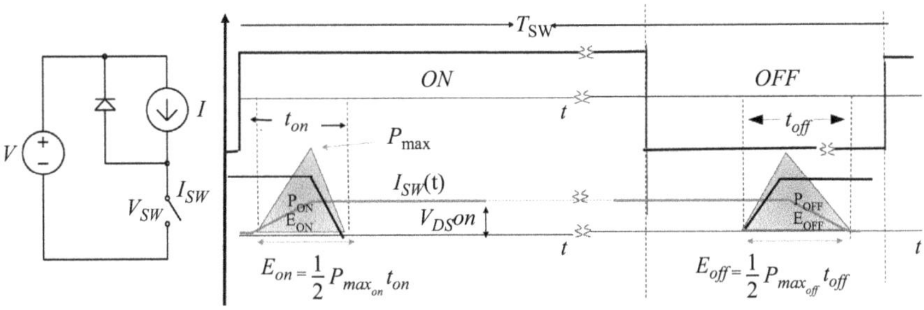

Figure 1.23 Switching losses for an ideal switch

1.15.2 Switching losses

Switching losses happen during the transition from open to closed and closed to open. The actual calculation is significantly more complex than the conduction losses because more parameters are involved. The switching in a real device is complicated by the parasitic elements that comprise the power switching devices. Figure 1.23 shows the switching losses for an ideal switch during the turn-on an turn-off process. Figure 1.24 is a linearization example obtained by processing the IGBT's characteristic curves from data sheet.

The following paragraph describes an approximated method for calculating the total energy dissipated in every switch according temperature, voltage, and current.

The energy released by switching is given as

$$E_t(T, V, I, \phi) = E_{on} + E_{off} \tag{1.42}$$

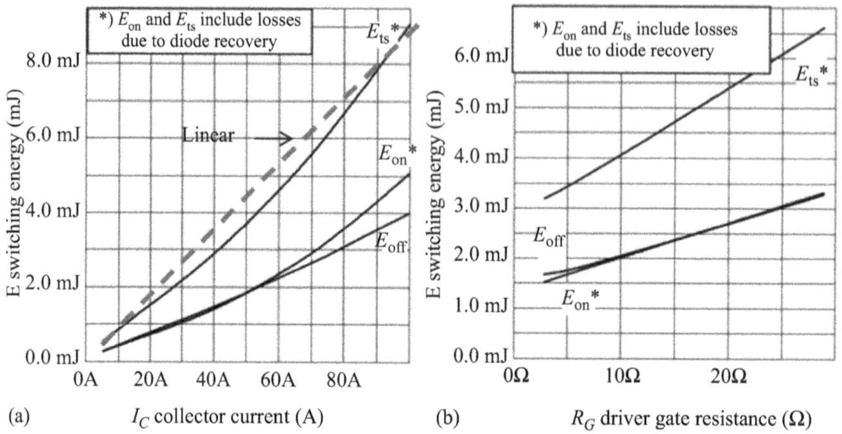

Figure 1.24 Example of IGBT curves from data sheet with linearization for switching losses: (a) typical switching energy losses as function of current (usually for inductive load, $T_{j_{max}}$, $V_{CE} = (2/3)V_N$, $V_G = 15\,V$ and minimal recommended R_G) and (b) energy losses a function of gate resistance

For constant T, V, I, and ϕ,

$$P_{SW\,loss} = E_t f_{SW} \tag{1.43}$$

To explore E_t dependence on current during switching, the other parameters are assumed to be constant:

$$E_{SW} = E_t(I_1) + E_t(I_2) + \cdots + E_t(I_n) \tag{1.44}$$

$$E_{SW}(n) = E_t(n)I(n) = \frac{E_t(I_P)}{I_p}I(n) \tag{1.45}$$

$$E_{SW} = \frac{E_t}{I_P}\sum_{k=1}^{n} I(k)$$

As a consequence, the power losses as a function of the current can be calculated as

$$P_{SW\,loss} = E_t f_{SW}\frac{I_{ave}}{I_P} \tag{1.46}$$

For a similar analysis as a voltage function

$$P_{SW\,loss} = E_t f_{SW}\frac{V_{CE\,ave}}{V_{CE\,max}} \tag{1.47}$$

Total average power is

$$P_{ave\,loss} = P_{SW\,loss} + P_{cond\,loss} \tag{1.48}$$

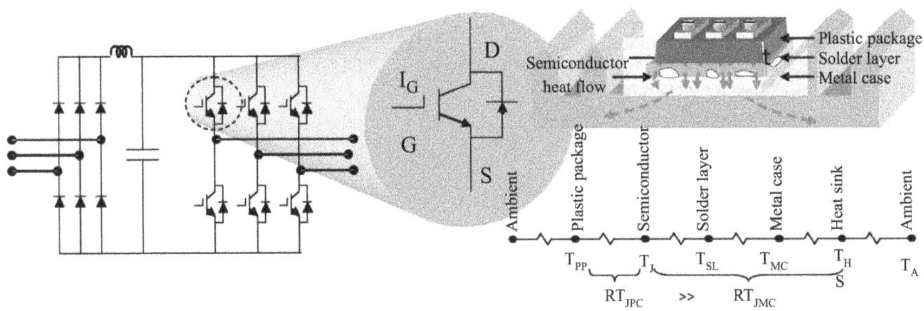

Figure 1.25 IGBT thermal model

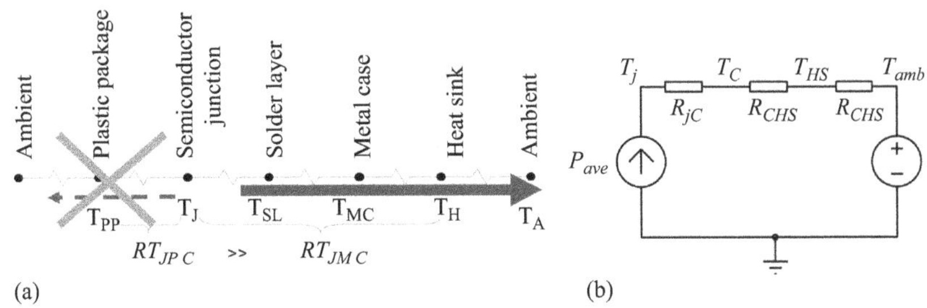

Figure 1.26 Simplified thermal model: (a) $R_{T_{JPC}} \gg T_{T_{JMC}}$ and (b) equivalent thermal circuit

1.16 Thermal consideration

The life of a power semiconductor is mainly defined by temperature of the junction T_j. The junction temperature is affected not only by the power losses but also by the ability to transfer heat to the environment. Consequently, in a power inverter, the calculation of the actual temperature relies on the estimation of the power losses and the understanding of the thermal model. Figure 1.25 shows how a power conductor transfers the power losses to the environment [9,38–41].

The thermal model can be simplified when thermal resistance through the plastic package between the transistor junction and the ambient is much greater than that through the solder layer, metal case, and heat sink to the ambient, $R_{T_{JPC}} \gg R_{T_{JMC}}$, as is shown in Figure 1.26. The junction temperature T_j can be obtained as [41]

$$T_j = P_{ave}\left[R_{JC} + R_{CHS} + R_{HSA}\right] + T_A \tag{1.49}$$

1.16.1 State modeling of the thermal system

The instantaneous computation of temperature can be performed using a detailed finite element approach or state-space variables modeling with lumped parameters

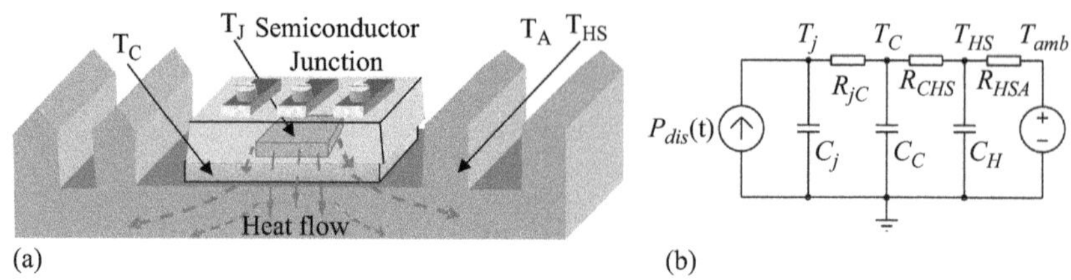

Figure 1.27 Dynamic thermal transistor model—junction–case temperature: (a) transistor and (b) expanded dynamic model

Table 1.3 Energy storage elements and state variables

Energy storage element	Energy storage relation	State variable
Dice (junction)	C_jT_j	T_j
Case	C_cT_c	T_c
Heatsink	C_hT_{hs}	T_{hs}

for thermal capacitance and resistances. Figure 1.27(b) presents a thermal model of the semiconductor with three state variables including the junction temperature T_j [41].

The state-space model of this system is defined in the form:

$$\begin{aligned}[\dot{x}] &= [A][x] + [B][u] \\ [y] &= [C][x] + [D][u]\end{aligned} \tag{1.50}$$

Energy storage elements and state variables are described in Table 1.3. The inputs to the system are the ambient temperature (T_{amb}) and the power dissipated at the junction (P_{in}).

Furthermore, thermal circuit equations can be derived as follows:

$$\begin{aligned}C_j\frac{dT_j}{dt} &= P_{in} - \frac{T_j - T_c}{R_{js}} \\ C_c\frac{dT_c}{dt} &= -\frac{T_c - T_j}{R_{jc}} - \frac{T_c - T_{hs}}{R_{chs}} \\ C_h\frac{dT_{hs}}{dt} &= -\frac{T_{hs} - T_c}{R_{chs}} - \frac{T_{hs} - T_{amb}}{R_{hsa}}\end{aligned} \tag{1.51}$$

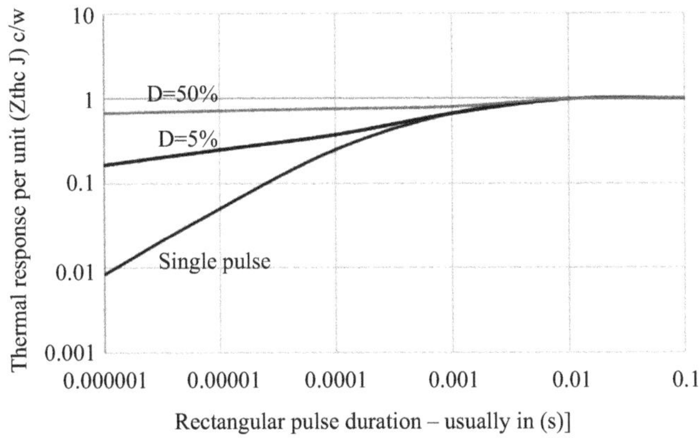

Figure 1.28 Typical per unit IGBT dynamic thermal response

Finally, the state-space system takes the form:

$$\begin{bmatrix} \dot{T}_j \\ \dot{T}_c \\ \dot{T}_{hs} \end{bmatrix} = \begin{bmatrix} -\frac{1}{C_j R_{js}} & \frac{1}{C_j R_{js}} & 0 \\ -\frac{1}{C_c R_{js}} & -\frac{1}{C_c}\left(\frac{1}{R_{js}} + \frac{1}{R_{chs}}\right) & -\frac{1}{C_c R_{chs}} \\ 0 & -\frac{1}{C_c R_{chs}} & -\frac{1}{C_h}\left(\frac{1}{R_{chs}} + \frac{1}{R_{hcs}}\right) \end{bmatrix} \begin{bmatrix} T_j \\ T_c \\ T_{hs} \end{bmatrix} + \cdots$$

$$\cdots + \begin{bmatrix} \frac{1}{C_j} & 0 \\ 0 & \frac{1}{C_h R_{hsa}} \end{bmatrix} \begin{bmatrix} P_{in} \\ T_{amb} \end{bmatrix} \tag{1.52}$$

Manufacturers of power semiconductors usually provide the plot for thermal impedance on the data sheet. This is obtained by determining the system response of the semiconductor model to rectangular pulses of different durations. The way the manufacture manual is provided to the designer is based on the curve shown in Figure 1.28 [41].

1.16.2 Thermal runaway

Thermal runaway is a catastrophic failure mode that occurs when the internal heat generated is larger than the ability of the available thermal path to release it. Consequently, the temperature increases extremely rapidly, generating an explosion of fire in order to release the trapped energy. The two factors that define thermal runaway are the power generated (power losses) and the thermal path. The design of reliable power drives should ensure that if the thermal path is clogged or blocked, thermal runaway does not occur. In addition, the designer should include diagnostics, monitoring, and protection systems to minimize the incidence of catastrophic failures caused by thermal runaway.

Thermal runaway mainly affects power FETs technology, but it extends to bipolar power technology because of its resistive characteristic for higher currents [42].

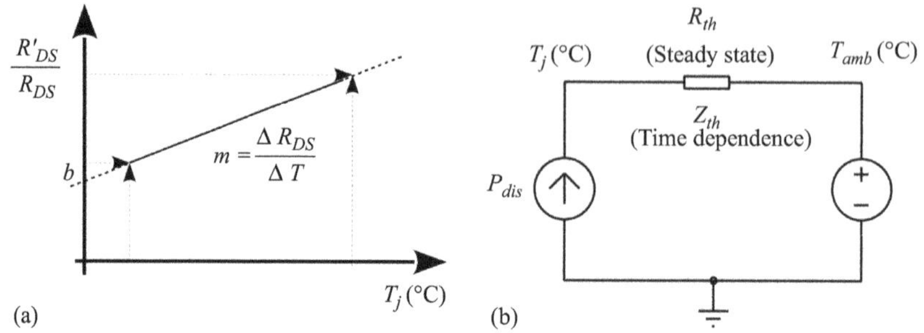

Figure 1.29 Voltage–current characteristic for variation in the semiconductor temperature: (a) resistance linearization and (b) thermal circuit

Increases in temperature produce different responses in conductors and semiconductors. In the former, a rise in temperature usually implies increases in resistance; however, in the semiconductors, the resistance decreases for higher temperatures. When semiconductors behave as conductor materials, their variation of temperature can be expressed as [43]

$$\rho(T) = \rho_0 \left[1 + \alpha(T - T_0)\right] \tag{1.53}$$

Applying the above equation to a MOSFET, we can obtain

$$R'_{DS} = R_{DS}(mT_J + b) \tag{1.54}$$

as is shown on Figure 1.29(a) and 1.29(b).

Thermal runaway is a phenomenon that happens when the heat generated is larger than the heat extracted. The following equation shows the variable dependence of this phenomenon:

$$P_{dis} = V_{to}I + R_{sd}I^2 = \frac{T_j - T_{amb}}{R_{th}} \tag{1.55}$$

From (1.55),

$$P_{dis} = V_{to}I + R_{ds}(25°\text{C})\left[mT_j + b\right]I^2 \tag{1.56}$$

and

$$T_j = P_{dis}R_{th} + T_{amb} \tag{1.57}$$

Replacing (1.57) in (1.56):

$$P_{dis}\left(1 - R_{ds}(25°\text{C})\, mR_{th}I^2\right) = V_{to}I + R_{ds}(25°\text{C})(mT_{amb} + b)\, I^2 \;\Rightarrow$$

$$P_{dis} = \frac{V_{to}I + R_{ds}(25°\text{C})(mT_{amb} + b)\, I^2}{1 - R_{ds}(25°\text{C})\, mR_{th}I^2} \tag{1.58}$$

Expression (1.58) is a positive feedback when $R_{ds}(25°\mathrm{C})\, mR_{th}I^2 \approx 1$, in this case $P_{dis} \rightarrow \infty$. The current for a thermal runaway is

$$I = \sqrt{\frac{1}{R_{ds}(25°\mathrm{C})\, mR_{th}}} \tag{1.59}$$

Equation (1.59) confirms that the thermal runaway only depends on the m temperature coefficient, the R_{ds} at 25°C, and thermal path resistance R_{th}.

References

[1] Renesas, ed., *Renesas semiconductor device reliability handbook*. Tokyo: Renesas, 2017.

[2] A. Papoulis and S. U. Pillai, *Probability, random variables and stochastic processes*. New York: Tata McGraw-Hill Education, 2002.

[3] A. E. Ginart, I. N. Ali, J. W. Goldin, *et al.*, "On-line intermittent connector anomaly detection," in *2011 IEEE AUTOTESTCON*, pp. 235–239, 2011.

[4] S. Arrhenius, "Über die dissociationswärme und den einfluss der temperatur auf den dissociationsgrad der elektrolyte," *Zeitschrift für physikalische Chemie*, vol. 4, no. 1, pp. 96–116, 1889.

[5] S. Arrhenius, "Über die reaktionsgeschwindigkeit bei der inversion von rohrzucker durch säuren," *Zeitschrift für physikalische Chemie*, vol. 4, no. 1, pp. 226–248, 1889.

[6] K. J. Laidler, *Chemical kinetics*. No. 544.4 LAI, 1987.

[7] S. Glasstone, K. J. Laidler, and H. Eyring, "The theory of rate processes; the kinetics of chemical reactions, viscosity, diffusion and electrochemical phenomena," Tech. Rep. McGraw-Hill Book Company, 1941.

[8] W. B. Nelson, *Accelerated testing: statistical models, test plans, and data analysis*, vol. 344. Hoboken, NJ: John Wiley & Sons, 2009.

[9] G. J. Vachtsevanos, F. Lewis, A. Hess, and B. Wu, *Intelligent fault diagnosis and prognosis for engineering systems*. Wiley Online Library, 2006.

[10] H. Eyring, "The activated complex in chemical reactions," *The Journal of Chemical Physics*, vol. 3, no. 2, pp. 107–115, 1935.

[11] S. S. Manson, "Behavior of materials under conditions of thermal stress," Tech. Rep. NACA-TR-1170. National Advisory Committee for Aeronautics, 1954.

[12] A. M. Trzynadlowski, *Power electronic converters and systems: frontiers and applications*. London: The Institution of Engineering and Technology, 2016.

[13] A. E. Ginart, D. W. Brown, P. W. Kalgren, and M. J. Roemer, "Online ringing characterization as a diagnostic technique for IGBTs in power drives," *IEEE Transactions on Instrumentation and Measurement*, vol. 58, pp. 2290–2299, Jul 2009.

[14] A. Schönung and H. Stemmler, "Geregelter drehstrom-umkehrantrieb mit gesteuertem umrichter nach dem unterschwingungsverfahren," *Bbc-Nachrichten*, vol. 46, no. 12, pp. 699–721, 1964.

[15] A. Ginart, R. Liou, A. Salazar, C. Restrepo, and M. Ernst, "Partial 5/3 level topology for solar grid-tie inverters," in *2014 IEEE Energy Conversion Congress and Exposition (ECCE)*, pp. 5736–5742, 2014. Permission for reproduction license number 4364861067712.

[16] T. Kawabata, T. Miyashita, and Y. Yamamoto, "Dead beat control of three phase PWM inverter," *IEEE Transactions on Power Electronics*, vol. 5, pp. 21–28, Jan 1990.

[17] R. Steele, *Delta modulation systems*. London: Pentech Press & Halsted Press, 1975.

[18] R. Baker and L. Bannister, "Electric power converter," *US Patent 3 867 643*, 1975.

[19] J. Rodriguez, J.-S. Lai, and F. Z. Peng, "Multilevel inverters: a survey of topologies, controls, and applications," *IEEE Transactions on Industrial Electronics*, vol. 49, pp. 724–738, 2002.

[20] C. I. Odeh and D. B. N. Nnadi, "Single-phase 9-level hybridised cascaded multilevel inverter," *IET Power Electronics*, vol. 6, pp. 468–477, 2013.

[21] R. Baker and L. H. Bannister, "Electric power converter," *United State Patent 3,867,643*, 1975.

[22] J. Rodriguez, S. Bernet, P. K. Steimer, and I. E. Lizama, "A survey on neutral-point-clamped inverters," *IEEE Transactions on Industrial Electronics*, vol. 57, pp. 2219–2230, 2010.

[23] P. Rodriguez, M. D. Bellar, R. S. M. noz Aguilar, S. Busquets-Monge, and F. Blaabjerg, "Multilevel-clamped multilevel converters (MLC^2)," *IEEE Transactions on Power Electronics*, vol. 27, pp. 1055–1060, 2012.

[24] P. Barbosa, P. Steimer, J. Steinke, M. Winkelnkemper, and N. Celanovic, "Active-neutral-point-clamped (ANPC) multilevel converter technology," in *2005 European Conference on Power Electronics and Applications*, 2005.

[25] S. S. Fazel, S. Bernet, D. Krug, and K. Jalili, "Design and comparison of 4-kV neutral-point-clamped, flying-capacitor, and series-connected H-bridge multilevel converters," *IEEE Transactions on Industry Applications*, vol. 43, pp. 1032–1040, 2007.

[26] A. Nabae, I. Takahashi, and H. Akagi, "A new neutral-point-clamped PWM inverter," *IEEE Transactions on Industry Applications*, vol. IA-17, pp. 518–523, 1981.

[27] Z. Li, P. Wang, Y. Li, and F. Gao, "A novel single-phase five-level inverter with coupled inductors," *IEEE Transactions on Power Electronics*, vol. 27, pp. 2716–2725, 2012.

[28] B. Jerusha, S. Vadhana, A. Mathy, and R. Ramaprabha, "Comparative study of hard switching and soft switching boost converter fed from a PV source," *International Journal of Engineering Research*, vol. 2, no. 5, pp. 328–331, 2013.

[29] M. H. Rashid, *Power electronics handbook*. Oxford: Butterworth-Heinemann, 2017.

[30] H. Pollock and J. O. Flower, "New method of power control for series-parallel load-resonant converters maintaining zero-current switching and unity

power factor operation," *IEEE Transactions on Power Electronics*, vol. 12, pp. 103–115, 1997.

[31] E. H. Kim and B. H. Kwon, "Zero-voltage- and zero-current-switching full-bridge converter with secondary resonance," *IEEE Transactions on Industrial Electronics*, vol. 57, pp. 1017–1025, 2010.

[32] D. M. Divan, "The resonant dc link converter-a new concept in static power conversion," *IEEE Transactions on Industry Applications*, vol. 25, pp. 317–325, 1989.

[33] Y. Murai and T. A. Lipo, "High frequency series resonant dc link power conversion," in *Conference Record of the 1988 IEEE Industry Applications Society Annual Meeting*, pp. 772–779, vol. 1, 1988.

[34] C. Hu, *Modern semiconductor devices for integrated circuits*, vol. 1. Prentice Hall, Upper Saddle River, NJ, 2010.

[35] S. Davis, "Enhancement-mode GAN MOSFET delivers impressive performance—A breakthrough in processing gallium nitride on a silicon substrate has produced enhancement-mode FETs with high conductivity and hyperfast switching," *Power Electronics Technology*, vol. 36, no. 3, p. 10, 2010.

[36] S. Hazra, A. De, L. Cheng, *et al.*, "High switching performance of 1700-V, 50-A SiC power MOSFET over Si IGBT/BiMOSFET for advanced power conversion applications," *IEEE Transactions on Power Electronics*, vol. 31, pp. 4742–4754, 2016.

[37] A. Volke, J. Wendt, and M. Hornkamp, *IGBT modules: technologies, driver and application*. Neubiberg: Infineon, 2012.

[38] A. R. Hefner, "A dynamic electro-thermal model for the IGBT," *IEEE Transactions on Industry Applications*, vol. 30, pp. 394–405, 1994.

[39] V. K. Khanna, *The insulated gate bipolar transistor: IGBT theory and design*. Piscataway, NJ: IEEE Press, 2003.

[40] Z. Luo, H. Ahn, and M. A. E. Nokali, "A thermal model for insulated gate bipolar transistor module," *IEEE Transactions on Power Electronics*, vol. 19, pp. 902–907, 2004.

[41] A. Ginart, I. N. Ali, I. Barlas, P. W. Kalgren, M. J. Roemer, and K. Goebel, "Observer based junction temperature estimator in thermoelectrical aging," in *2010 IEEE Aerospace Conference*, pp. 1–9, 2010. Permission for reproduction license number 4365081013314.

[42] A. Ferrara, P. Steeneken, B. Boksteen, A. Heringa, A. Scholten, J. Schmitz, and R. J. E. Hueting, "Physics-based stability analysis of MOS transistors," *Solid-State Electronics*, vol. 113, pp. 28–34, 2015.

[43] D. Wolpert and P. Ampadu, "Temperature effects in semiconductors," *Managing temperature effects in nanoscale adaptive systems*. pp. 15–33, New York, NY: Springer, 2012.

Chapter 2

Early diagnosis in power semiconductors: MOSFET, IGBT, emerging materials (SiC and GaNs)

Antonio Ginart[1], José M. Aller[2,3], and George J. Vachtsevanos[4]

2.1 Introduction

The reliability of a power converter depends mainly on the endurance of its main component, the power semiconductor. Therefore, particular attention is paid in this chapter to understand the models that allow the identification of features and early indicators of problems that establish the groundwork for early fault diagnosis in inverters.

The two major switch technologies that control the market are field-effect transistors (FETs) and insulated gate bipolar transistors (IGBTs), which are bipolar devices integrating concepts from bipolar junction transistors (BJTs; IGBTs) which are bipolar devices integrating concepts from BJTs and FETs [1]. From the beginning of power electronics, silicon-based semiconductors have been the undisputed king, but modern silicon-based power semiconductors are being challenged by silicon carbide (SiC) and more recently by gallium nitride (GaN). Regardless of the semiconductor material, early diagnostics in semiconductors are based not only on the understanding of the failure mechanisms but also on the "*parasitic structures*" that are intrinsically associated with the fabrication of the device. The identification and characterization of these "*parasitic or associated structures*" allow for the development of unified models with general characteristics across different materials and structures of power semiconductors. The aging effects are, in general, reflected in the parasitic structures early in the process of degradation, creating an ideal approach for understanding failure propagation. Fortunately, semiconductor devices share a similar structure, and a standard model across all power semiconductors is used in this chapter to understand aging and to enable early diagnostics.

[1]Smart Wires, USA
[2]Electrical Engineering Department, Universidad Politécnica Salesiana – Sede Cuenca, Ecuador
[3]Electrical Engineering Department, Simon Bolivar University, Venezuela
[4]School of Electrical and Computer Engineering, Georgia Institute of Technology, USA

Table 2.1 Electrical properties for power semiconductor material [3]

Electric property	Si	SiC(4H)	GaN	Ga_2O_3
Band gap energy (eV)	1.12	3.28	3.39	4.8
Critical electrical field (MV/cm)	0.29	2.5	3.0	3.8
Electron mobility (cm^2/V s)	1,200	800	900	300
Intrinsic carrier conc. (1/cm^3)	1.5×10^{-10}	0.82×10^{-8}	1.9×10^{-10}	1.79×10^{-22}
Relative permittivity (ϵ/ϵ_0)	11.8	9.0	9.0	10
Saturation velocity (cm/s)	1.0×10^7	2.0×10^7	2.5×10^7	–
Thermal conductivity (W/mK)	130	360–490	150–200	10.9–27

Metal–oxide–semiconductor FETs (MOSFETs) based in silicon are used mainly in applications rated less than 200 V and nowadays have been used in some applications above 1 kV; SiC power MOSFETs are a mature technology, and they compete with Si IGBTs in the 600 V–2 kV market. New technological developments are providing MOSFETs that target the 15 kV medium-voltage application range. GaN-based MOSFETs are mostly confined under 600 V, whereas bipolar GaN-based devices have been relatively less successful.

IGBTs are typically used for voltages larger than 100 V up to a few kV due to their superior robustness and low cost [1]. The high-power MOSFET and IGBT device structures are radically different from the highly integrated low-power devices because they use different manufacturing processes and technologies to support high electric fields associated with large power; thus, vertical device construction is preferred with a large-scale integration of devices in parallel. The geometric disposition connecting individual devices in shunt create well-known devices such as the HexFET [2]. MOSFETs are ideal semiconductors for paralleling due their positive temperature resistance. Modern IGBTs have a similar characteristic positive temperature coefficient that affords high-current operation allowing paralleling.

From the material semiconductor point of view, Table 2.1 summarizes the electrical properties of the Si, SiC_4H, and GaNs that are commercially available in the market. In addition, Ga_2O_3 is incorporated into Table 2.1 due to its potential to become the next trend in power devices [3,4].

2.1.1 Power device stress factors

Industry reliability practices have led to the classification of vertical power devices stress and failure factors in two distinct categories: silicon die defects and die packaging defects [5–7]. The die defects in MOSFETs and IGBTs are generally categorized with regard to the oxide region under field distortion defects and oxide defects. Field distortion defects are a result of the electric field in the power devices getting distorted [8]. This distortion is caused when a high voltage is applied to the device in the presence of polar molecules. A polar molecule is such that one end of the molecule is positively charged, and the other end is negatively charged (e.g., water and atmospheric ionic contaminants). Oxide defects are the second most common

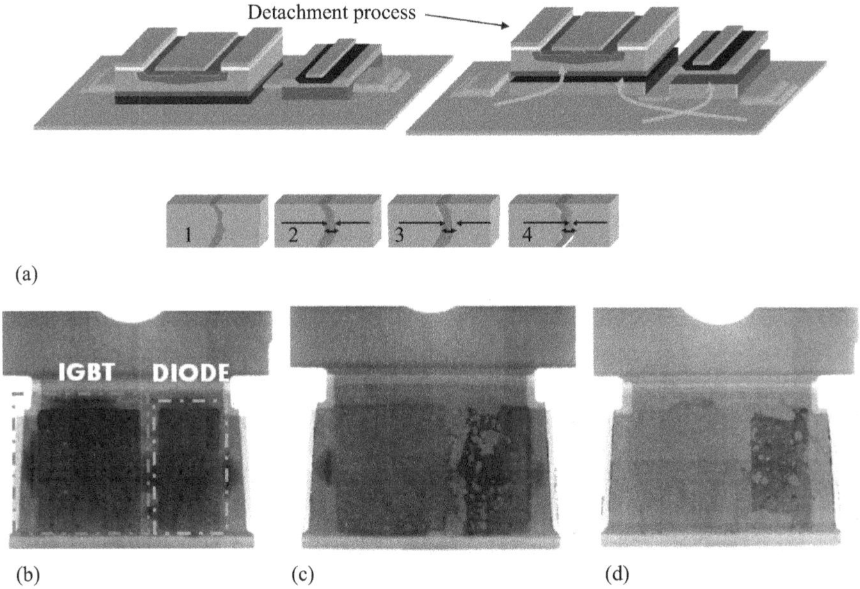

Figure 2.1 Die attach defects: (a) detachment process with void formation, (b) healthy, (c) low degraded, and (d) medium degraded

failures associated with die defects. Defects in the device's layer of gate oxide can cause failures such as short circuiting from gate to source. These defects are often enabled by thermal damage due to thermal cycling, which in turn leads to more trap generation. These traps cause more conduction and eventually the short circuiting.

Die packaging defects fall under several categories: die attach damage, wire bond fatigue, and metal corrosion [9]. Die attach defects are caused by thermal differentials created between the die and the headers of the devices [8]. The ultimate result is the creation of voids and cracks in the die attach region, leading to degradation in the on-resistance and thermal resistance of the device. Wire bond fatigue is also thermally induced and results in the separation of the wire bond of the device's dice. Finally, metal corrosion occurs due to intrusion of water particles into the device packaging through the atmosphere. Naturally, the introduction of water in the metallic contact area leads to corrosion and that eventually results in device degradation in the form of on-state resistance shift.

Figure 2.1 shows a detachment process with void formation, from a healthy to a medium degraded state.

2.1.2 Silicon power MOSFET structure and parasitics

Parasitics play a fundamental role in understanding and modeling aging in semiconductors; consequently, in the following section, power semiconductor structures will be presented, highlighting the parasitic components that appear due to manufacturing.

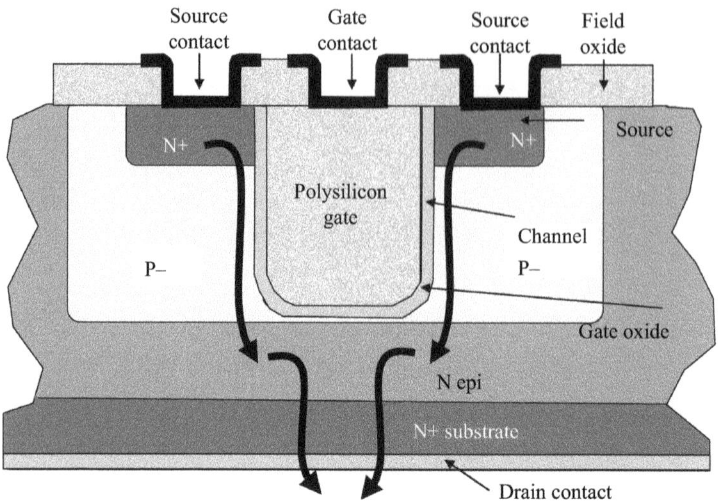

Figure 2.2 Structure of a commercial vertical power MOSFET

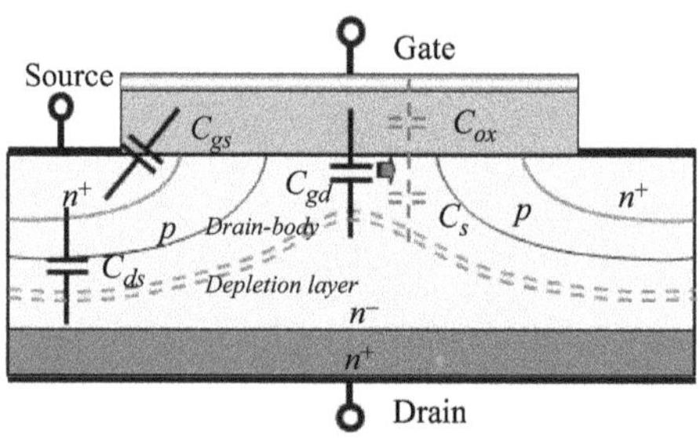

Figure 2.3 Parasitics component of a vertical power MOSFET

A general structure of a vertical power MOSFET device is shown in Figure 2.2 [8]. A power device is constructed by forming a mesh of vertical devices positioned in parallel to produce high power. Therefore, the structure shown in Figure 2.2 is connected in parallel in order to be integrated into a power MOSFET.

Parasitic components define the properties and behavior of a semiconductor power device. These properties change over time, providing essential tools for detecting early precursors to failures. Figure 2.3 shows a diagram with the material that creates the oxide and the disposition of the main three parasitic capacitors that are characteristic of a power semiconductor: C_{gs} (gate-to-source), C_{ds} (drain-to-source), and C_{gd} (gate-to-drain) [8].

2.1.3 SiC power MOSFET structure and parasitics

Commercial silicon carbide power devices are mature technologies and are generally based on a 4H–SiC structure. Figure 2.4 shows the basic configuration and the

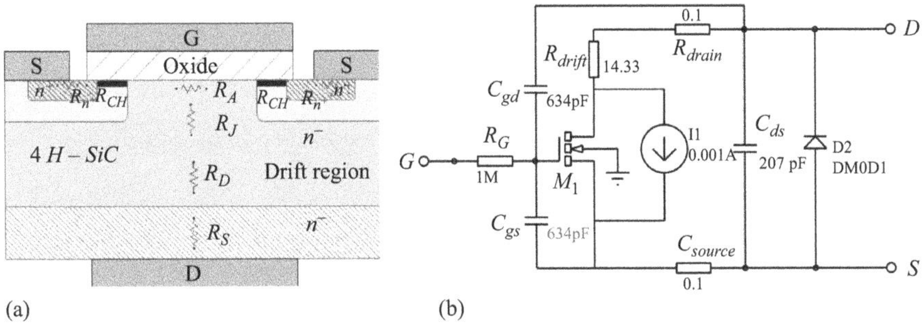

Figure 2.4 Parasitic component of power SiC_4H: (a) power SiC_4H structure and (b) power SiC_4H equivalent circuit

respective spice model presented by Lotfi *et al.* [2,10]. There is a substantial similitude between the silicon power MOSFET and SiC structures, with the difference that in the SiC, the epilayer is substituted by the 4H–SiC structure. The aging and reliability issue is similar as well; it is mainly confined to the electrical side to V_{th} shift due to impurities and hot carriers trapped in the interface of SiC and SiO_2 [11]. From the mechanical side, thermocycling and power cycling show a similar effect to that in power modules, increasing the thermal resistance of the wire bonds and also increasing thermal resistances by dice detachment. When dice detachment and void formation happen, for the same level of the current will produce higher dice temperature and consequently higher voltage drop and Ron drain-source. The results are shown by Luo [12].

2.1.4 GaNs structure and parasitics

Commercial power GaNs technology is different from vertical power devices (SiC and silicon) because its structure is based on lateral arrangements, limiting its application to single devices of lower than 1 kV and 100 A. It has two main representations, the cascade device and the e-GaNs [13]. The cascade is produced by Transphorm[1] [14] and consists in a combination of a JFET (MIS-HEMT) and a low-voltage silicon-based MOSFET (LV MOSFET) assembled in a single package up to 900 V. The parasitic structure is the combination of these two elements and is shown in Figure 2.5. The GaNs system is the main manufacturer of eGANs with dice from EPC. eGANs have a structure similar to that of the silicon n-type lateral MOSFET. The highly conductive (high mobility) channel is formed by growing AlGaN on top of the GaN layer (high resistance). The interface between AlGaN/GaN creates a zone with ultra-high mobility electrons (2DEG). When bias is applied, the gate forms a depletion region that blocks the conduction of the channel [15,16].

[1] https://www.transphormusa.com/company/

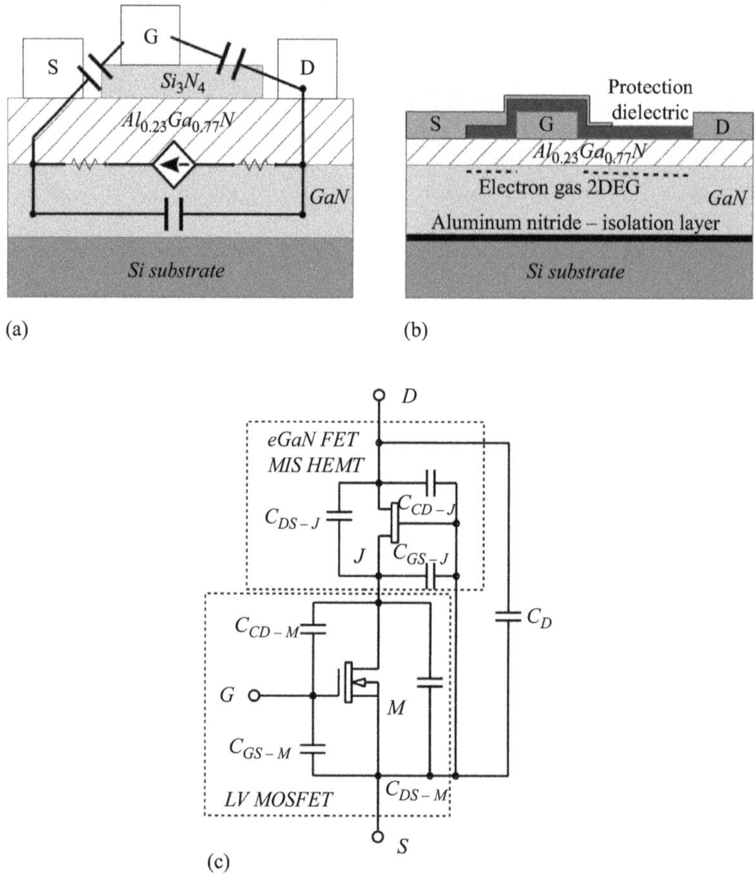

Figure 2.5 GaNs structure: (a) MIS-HEMT, (b) high mobility GaNs structure AlGaN/GaN (eGans), and (c) LV MOSFET parasitic component

2.1.5 IGBT structure and latch-up

A power semiconductor is arranged in such a way that the electrical connection emerges from the upper part of the dice. The bottom is metalized to the plate that transfers the heat generated over the junction to the heat sink. Figure 2.6 shows the two components (fast diode and n-type IGBT) integrated into the power device and encapsulated in the TO-220AB package corresponding to the IGBT IRG4BC30KD from International Rectifier, now Infineon. It also shows the X-ray image of the IGBT with a freewheeling diode as part of the device package. From the package point of view, two primary failure modes are well known: in first, the wire bond is detached from the dice, while the second is detachment from the substrate; processing of the first is directly related to the power cycling while the second relates to thermocycling.

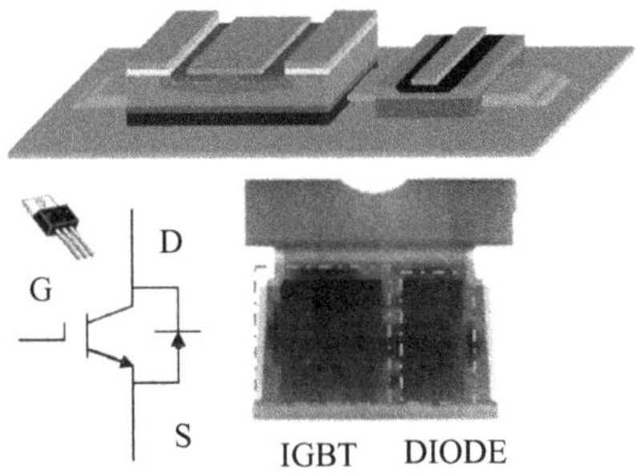

Figure 2.6 IGBTs packages TO-220

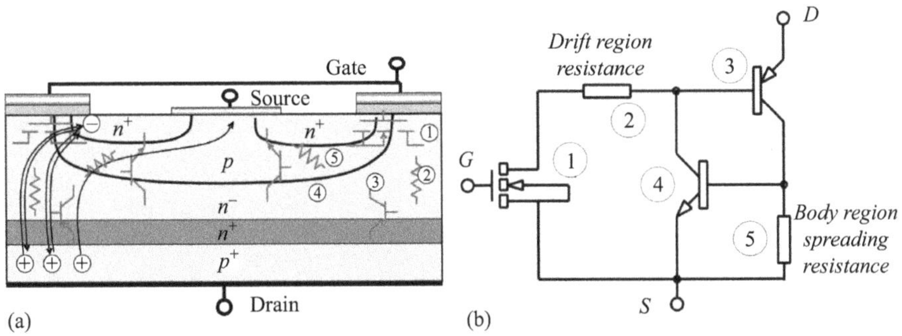

Figure 2.7 IGBTs structure: (a) IGBT internal structure and (b) equivalent circuit

The parasitic structure of the IGBT is similar but more complex than the vertical MOSFET shown in the previous section. Figure 2.7 shows a detail of the parasitic structure and the circuit representation of the IGBT. In addition to the PNP BJT power transistor market as ③, in Figure 2.7(a), a parasitic NPN market as ④ is formed. This transistor could get biased forward due to the voltage drop in the body region speeding resistances. In this case, the IGBTs behaves as silicon-controlled rectifiers (SCRs), losing the ability to turn off by gate command. This phenomenon is called IGBTs "latch-up" [17]. The voltage drop is proportional to the current and the body resistance. The trigger of "latch-up" is usually set by the manufacturer to at least six times the nominal current; moderns IGBTs have latch-up triggers that are even higher.

The latch-up phenomena are highly dependent on the body region speeding resistance, which grows significantly based on temperature. The positive thermal coefficient of the body region resistance is desirable for enabling IGBTs in paralleling operation but have an adverse effect for latching, producing a trade-off design between these two characteristics. To illustrate temperature dependence of the body region for a different level of doping in semiconductor, Figure 2.8(a) shows the variation of resistivity vs. temperature for different doping levels. The effect of temperature is depicted in Figure 2.8(b), where resistivity increases up to four times when the temperature increases from 25 to 300°C; consequently, a high temperature can reduce

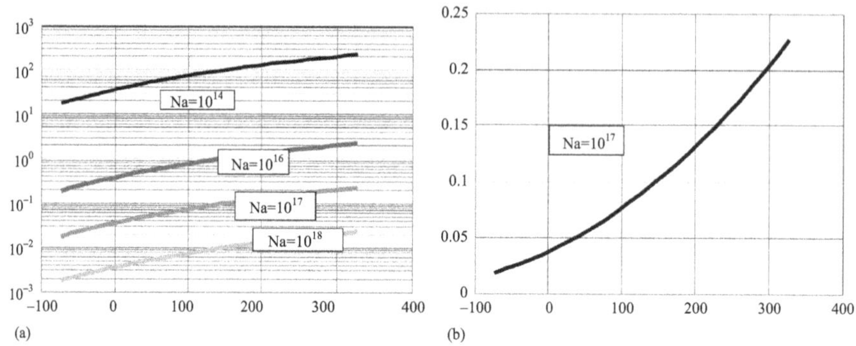

Figure 2.8 Resistivity vs. temperature dependence for different doping levels for n-type semiconductors: (a) variation for different doping levels and (b) detail of resistivity variation from −75 *to* 300°C

by several times the required magnitude of current required to trigger the latching effect [8].

2.2 Switching process in semiconductors

The switching process define the power semiconductor. Figure 2.9 shows how the different parasitic capacitances influence the switching turn-on process [17]. Figure 2.9(a)–2.9(d) shows the equivalent circuit for the MOSFET used to estimate the characteristic voltage and current in the device as the switching process progresses. Figure 2.9(a) and 2.9(b) shows how the power transistor begins to switch with the current that is initially circulating through the freewheeling diode D_F. During these stages, the influence of C_{gd} (gate-to-drain) and C_d (drain-to-source) with R_G, the gate resistance, is very important. In the final stages, as shown in Figure 2.9(c) and 2.9(d) R_{DS}, drain-to-source resistance, together with C_{gd} defines the process. Figure 2.9 summarizes the influences of the capacitances on the switching time during on-state, and how the variation in the values of the parasitic parameters will affect the time spent in the different stages. The gate drive voltage signal is a step response; the other gate-to-source voltage (V_{GS}) and gate and drain currents, I_G and I_D, are also shown. It should be noted that Figure 2.9(a)–2.9(d) is correlated with stages ①–④ in Figure 2.10. Figures 2.9 and 2.10 show the effect of parasitic capacitance on the characteristic waveforms of the device.

The process of switching is intrinsically related with the gate driver and its interaction with the capacitive components of the power semiconductor. This characteristic holds for the entire power semiconductor class regardless if they are based in silicon or other materials such as Sic or GaNs.

Given the influences of internal parasitic elements of power semiconductors, it is then important to classify the effects of aging in these parasitic elements. Figure 2.11 shows a schematic of how aging can be represented in changes of parasitic capacitance over time, using the model of the power MOSFET with the relevant parasitic elements

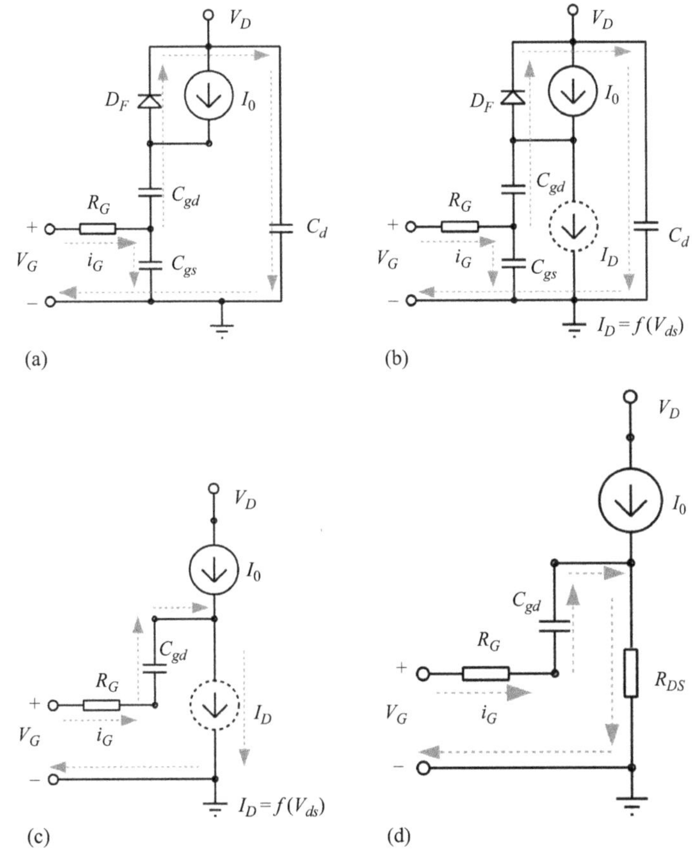

Figure 2.9 Capacitance interaction during switch ON

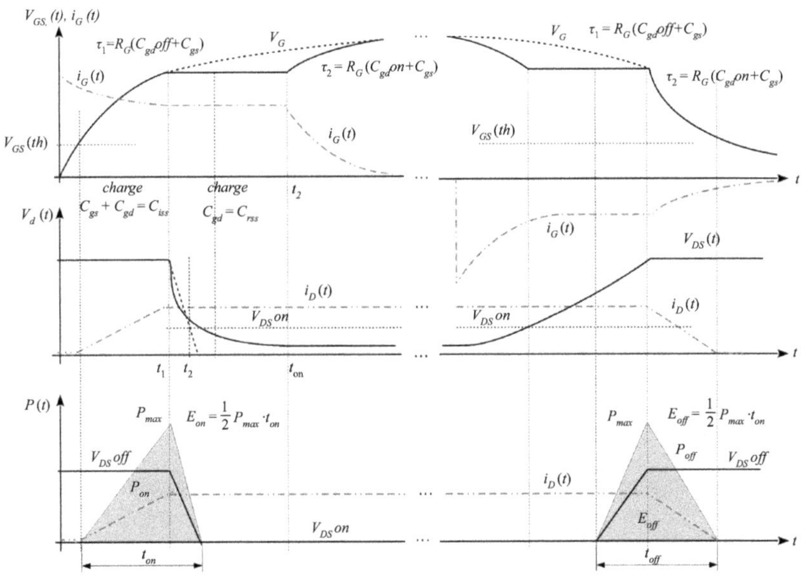

Figure 2.10 Effects of parasitics capacitance as a function of time

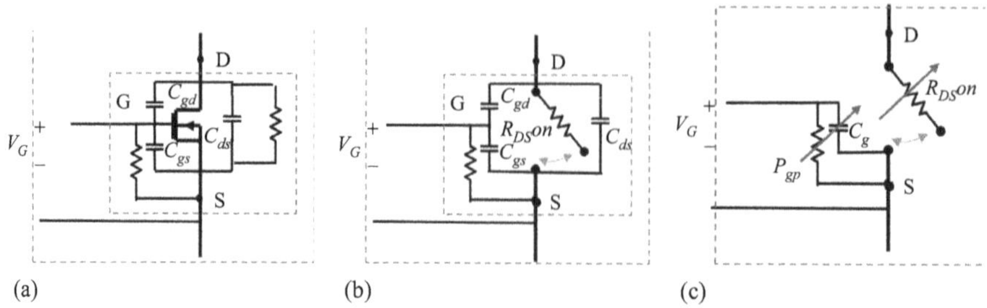

Figure 2.11 Power device simplified circuit aging

during switch-on process. Figure 2.11(a)–2.11(c) shows the model's simplification by lumping the components together. Increasing the power losses during switching can be represented as regular increases of the value of R_G. Moreover, the changes in the capacitance by ion movement or new trap charges can be accounted for as changes in C_g over time. Figure 2.11(c) is the model considered in this section for evaluation; specifically the Cg is the parasitic element modeled and monitored in the aged power device to characterize the aging effects. Other works [18] have used on-resistance R_{DS} as an indicator of device aging.

2.2.1 Field distortion acceleration model

Acceleration models have been used extensively to represent equipment life reduction. Representative examples include the Arrhenius model for temperature [19]:

$$L(T) = AFe^{(E/kT)} \tag{2.1}$$

where $L(T)$ is the life under thermal conditions; A, B are experimental constants; E is the activation energy; T is the temperature in K; k is the Boltzmann's constant 8.6171×10^{-5} eV/°C.

The acceleration factor, AF, is determined generally from the Arrhenius model (2.1):

$$AF = e^{((E_a/k)((1/T_1)-(1/T_2)))} \tag{2.2}$$

Crook [20] proposed the following expression in the case of two aging acceleration factors (temperature and voltage):

$$AF = e^{((E_a/k)((1/T_1)-(1/T_2)))}e^{-((E_2-E_1)/E_c)} \tag{2.3}$$

where E_c has been experimentally shown to equal 0.062 MV/cm [20].

The power law used commonly in determining the life expectancy of power devices based on the data presented in [21] is written as

$$AF = A \times 10^{CT} \times 10^{B \cdot V_G + a} \tag{2.4}$$

Several manufacturers use gate voltages and temperatures for aging estimation. Based on the data in [21], the values for the IRF4XX series are computed to be $A = -17.2$, $B = 0.4$, $C = -0.15$, and the respective curves are shown in Figure 2.12.

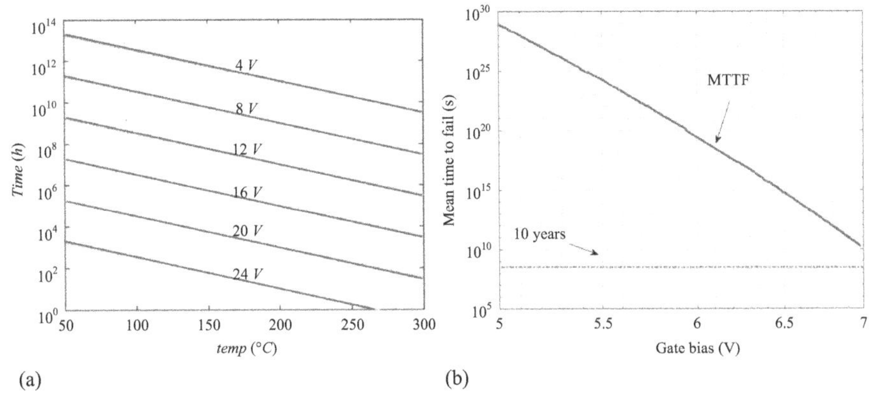

Figure 2.12 Power MOSFET silicon life expectancy under power law: (a) IRF4xx life expectancy and (b) HTGB voltage acceleration stress test—mean time to fail vs. gate voltage (EPC2016C) @ 150°C

The thermal activation energy of a failure mechanism is determined by testing at a minimum of two different temperature stress levels, as shown in Figure 2.12. Figure 2.12(a) shows the power MOSFET life expectancy using the power law shown in (2.4). It can be seen that the life expectancy of MOSFET devices changes with varying bias voltages and is changing uniformly, as the operating temperature changes. In lab testing and field operations, it has been shown that the device's failure rate accelerates operated in higher temperature regions. The same principle applies to eGANs FETs. Figure 2.12(b) shows how increased life expectancy has been obtained for GaNs materials [22].

2.3 Relevant indicators in power semiconductors

The following section discusses the most used features in early diagnostics of power semiconductors and their relation to the physics fundamentals associated with the relevant phenomena. Modern semiconductors share significant similarities among the power structures. Consequently, the early indicators and fault progression are shared across power devices in most cases. Examples and research in the area are substantial, and for this reason, representative examples for different semiconductor technologies will be chosen to explain the fundamental relations and why they become early indicators of health.

Aging in semiconductors usually has multiple failure mechanisms that develop simultaneously before the predominant mechanism becomes evident and finally leads to device failure. In vertical device structures, the main failure mechanisms that have been reported are as follows:

- Changes in V_{th} and parasitic capacitance [21];
- Ringing characterization and turn-on delay;

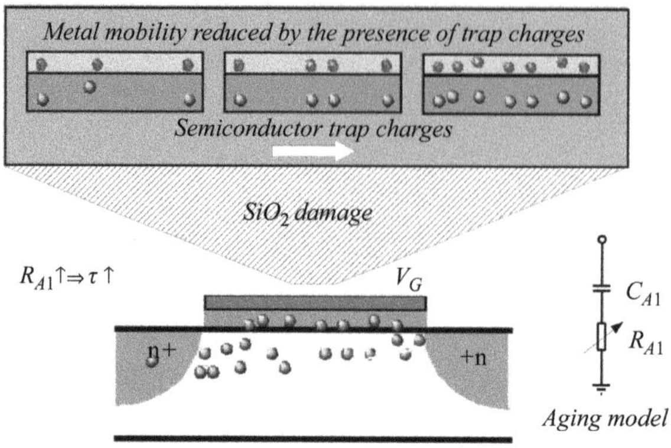

Figure 2.13 Metal mobility reduced by the presence of trap charges

- Wire bond fatigue and void formation [22–26]; and
- Increases in junction temperature.

2.3.1 Voltage V_{th} and capacitance shift

Aging could shift in threshold voltage (V_{th}) and capacitance parameters in the structure of FET devices, including IGBTs, primarily due to the mobility of impurities, especially $Na + Ion$ [11]. This phenomenon was a vital reliability issue in the early stages of MOSFET fabrication because of the early stage fabrication process. As the manufacturing technology evolved and purer silicon dioxide material could be grown on devices, less importance was placed on ion migration loss in reliability research for large-scale integration [27–29]. On the other hand, for power devices fabricated with new materials and vertical technology, oxide contamination and field distortion remain as a significant type of failures [30] due to the process itself and the large fields associated with the operation.

The capacitance mainly changes by the mobility of trap charges during fabrication and additional traps generated during operation as shown in Figure 2.13. These charges are primarily created from bias voltages and temperature. Regardless of the origins of the impurities, their mobility produces a variation of capacitances that can be used as an indicator of aging in the semiconductor devices [21].

2.3.1.1 Fundamental equations and modeling of V_{th} displacement and capacitance change by aging

Ginart *et al.* [25] explain how the initial trap charge is formed by hot ion as

$$\Delta\Phi_{ox} = \frac{K_s}{K_0}x_0E_s - \frac{1}{K_0E_s}\int_0^{x_0} x\rho_{0x}(x)dx \tag{2.5}$$

where Φ_{ox} is the potential in the SiO_2; K_s is the semiconductor dielectric constant; K_0 is the oxide dielectric constant; x_0 is the oxide thickness; E_s is the electric field in the semiconductor at the oxide-semiconductor interface; ρ_{0x} is the charge density in the oxide.

This implies that for the same potential in the silicon, Φ_s,

$$\Delta\Phi_G = -\frac{1}{K_0 E_s}\int_0^{x_0} x\rho_{0x}(x)dx \tag{2.6}$$

Finally, computing the capacitance for a nonideal silicon, SiC and GaNs power semiconductor:

$$C = \frac{C_0}{1 + (K_0 W_t / K_s x_0)} \tag{2.7}$$

where W_t is the depletion width calculated by

$$W_t = \sqrt{\frac{2K_s E_s}{qN_A}\Phi_{0x}} \tag{2.8}$$

where Φ_{0x} is the oxide potential related to the semiconductor's doping concentration, and C_0 is the oxide capacitance computed by

$$C_0 = \frac{K_0 E_0 A_G}{x_0} \tag{2.9}$$

and A_G is the gate area.

Following the Arrhenius equation shown in (2.4), the diffusion-like movement of trap charge impurities caused by stress-induced aging is modeled as time moves forward. The diffusion process mobilizes the impurities in an exponential fashion as a function of temperature and the bias voltage, V_G, thus changing the oxide potential as defined in (2.5). Therefore, the change in the oxide potential leads to the depletion width changing, which ultimately produces the shift in the device capacitance with respect to the voltage. Figure 2.14 shows the MATLAB® simulation results of the mobility of the trap charge impurities in the power devices and the corresponding shift in the capacitance. The simulation results are shown for the timescale of 1, 10, and 30 years. The temperature was kept constant at 100°C along with the initial trap charges, and the gate voltage V_G was set at 8 V. The impurities, represented per m^2, are shown to migrate over time from being uniformly distributed across the device (year *one*) and aggregating on one end by year 30. Correspondingly, the capacitance starts varying, and the shift over time is becoming obvious. The initial capacitance is shown in circles, whereas the simulated capacitance at the respective time is shown in a solid line, illustrating a clear shift in capacitance over time as the migration of impurities starts accelerating.

Figure 2.15 shows the results of the shifting of the IGBT transistor capacitance as a function of the gate voltage. The normalized results show the shift in capacitance of the device before and after the device underwent the accelerated aging described in the previous section. As can be seen, the shifting of the device produces similar results to those obtained in the simulation. It should be noted that since MOSFETs and

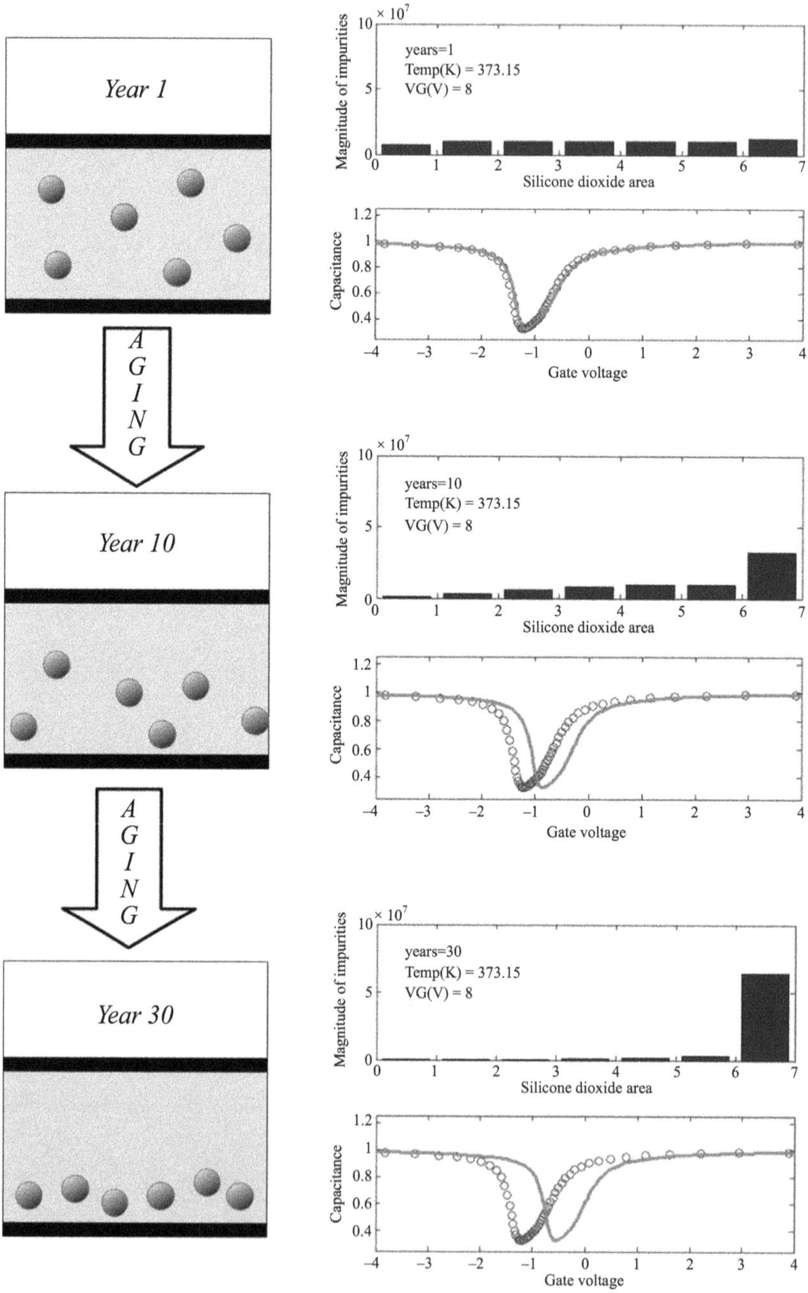

Figure 2.14 Migration of trap charges and capacitance variation with 100°C and $V_G = 8\ V$

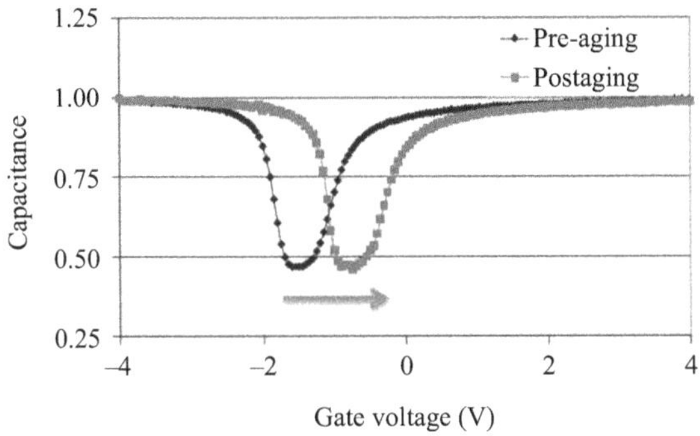

Figure 2.15 Capacitance shift of IGBT device before and after accelerated aging

Table 2.2 Parasitic components specification in a data sheet

IGBT			
Internal emitter inductance (L_E)	7.5 nH		
Output capacitance (C_{oes})	110 pF	CGE = 900 pF	$VGE = 0$; VCC = 30 V
Input capacitance (C_{ies})	920 pF	CCE = 83 pF	$f = 1$ MHz
Reverse transfer capacitance (C_{res})	27 pF	CGC = 27 pF	See Figure 2.16
DIODE			
Reverse recovery time	42 ns	60 ns (*max*)	$T_j = 25°C$
$IF = 12$ A; $VR = 200$ V; $(di/dt) = 200\ \mu$A	80 ns	42 ns (*max*)	$T_j = 125°C$

IGBTs have a similar gate structure, as describe earlier, the gate parasitic capacitance that has been modeled and tested would have similar results for both devices.

2.3.2 Ringing characterization and turn-on delay

The characterization of oscillation and rising time during switching have been used to determine the aging of the power device [25,31–33]. The reasons why the variation in the ringing and turn-on time represents aging are related to changes in the parasitic components. This section explains the physics of failure and parameters variations associated with this phenomenon. Table 2.2 shows an example of how the parasitic components are specified in a manufacturer's data sheet.

Figure 2.16(a) shows the typical model of the parasitic capacitances in a power device formed by an oxide insulation layer or other insulation material, and a highly doped semiconductor layer. The three main parasitic capacitors have a high dependence on the voltage present in the terminal of the power device (V_{CE} or V_{DS}) as shown in Figure 2.16(b). Figure 2.16(c) shows the parasitic components and their response simulated in Spice/ICAP. Note that a resistance in parallel has been added to simulate the losses. This particular model was implemented in parallel, but modeling with resistances in series is possible also equivalent series resistance (ESR).

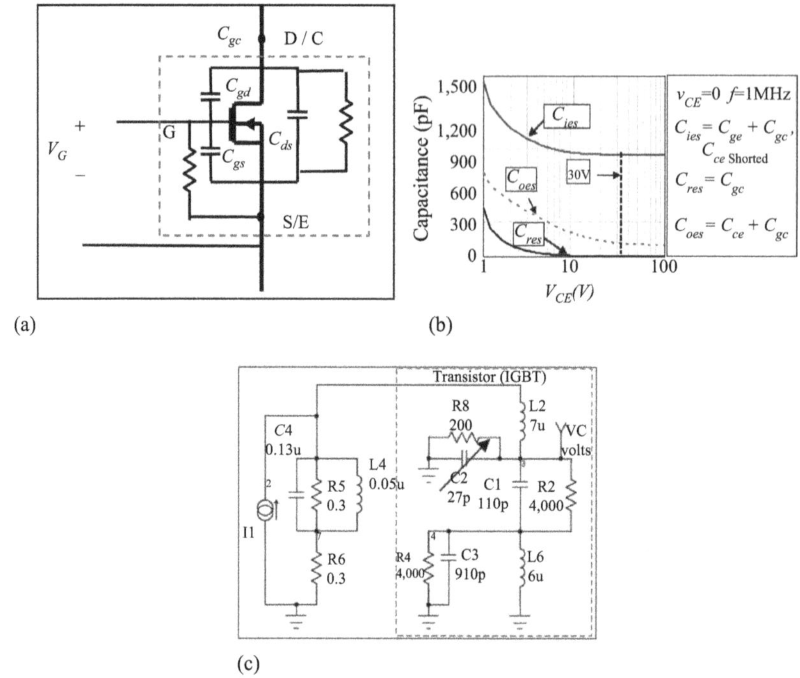

Figure 2.16 Capacitance shift of IGBT device before and after accelerated aging: (a) parasitic capacitances, (b) capacitance voltage dependence, (c) SPICE simulation

During aging, the difference in capacitance value manifests as changes in frequency, and its losses as a damping oscillation, as shown in Ginart *et al.* [31]. To prove this concept and other feature aging, a high-speed aging method is applied to IGBTs and power MOSFETs as proposed in [32]. Thermoelectrically aging is very similar to the power cycling method that consists of levering the high heating properties of the self-applied current when the cooling features (heat sinks) are removed. The differences between power cycling and thermoelectrical aging are that the later includes feedback over its owns temperature, which opens the circuit to alleviate the thermal stress, consequently failing to open the power switching devices, which proves there is a latch-up, leading to the loss of controllability.

The process described in detail in [25,31–33] is summarized as follows:

1. **Initialization:** The reference temperature of the junction is initially set at $T_{set}(1) = 30\,\text{K}$ above the maximum operational junction temperature defined by the manufacturer.
2. **Data acquisition and feedback control:** Measurements for $T_{c(front)}$, $T_{c(back)}$, V_{ge}, V_{ce}, I_c, and I_g are acquired using sensors connected to the data acquisition (DAQ) computer. Estimates for T_j are computed from measurements of $T_{c(front)}$ and $T_{c(back)}$ using the thermal model. V_{ge}, I_c, V_{ce}, and I_g are acquired to detect a latch-up event.

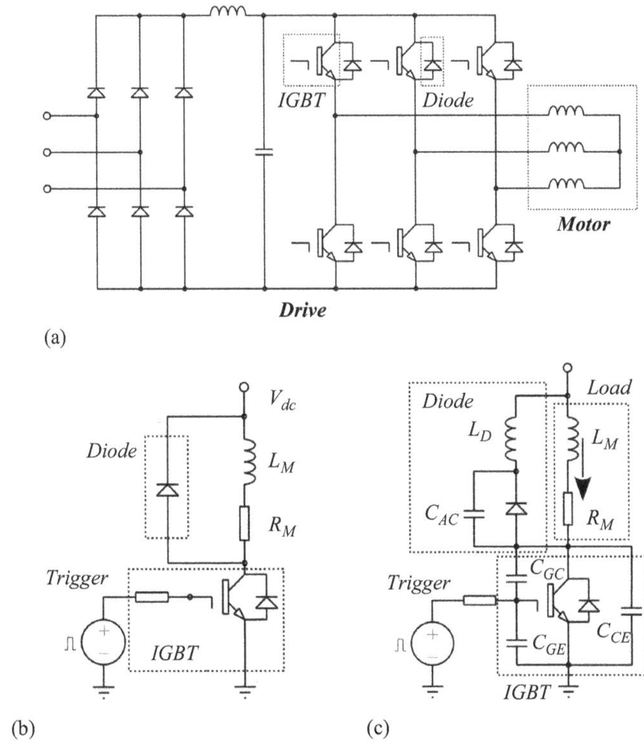

Figure 2.17 Correspondence between components of the standard system and the ringing evaluation platform: (a) power drive schematic, (b) simplified evaluation test-bed, and (c) transistor model, parasitic and motor winding

3. **Latch-up:** When a latch-up condition occurs, the transistor loses controllability and remains in a permanent on-state. Due to latching, the transistor cannot be successfully turned off, and the temperature continues to rise.
4. **Latch-up recovery:** Once a latch event is detected, the transistor is disconnected to allow the IGBT to recover. During this phase, the transistor is given sufficient time to cool, allowing for a reduction in T_j.
5. **Repeat:** The aging of the transistor is performed by a succession of latching. The next latching generally occurs at a lower ΔT_{set}, which typically ranges from 10 to 20 K. The reduction in temperature is a testament that degradation is taking place in the semiconductor. When applying thermoelectrical aging methodology, it takes a few cycles to generate a hard failure as a result of thermal runaway.

Ginart *et al.* proposed a simplified platform to study the effects of ringing resulting from aged or degraded transistor in a power drive system [25,31,32]. The correspondence between components of the standard system (denoted in the dashed frame) and the ringing evaluation platform is shown in Figure 2.17. The simplified

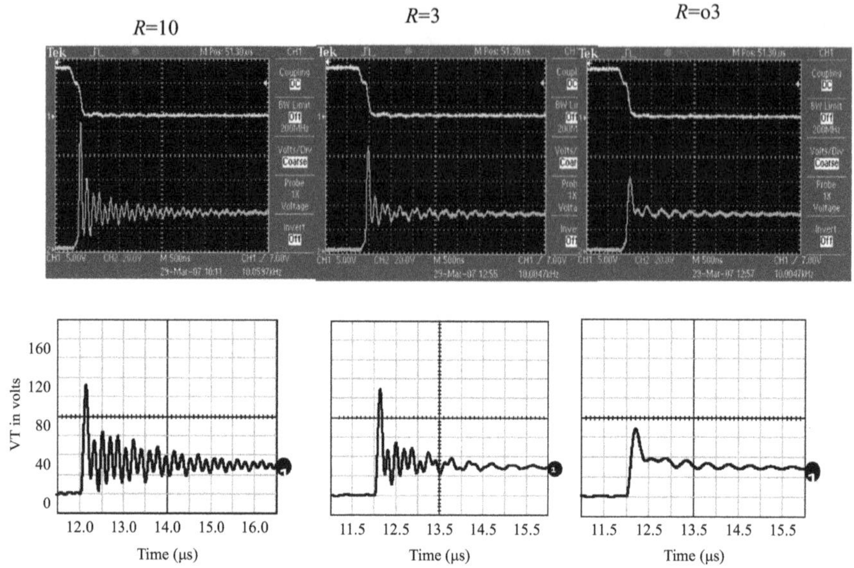

Figure 2.18 Capacitance shift of IGBT device before and after accelerated aging

diagram, shown in Figure 2.17(b), identifies the three main components used to study the effects of ringing in power drives: a single-phase of the three-phase motor winding, a freewheeling diode, and an IGBT transistor. Figure 2.17(c) details the main parametric elements of the power devices involved in the switching.

The IGBT is typically modeled with the three terminal capacitances C_{GE}, C_{GC}, C_{CE} , and one parasitic inductance-capacitance (L_E). In the case of FETs, the devices are named C_{GS}, C_{GD}, C_{DS}, correspondingly. In the diode, only two parametric components are highlighted: the cathode–anode capacitance (C_{AC}) and the parasitic diode inductance (L_D). The motor is represented by its stator inductance and resistance. Note, the winding capacitance and coil ground capacitance are omitted since the main interest is on the power device.

Figure 2.18 depicts an example of how the parasitic elements change in the model in order to match the ringing observed in the experimental results for different latching stages. The match of the ringing shapes for different aging stages in the physical model (Figure 2.19) is achieved with the variation of only one parameter, R_8, which controls the increase in losses of capacitance C_{GD}. R_8 variations are 10,000, 2,000, and 200 Ω, respectively (see Figure 2.16(c)).

It is remarkable how the experimental ringing aging in IGBTs can be reproduced using the modeling described above. However, it is not the only combination of parameter changes that can reproduce this phenomenon. Brown *et al.* [33] continued the work in this area, including ringing and evaluation with a seeded fault in full three-phase power drive systems. They developed a theoretical model that explains how C_{res}

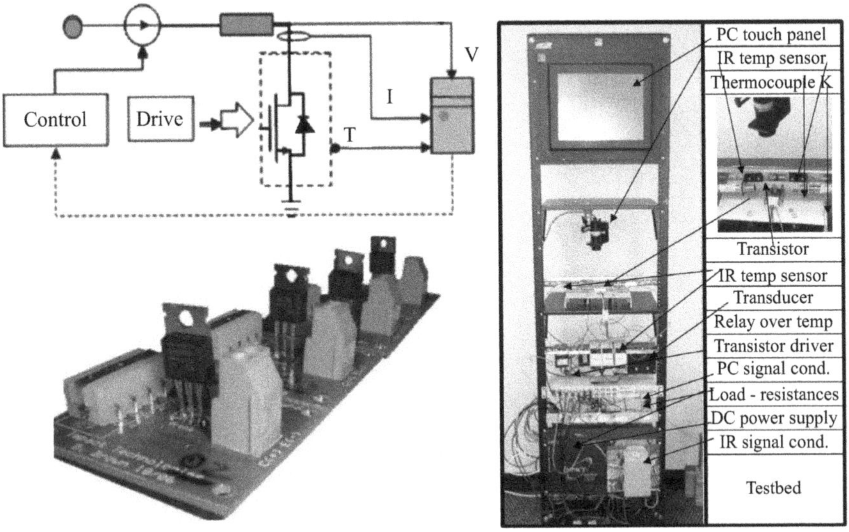

Figure 2.19 Diagram of the thermoelectrical accelerated aging system and IGBT test board and aging system

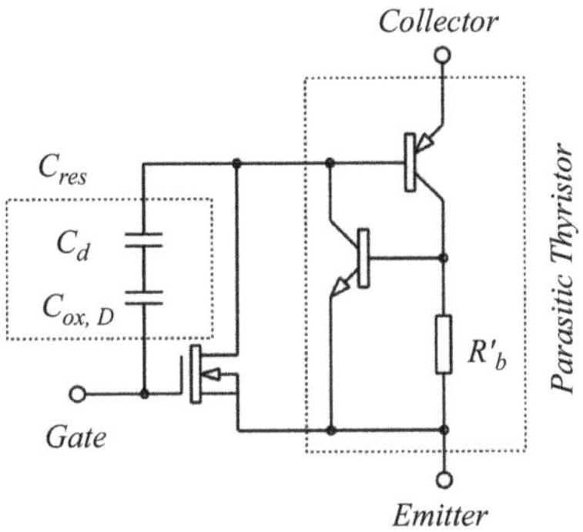

Figure 2.20 IGBT model [33]

changes based on the junction temperature (see Figure 2.20). Their modeling approach is based on (2.10), that expresses the variation of the intrinsic carrier concentration in function of temperature. In this case, the shift in temperature is 100°C (from 275 to 375 K) and the intrinsic carrier concentration (n_i) for silicon at 300 K is $1.01 \times 10^{10}\text{cm}^{-3}$.

$$n_i(T_j) = C_0 \left(\frac{T_j}{300}\right)^2 e^{-(a_0/T_j)} \tag{2.10}$$

where $a_0 = 6.884 \times 10^3$ K and $C_0 = 9.38 \times 10^{19}\text{cm}^{-3}$.

The partial derivative of n_i with respect to T_j can be computed as

$$\frac{\partial n_i}{\partial T_j} = n_i T_j^{-1}(2 + a_0 T_j^{-1}) \tag{2.11}$$

The surface potential, Φ_s, can be computed based on the relative number of charge carriers between the oxide interface and the epilayer, as

$$\Phi_s(T_j, n_i) = \frac{k_T T_j}{q} \ln\left(\frac{n_s}{n_i}\right) \tag{2.12}$$

where n_s is the electron concentration near the surface of the epilayer/gate oxide boundary.
The effect of T_j on Φ_s can be found by taking the partial derivative of Φ_s with respect to n_i:

$$\frac{\partial \Phi_s}{\partial T_j} = \frac{\partial \Phi_s}{\partial n_i} \cdot \frac{\partial n_i}{\partial T_j} = -\frac{k_T}{q}(2 + a_0 T_j^{-1}) \tag{2.13}$$

The reverse transfer capacitance, C_{res}, can be calculated as the sum of two series capacitances. The first one is the oxide capacitance $C_{ox,D}$, constituted by the gate electrode, the silicon dioxide, and the top of the epilayer. The second term is the depletion region capacitance, C_d.

$$C_{res}^{-1} = C_{ox,D}^{-1} + C_d^{-1} \tag{2.14}$$

$$C_d^{-1} = C_{d0}^{-1}\sqrt{1 - \left[\frac{2k_T a_0 C_{d0}^2}{q^2 \varepsilon_s N_d} \ln\left(\frac{T_j}{300}\right)\right]} \tag{2.15}$$

Substituting this expression for C_d into (2.14) produces an analytical expression for C_{res}:

$$C_{res} = \left[k_1 + k_2\sqrt{1 - k_3 \ln\left(\frac{T_j}{300}\right)}\right] \tag{2.16}$$

where

$$k_1 = C_{ox,D}^{-1}, \quad k_2 = C_{do}^{-1}, \quad \text{and } k_3 = (2k_T a_0 C_{d0}^2 / q^2 \epsilon_s N_d)$$

The modeling used the computed values of t_{off} from the experimental data collected during seeded fault testing. As shown in Figure 2.21, a total of 160 values of t_{off} were obtained for each transistor in order to show statistical difference. Attenuation in the modeling was similar to that observed with the ringing increase during the turn-on time.

2.3.3 *Detachment and wire bond fatigue*

Detachment reduces the capability of releasing heat, elevating the junction temperature of the dice. As discussed in the previous section, in power devices, R_{DS} increases proportional to T_j, as observed in FET technologies. A similar effect happens on IGBTs (BJT) with their resistive part: V_{TO} is the only element reduced when

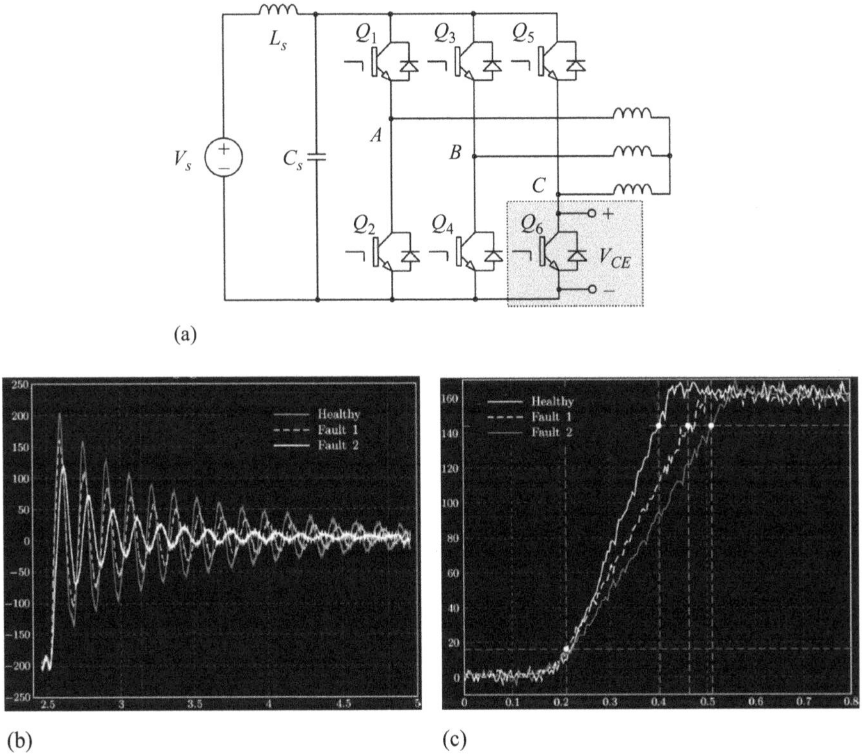

Figure 2.21 Seeded time-series plots comparing the ringing and time delay beet wing aged IGBTs and healthy: (a) inverter bridge, (b) ringing, and (c) time delay. IEEE License Number 4362891244959. Reproduced, with permission, from Reference [33]

temperature increases, but its total impact in the voltage drop is low because it is only significant at minimal current, as explained in Chapter 1. This behavior is present across different technologies, silicon, SiC, and GaNs. Several authors use this phenomenon to estimate the actual junction temperature [26,34–37].

2.3.3.1 Detachment in GANs

In GaNs, R_{ds} is affected by detachment similar to that of silicon and SiC devices. Song *et al.* [34] aged three power devices using thermocycling, and the conclusions of their work are shown in Figure 2.22. Their work verify R_{ds} *max* is increasing during aging. The explanation is similar to the other power devices. The same power is generated by the losses, and it does not change. However, the effective thermal path deteriorates when compared with the thermal path prior to aging, resulting in a higher junction temperature during operation and, consequently, a higher electrical resistance between terminals. Figure 2.22 shows a similar trend for the tree samples. Initially, ΔT_j and $R_{DS\,max}$ (max R_{ds}) remain almost constant up to a large number of thermal cycles, and

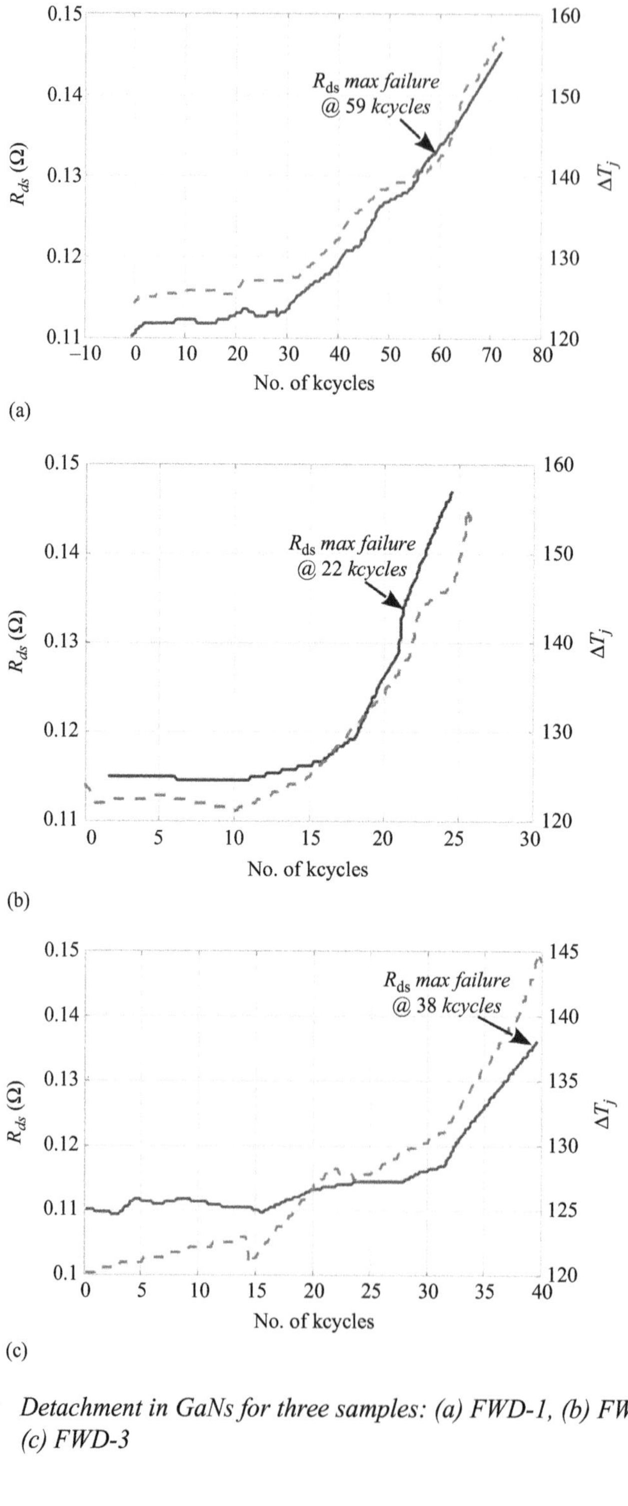

Figure 2.22 Detachment in GaNs for three samples: (a) FWD-1, (b) FWD-2, and (c) FWD-3

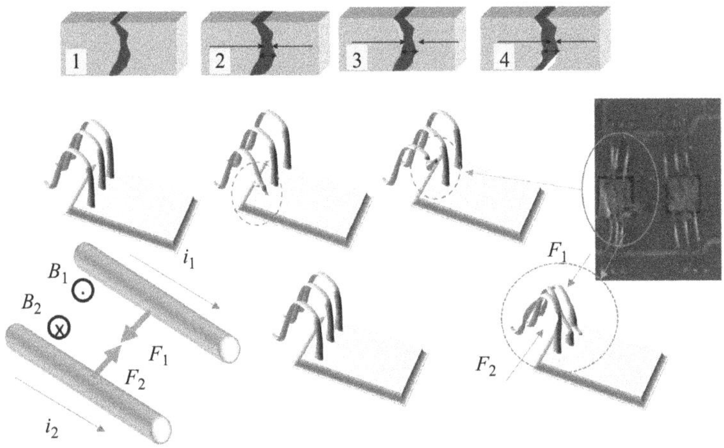

Figure 2.23 Wire bound fatigue and Lorentz's force

then they gradually increase after 15 or 30 cycles. Their failures occur between 22 and 59 cycles with $R_{D\,max}$ exceeding $R_{DS\,max} + 20\%$. Generally, the on-state resistance changes correspondingly with the temperature rise [38].

2.3.3.2 Wire bond and solder bond fatigue

In a similar way that the thermal path deteriorated in the void formation, the wire and the wire bond affected the electrical interconnection as they aged. While in void formation, the preferred method of stress is thermocycling; in wire bond fatigue, the preferred method of stress is power cycling. Power cycling [39] produces thermal stress by applying current to a power semiconductor device without thermal dissipation (heat sink). Consequently, detachment and void formation can occur when some layers of the semiconductor become detached from the packaging [23–26], but power cycling more directly stresses the wires and wire bonds. For large currents, the wires and wire bonds experience Lorenz forces together with thermal expansion, which act by pulling the wires together and at the same time stressing or pulling away their bond (see Figure 2.23).

The following equation shows the forces that are applied in the interconnection wires while current passes through

$$F = \frac{I}{c}\int dl \times B \tag{2.17}$$

The effect of wire bond fatigue is common to all semiconductors, as it relates to the interconnection and the solder method. Wire bond fatigue produces an effect similar to that of detachment or void formation, but for entirely different reasons. In both cases, the net result is increased R_{DS} with the same amount of current circulating, but in wire bond fatigue, the current circulation takes place over a smaller number of wires than before, or through a connection with higher losses. The R_{DS} increase

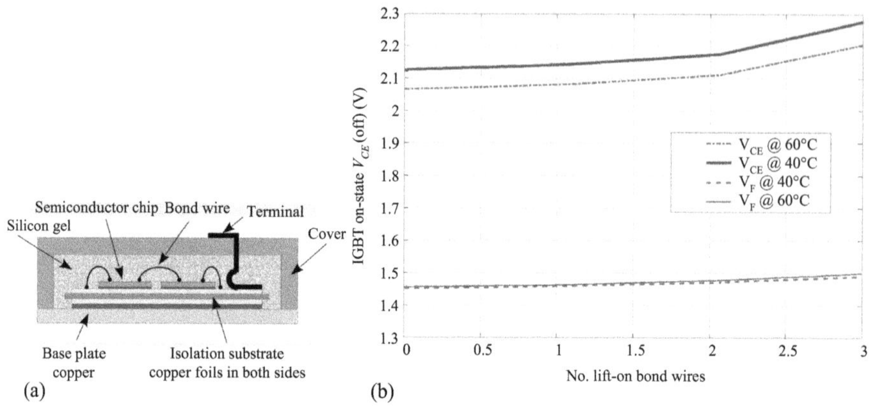

Figure 2.24 Monitoring to estimate aging of bond wire-IGBT modules stressed by power cycling[39]: (a) IGBT structure and (b) IGBT on-state voltage V_{CE}

is mainly caused by electrical effects rather than by reduction in the thermal path. It manifests as a larger voltage drop due to a more electrically resistive path, but it is less related to the increases in the junction temperature, which is the leading cause of detachment or void formation. Smet *et al.* [39] present a method for estimating the wire bond fatigue in silicon carbide bases in the increment of V_{CE}. The summarized results are shown in Figure 2.24.

2.3.3.3 Present and future in life estimation models

In modern power semiconductors, the power modules manufactured attempt to increase reliability using an ultrasonic wire bonding of the semiconductors. In addition, new methods of bonding are being developed for increased reliability in the new generation of power semiconductors, as shown in Figure 2.25, where the elimination of the layer of solder is displayed as one of the new approaches.

State-of-the-art life estimation models can predict reliability, but there are still significant ways for these models to evolve. An example of a modern life model limitation was exposed by testing the prediction done by several Coffin–Manson methods, which found up to six times the variance among the different methods in life estimation prediction [41]. Table 2.3 shows the methods compared [41–44]. Chapter 8 of this book discusses prognostic tools that can predict accurately the component's remaining useful life [45].

2.3.4 Junction temperature of power semiconductor

The junction temperature measurement of a power semiconductor device is instrumental in two main ways. The first one is for thermal management of power modules in order to establish a safe operation of the devices. The second use is related to the estimation of the aging of components for reliability reasons.

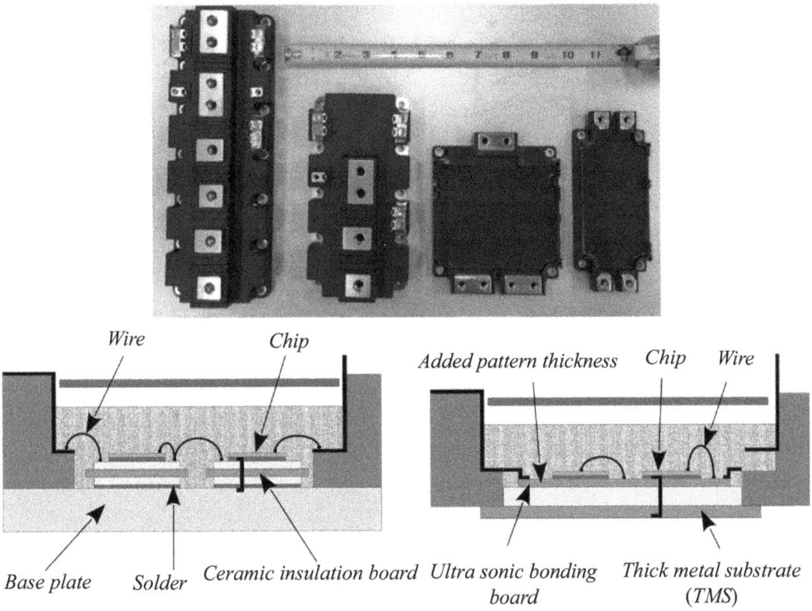

Figure 2.25 Manufacture improvement between sixth and seventh generation of IGBTs modules [40]

Table 2.3 The commonly employed cycle to failure models (analytical models)

No.	Model name	Analytical life time	Major impacts
1	Coffin–Manson [11]	$N_f = A\Delta T^{-n}$	only amplitude of temperature
2	General Coffin–Manson [7]	$N_f = A(\Delta T - \Delta T_0)^{-n}$	Model 1 + elastic region
3	Modified Coffin–Manson [16]	$N_f = A\Delta T^{-n} e^{(E_a/k_B T_m)}$	Model 1 + absolute temperature
4	General modified Coffin–Manson	$N_f = A(\Delta T - \Delta T_0)^{-n} e^{(E_a/k_B T_m)}$	Model 1 + abs. temp + elastic region
5	Bayerer [12]	$N_f = A\Delta T^{-n} e^{(\beta_2/T_{min})} t_{on}^{\beta_3} I^{\beta_4} V^{\beta_5} D^{\beta_6}$	Model 3 + power on time
6	General Bayerer's	$N_f = A(\Delta T - \Delta T_0)^{-n} e^{(\beta_2/T_{min})} \times t_{on}^{\beta_3} I^{\beta_4} V^{\beta_5} D^{\beta_6}$	Model 3 + power on time + elastic region

The three primary methods currently used to evaluate the temperature of power semiconductor devices are as follows [37]:

- Optical methods:
  - Infrared sensor and
  - Infrared cameras or similar.

- Physical contact methods:
  - Perform with the use of thermal probes (thermistors or thermocouples),
  - Thermocouples mainly J and K, and
  - RTD.
- Electrical methods:
  - They are based on the relation present between the electrical parameter and the actual junction temperature.
- Combined methods:
  - Use a combination of the first two with the electrical method, as in the case of the observer.

2.3.4.1 Observer based for junction temperature estimation

In the preceding sections, it was shown that the junction temperature is fundamental in determining the aging and remaining life of power semiconductor devices. Unfortunately, it is not easy to directly measure these variables. In [46], the use of an observer that is based on the operational condition and a validated thermal model that can estimate the junction temperature in real time is proposed. In this specific case, a junction temperature is estimated for an SiC device.

To develop the observer, a thermal model with four state variables is used, of which three are the measured temperature of the case (T_{CE}), the temperature of the drain (T_C), and the temperature of the source (T_s). The junction temperature (T_j) is the four state variable, which is the state variable to be estimated [47]. The input of the system is the power losses and the ambient temperature (T_{amb}) from a complete definition of the system. To compute the instantaneous power losses (P_{in}) from the transistor, measurement of the voltage between the drain and source (V_{DS}) and the I_{DC} current is both performed. The simplified model of the system is shown in Figure 2.26, while Figure 2.26(b) shows the physical location of the measurable temperatures. Figure 2.26(c) shows a simple model, while Figure 2.26(d) shows a more complete system that allows for a more accurate representation of the transistor.

Energy storage elements and state variables are described in Table 2.4. The inputs to the system are the ambient temperature (T_{amb}) and the power dissipated at the junction (P_{in}).

Furthermore, thermal circuit equations can be derived as follows:

$$\begin{aligned}
C_j\frac{\partial T_j}{\partial t} &= P_{in} - \frac{T_j - T_s}{R_{js}}\\
C_s\frac{\partial T_s}{\partial t} &= -\frac{T_s - T_j}{R_{js}} - \frac{T_s - T_c}{R_{sc}}\\
C_c\frac{\partial T_c}{\partial t} &= -\frac{T_c - T_s}{R_{sc}} - \frac{T_c}{R_{ca}} - \frac{T_{amb}}{R_{ca}}\\
C_{ce}\frac{\partial T_{ce}}{\partial t} &= P_{in} - \frac{T_{ce} - T_j}{R_{ce}} - \frac{T_{ce}}{R_{cea}} - \frac{T_{amb}}{R_{cea}}
\end{aligned} \tag{2.18}$$

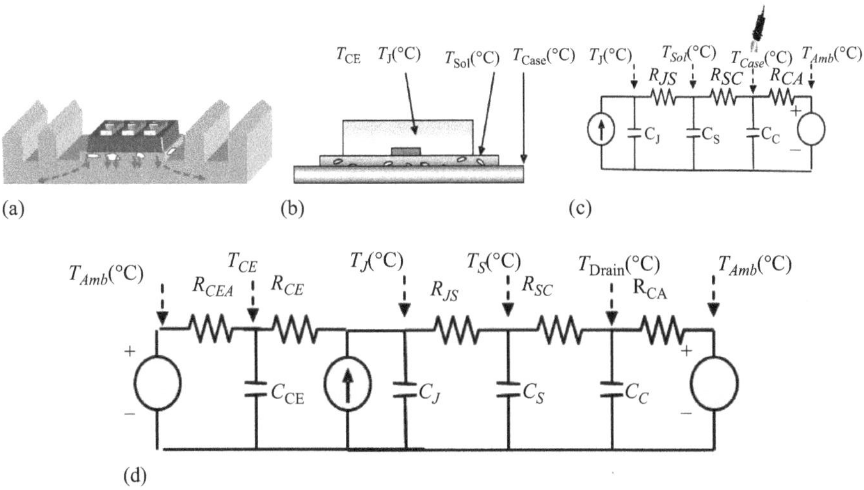

Figure 2.26 Dynamic model of the junction–case temperature [47]: (a) transistor, (b) transistor profile, (c) temperature model, and (d) expanded dynamic model of the junction–case temperature

Table 2.4 Energy storage elements and state variables

Energy storage element	Energy storage relation	State variable
Dice (junction)	$C_j T_j$	T_j
Source	$C_s T_s$	T_s
Drain	$C_c T_c$	T_c
Ceramic case	$C_{ce} T_{ce}$	T_{ce}

Finally, the state space system takes the form of

$$\begin{bmatrix} \dot{T}_j \\ \dot{T}_s \\ \dot{T}_c \\ \dot{T}_{ce} \end{bmatrix} = \begin{bmatrix} -\frac{1}{C_j R_{js}} & \frac{1}{C_j R_{js}} & 0 & 0 \\ \frac{1}{C_s R_{js}} & -\frac{1}{C_s}\left(\frac{1}{R_{js}} + \frac{1}{R_{sc}}\right) & \frac{1}{C_s R_{sc}} & 0 \\ 0 & \frac{1}{C_c R_{ca}} & -\frac{1}{C_c}\left(\frac{1}{R_{sc}} + \frac{1}{R_{ca}}\right) & 0 \\ \frac{1}{C_{ce} R_{ce}} & 0 & 0 & -\frac{1}{C_{ce}}\left(\frac{1}{R_{ce}} + \frac{1}{R_{sc}}\right) \end{bmatrix} \begin{bmatrix} T_j \\ T_s \\ T_c \\ T_{ce} \end{bmatrix}$$

$$+ \cdots + \begin{bmatrix} \frac{a}{C_j} & 0 \\ 0 & 0 \\ 0 & \frac{1}{C_s R_{ca}} \\ 0 & \frac{1}{C_{ce} R_{cea}} \end{bmatrix} \begin{bmatrix} P_{in} \\ T_{amb} \end{bmatrix} \tag{2.19}$$

2.3.4.2 Device selection observer

The diagram of the observer control block is shown in Figure 2.27(a).

Assuming that the state x is approximately the same as the estimated state $\tilde{x}$,

$$\begin{aligned}\dot{\tilde{x}} &= A\tilde{x} + Bu + L(y - \tilde{y}) \\ \dot{\tilde{x}} &= A\tilde{x} + Bu + L(y - C\tilde{x})\end{aligned} \tag{2.20}$$

an observer error equation can be derived as

$$\dot{x} - \dot{\tilde{x}} = Ax - LC(x - \tilde{x}) \tag{2.21}$$

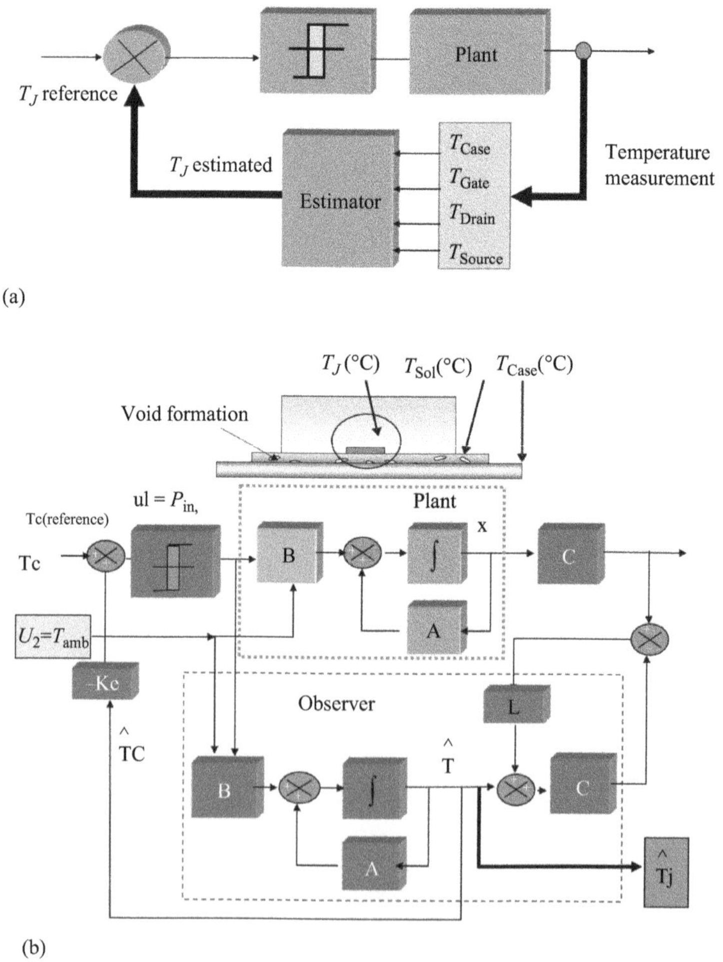

Figure 2.27 Transistor temperature observer system [47]: (a) block diagram and (b) observer implementation

Then,

$$\dot{e} = (A - LC)e \tag{2.22}$$

If the matrix L is such as that error converges to zero with an adequate speed of convergence, this criterion determines the value of L.

The general structure to solving this problem after checking for controllability is finding a simple matrix L that provides eigenvalues in the left-hand plane to ensure convergence to zero in the error matrix, which is shown as

$$\det\{sI - (A - LC)\} = (s - \mu_1)(s - \mu_2)\cdots(s - \mu_n) \tag{2.23}$$

Multiple solutions are possible. In order to narrow the set of solutions adequate for this problem, a systematic procedure is undertaken as follows: The matrix for $C = I$, the identity matrix, L may have a solution independent of the error of the junction temperature (T_j). Held *et al.* [46] propose this simplified solution L as

$$L = \begin{bmatrix} 0 & \alpha & 0 & 0 \\ 0 & \alpha & 0 & 0 \\ 0 & 0 & \beta & 0 \\ 0 & 0 & 0 & \phi \end{bmatrix} \tag{2.24}$$

Based on (2.24), Figure 2.28(a) shows the temperatures that is estimated based on the experimental data from the accelerated aging test. The power computed from the test bench and the input power is shown in Figure 2.28(b).

$$P_{in}(t) = V_{DS}(t) \cdot I_D(t) \tag{2.25}$$

Figure 2.28(a) shows the experimental results for the drain, gate, source, and case (ceramic) temperatures extracted from the test bed. The results are in very close

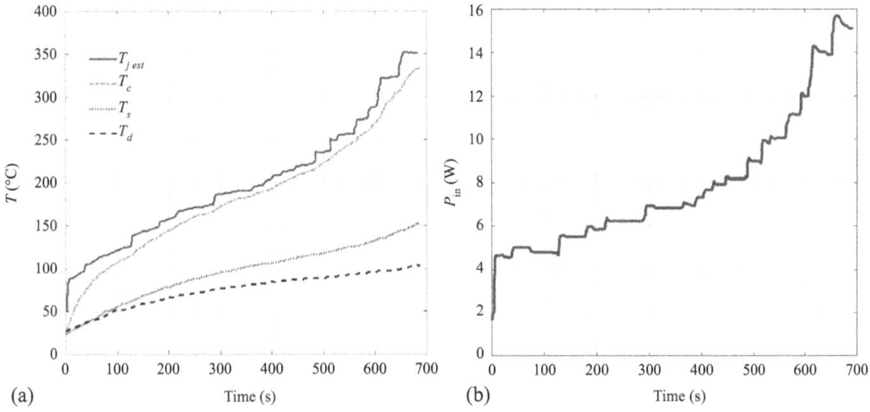

Figure 2.28 Thermal model from experimental data [47]: (a) estimated temperatures T_j, T_c, T_s, and T_d and (b) input power from experimental data

agreement with the model-based results. It should be noted that both model-based theoretical temperatures and the actual experimental results have the same input power as shown in Figure 2.28(b).

References

[1] V. K. Khanna, *Insulated gate bipolar transistor IGBT theory and design*. Piscataway, NJ: IEEE Press, 2004.

[2] M. Lotfi, D. Zohir, and S. Toufik, "A new spice macromodel of 4H – SiC vertical double implanted MOSFET (DIMOS)," *Microelectronics and Solid State Electronics*, vol. 1, no. 5, pp. 111–117, 2012.

[3] I. Lee, A. Kumar, K. Zeng, U. Singisetti, and X. Yao, "Modeling and power loss evaluation of ultra wide band gap Ga_2O_3 device for high power applications," in *2017 IEEE Energy Conversion Congress and Exposition (ECCE)*, pp. 4377–4382, 2017.

[4] M. Östling, R. Ghandi, and C. M. Zetterling, "SiC power devices 2014; present status, applications and future perspective," in *2011 IEEE 23rd International Symposium on Power Semiconductor Devices and ICs*, pp. 10–15, 2011.

[5] M. Ciappa, "Selected failure mechanisms of modern power modules," *Microelectronics Reliability*, vol. 42, no. 4–5, pp. 653–667, 2002.

[6] D. K. Schroder, *Semiconductor material and device characterization*. Hoboken, NJ: Wiley-IEEE Press, 2006.

[7] S. Clemente, B. Pelly, and A. Isidori, "Understanding HEXFET switching performance," *International Rectifier, Application Note*, vol. 947, pp. 1–14, 1993.

[8] A. E. Ginart, I. N. Ali, J. R. Celaya, P. W. Kalgren, S. D. Poll, and M. J. Roemer, "Modeling SiO_2 ion impurities aging in insulated gate power devices under temperature and voltage stress," tech. rep., IMPACT TECHNOLOGIES LLC ROCHESTER NY, 2010. This is an open-access article distributed under the terms of the Creative Commons Attribution 3.0 United States License.

[9] V. R. Manikam and K. Y. Cheong, "Die attach materials for high temperature applications: A review," *IEEE Transactions on Components, Packaging and Manufacturing Technology*, vol. 1, no. 4, pp. 457–478, 2011.

[10] M. Lotfi and S. Toufik, "The electro-thermal sub circuit model for power MOSFETs," in *2012 24th International Conference on Microelectronics (ICM)*, pp. 1–4, 2012.

[11] D. R. Kerr, J. S. Logan, P. J. Burkhardt, and W. A. Pliskin, "Stabilization of SiO_2 passivation layers with P_2O_5," *IBM Journal of Research and Development*, vol. 8, no. 4, pp. 376–384, 1964.

[12] H. Luo, F. Iannuzzo, F. Blaabjerg, M. Turnaturi, and E. Mattiuzzo, "Aging precursors and degradation effects of SiC-MOSFET modules under highly accelerated power cycling conditions," in *2017 IEEE Energy Conversion Congress and Exposition (ECCE)*, pp. 2506–2511, 2017.

[13] B. W. Williams, "Power electronics: principles and elements," *Devices, Drivers, Applications and Passive Components*. Barry W Williams, p. 1432, 2006.

[14] K. Smith and R. Barr, *Reliability lifecycle of 'hboxGaN power devices*. Goleta, CA: Transphorm, 2017.

[15] S. Ghosh, S. A. Ahsan, A. Dasgupta, S. Khandelwal, and Y. S. Chauhan, "GaN HEMT modeling for power and RF applications using ASM-HEMT," in *Emerging Electronics (ICEE), 2016 3rd International Conference on*, pp. 1–4, IEEE, 2016.

[16] C.-C. Wu and S.-L. Jeng, "Simulation model development for packaged cascode gallium nitride field-effect transistors," *Crystals*, vol. 7, no. 8, p. 250, 2017.

[17] N. Mohan, T. M. Undeland, and W. P. Robbins, *Power electronics: Converters, applications and design*. Hoboken, NJ: John Wiley & Sons, Inc., 2002.

[18] J. R. Celaya, N. Patil, S. Saha, P. Wysocki, and K. Goebel, *Towards accelerated aging methodologies and health management of power MOSFETs (technical brief)*. 2009.

[19] S. Arrhenius, "Über die reaktionsgeschwindigkeit bei der inversion von rohrzucker durch säuren," *Zeitschrift fur physikalische Chemie*, vol. 4, no. 1, pp. 226–248, 1889.

[20] D. L. Crook, "Method of determining reliability screens for time dependent dielectric breakdown," in *Reliability Physics Symposium, 1979. 17th Annual*, pp. 1–7, IEEE, 1979.

[21] R. Pearce, S. Brown, and D. Grant, "Measuring HEXFET characteristics," *International Rectifier Application Note*. http://www. irf. com, El Segundo, CA, 1993.

[22] C. Jakubiec, R. Strittmatter, and C. Zhou, EPC eGaN FETs reliability testing: Phase 8. El Segundo, CA: Efficient Power Conversion Corporation, 2016.

[23] B. Saha, J. R. Celaya, P. F. Wysocki, and K. F. Goebel, "Towards prognostics for electronics components," in *Aerospace Conference, 2009 IEEE*, pp. 1–7, IEEE, 2009.

[24] G. Sonnenfeld, K. Goebel, and J. R. Celaya, "An agile accelerated aging, characterization and scenario simulation system for gate controlled power transistors," in *AUTOTESTCON, 2008 IEEE*, pp. 208–215, IEEE, 2008.

[25] A. E. Ginart, D. Brown, P. W. Kalgren, and M. J. Roemer, "On-line ringing characterization as a PHM technique for power drives and electrical machinery," in *2007 IEEE Autotestcon*, pp. 654–659, 2007. Permission for reproduction license number 4366051433153.

[26] P. Srinivasan, *Reliability of solder die attaches for a high power application*. PhD thesis, University of Maryland, College Park, 2000.

[27] E. Miranda and J. Sune, "Electron transport through broken down ultra-thin SiO_2 layers in MOS devices," *Microelectronics Reliability*, vol. 44, no. 1, pp. 1–23, 2004.

[28] A. Feinberg, P. Ersland, V. Kapar, and A. Widom, "On aging of key transistor device parameters," *Annual Technical Meeting-Institute of Environmental Sciences and Technology*, vol. 46, pp. 231–236, 2000.

[29] A. Ferrara, P. Steeneken, B. Boksteen, *et al.*, "Physics-based stability analysis of MOS transistors," *Solid-State Electronics*, vol. 113, pp. 28–34, 2015.

[30] S. Clemente and K. Teasdale, "Understanding and using power MOSFET reliability data," in *HEXFET Power MOSFETs Designer Manual Application notes and reliability Data. International Rectifier*, vol. 1, 1993.

[31] A. Ginart, M. J. Roemer, P. W. Kalgren, and K. Goebel, "Modeling aging effects of IGBTs in power drives by ringing characterization," in *2008 International Conference on Prognostics and Health Management*, pp. 1–7, 2008.

[32] A. E. Ginart, D. W. Brown, P. W. Kalgren, and M. J. Roemer, "Online ringing characterization as a diagnostic technique for IGBTs in power drives," *IEEE Transactions on Instrumentation and Measurement*, vol. 58, pp. 2290–2299, 2009.

[33] D. W. Brown, M. Abbas, A. Ginart, I. N. Ali, P. W. Kalgren, and G. J. Vachtsevanos, "Turn-off time as an early indicator of insulated gate bipolar transistor latch-up," *IEEE Transactions on Power Electronics*, vol. 27, pp. 479–489, 2012. Permission for reproduction license number 4366060736111.

[34] S. Song, S. Munk-Nielsen, C. Uhrenfeldt, and K. Pedersen, "Power cycling test of a 650 V discrete GaN-on-Si power device with a laminated packaging embedding technology," in *2017 IEEE Energy Conversion Congress and Exposition (ECCE)*, pp. 2540–2545, 2017.

[35] H. Chen, B. Ji, V. Pickert, and W. Cao, "Real-time temperature estimation for power MOSFETs considering thermal aging effects," *IEEE Transactions on Device and Materials Reliability*, vol. 14, pp. 220–228, 2014.

[36] F. Stella, G. Pellegrino, E. Armando, and D. Daprà, "On-line temperature estimation of sic power MOSFET modules through on-state resistance mapping," in *2017 IEEE Energy Conversion Congress and Exposition (ECCE)*, pp. 5907–5914, 2017.

[37] Y. Avenas, L. Dupont, and Z. Khatir, "Temperature measurement of power semiconductor devices by thermo-sensitive electrical parameters – A review," *IEEE Transactions on Power Electronics*, vol. 27, pp. 3081–3092, 2012.

[38] G. S. Inc., *"gs66508p," data sheet.* 2016.

[39] V. Smet, F. Forest, J. J. Huselstein, A. Rashed, and F. Richardeau, "Evaluation of v_{rmce} monitoring as a real-time method to estimate aging of bond wire-IGBT modules stressed by power cycling," *IEEE Transactions on Industrial Electronics*, vol. 60, pp. 2760–2770, 2013.

[40] M. Electric, *Power modules.* Tokyo: Mitsubishi Electric Corporation, 2018.

[41] Y. Zhang, H. Wang, Z. Wang, Y. Yang, and F. Blaabjerg, "Impact of lifetime model selections on the reliability prediction of IGBT modules in modular multilevel converters," in *2017 IEEE Energy Conversion Congress and Exposition (ECCE)*, pp. 4202–4207, 2017.

[42] B. Ji, X. Song, W. Cao, *et al.*, "In situ diagnostics and prognostics of solder fatigue in IGBT modules for electric vehicle drives," *IEEE Transactions on Power Electronics*, vol. 30, pp. 1535–1543, 2015.

[43] W. Kexin, D. Mingxing, X. Linlin, and L. Jian, "Study of bonding wire failure effects on external measurable signals of IGBT module," *IEEE Transactions on Device and Materials Reliability*, vol. 14, pp. 83–89, 2014.

[44] H. Oh, B. Han, P. McCluskey, C. Han, and B. D. Youn, "Physics-of-failure, condition monitoring, and prognostics of insulated gate bipolar transistor modules: A review," *IEEE Transactions on Power Electronics*, vol. 30, pp. 2413–2426, 2015.

[45] G. J. Vachtsevanos, F. Lewis, A. Hess, and B. Wu, *Intelligent fault diagnosis and prognosis for engineering systems*. Hoboken, N.J: Wiley Online Library, 2006.

[46] M. Held, P. Jacob, G. Nicoletti, P. Scacco, and M. H. Poech, "Fast power cycling test of IGBT modules in traction application," in *Proceedings of Second International Conference on Power Electronics and Drive Systems*, vol. 1, pp. 425–430, 1997.

[47] A. Ginart, I. N. Ali, I. Barlas, P. W. Kalgren, M. J. Roemer, and K. Goebel, "Observer based junction temperature estimator in thermoelectrical aging," in *2010 IEEE Aerospace Conference*, pp. 1–9, 2010. Permission for reproduction license number 4365081013314.

Chapter 3
Early diagnosis in DC-link capacitors: electrolytic and films

Chetan Kulkarni[1], José Celaya[2], and Kai Goebel[3]

3.1 Introduction

The development of prognostics methodologies for electronics has become more important as more electrical systems are being used in a wide range of application fields including aeronautics, maritime, and automotive. Implementation of prognostics methodologies for electronics presents challenges due to the great variety of components being used in a system in addition to a general lack of understanding of how electronic components fail. Traditional reliability techniques in electronics tend to focus on understanding the time to failure for a batch of components of the same type making statistical estimates from accumulated data aggregates by resulting in, for example, the mean time between failure estimate. There is now a drive to understand, in more depth, how faults progress as a function of loading and environmental conditions. Just until recently, it was believed that there were no precursor to failure indications for electronics components and systems. That is now understood to be incorrect, since electronics systems, much like mechanical systems, undergo a measurable degradation process from which one can in principle derive features that can be used to provide early warnings to failure. By modeling the degradation process, one can potentially predict the remaining useful life (RUL) as a function of future use and environmental conditions.

Almost all critical and complex systems today consist of embedded electronic modules for monitoring, control, and enhanced functionality. However, it has been found that these modules are often also the first elements in the system to fail, thus reducing overall system reliability [1,2]. Studying and analyzing the performance degradation of such critical electronic systems enhances operational safety, reliability, assured system performance, avoids catastrophic failures, and reduces maintenance costs over the life of the system [3,4]. In addition, an understanding of the deterioration mechanisms often attributed to adverse storage and operating conditions helps to

[1]SGT, Inc., NASA Ames Research Center, United States
[2]Schlumberger, Software Technology Innovation Center, United States
[3]NASA Ames Research Center, United States

systematically track the changes in system behavior and performance, develop the capability to anticipate failures, and predict the RUL of the electronic systems long before the components actually fail.

While electrolytic capacitors suffer from low reliability, they have high criticality in advanced electrical and electronic systems. In addition to this, degradation at component level could lead to cascading faults at subsystem and system level. This makes them good candidates for health-management implementation. In order to mitigate the effects of capacitor failures in critical safety systems, we introduce here a condition-based prognostics methodology. This methodology provides ability to identify and monitor degradation effects and to estimate RUL of the components periodically. This method will further allow for prognostics-based decision-making for optimal maintenance scheduling of the system or for implementation of mitigation strategies in the case of contingencies during operation. In the next section, we discuss in brief earlier work on prognostics for capacitors at both component and system level.

The term "diagnostics" relates to the ability to detect and isolate faults or failures in a system. "Prognostics," on the other hand, is the process of predicting the health condition and RUL of components based on current state, and current and future operating conditions. Prognostics and health management (PHM) methods combine sensing, data collection, interpretation of environmental, operational, and performance-related parameters to compute current system state, infer component health, and anticipate damage propagation due to degradation of components. This makes them one of the key enablers for predicting system-level performance and making decisions on when to perform condition-based maintenance, thus providing means for lowering life cycle costs in critical systems. PHM methodologies can be implemented using a variety of techniques that combine trend analysis and parameter estimation methods to model performance degradation as a function of duration of use and history of operating conditions. Systematically developed PHM methods advance the goal of intelligent decision-making that in turn allows making better strategic and business-based decisions.

The prognostics methodology shown here is based on a Bayesian tracking framework and dynamic degradation models. A validation methodology is presented to assess the validity of the method using available run-to-failure data. The approach periodically estimates RUL.

Degradation in electronic components is harder to detect and analyze compared to mechanical and structural material, because relevant sensor measurements are often times not taken and accurate physics-based models are not readily available for most electronic components [1]. Prognostics failure predictions have to be accurate to support decision-making for operational decisions including repair and replacement. The dynamic degradation models need to be compared with real data that will be used to refine, verify, and validate the accurate time-varying functions that define how parameters change during system operation. This also addresses the issue that component parameters can vary from one manufacturer to another, and between batches manufactured by the same manufacturer [5]. This motivates the need for developing general, first principles approaches for modeling degradation processes and predicting RUL. Data-driven methods can supplement the general methodology to derive capacitor degradation models for specific types of capacitors.

3.1.1 Research challenges

Physics-based degradation modeling (PBDM) considers the primary mechanical, chemical, electrical, and thermal processes that govern the behavior of a component. It also considers the effect of stressors (e.g., higher than rated temperatures, voltage spikes) on component performance. The methodology establishes the cumulative damage due to various failure mechanisms as a function of the state of the component, the relevant physical processes that govern component behavior, and its operating conditions. A failure mode is defined by the causes for the failure [6]. Failure thresholds can also be defined by the way in which a component, subsystem, or system fails to meet or deliver the intended functionality and behaviors. A component can have multiple failure modes, each of which can be derived from the underlying physics-based degradation model associated with the component [1]. In some cases, a theoretical physics of failure model can be complex and difficult to derive; under such conditions, empirical methods using selective accelerated life testing to derive the failure mode effects [6] are developed and implemented.

Implementation of such approaches to electronics enhances the ability to detect anomalies in the device/component early in its life cycle. This helps in tracking the degradation due to aging or operational use and mitigates the risk of sudden failure for critical applications.

Developing physics-based models which incorporate underline degradation mechanism is the key to generating accurate prediction models and RUL of components. However, a number of important challenges remain before these approaches can be applied widely in real-world monitoring, behavior analysis, performance-driven predictions, and decision-making. The key issues that are currently being addressed in the PHM community include system parameter monitoring, selection of precursors to failures, combining this with parameter monitoring to determine the type of sensors and their placement to ensure tracking of relevant system behavior and operating conditions, *in-situ* data acquisition, feature extraction for refining the degradation models, and behavior prediction from the collected data [7,8]. Challenges remain to develop methodologies for predicting the overall system performance, and also to understand cascading failures effects in subsystems and systems.

3.1.2 Organization

Section 3.2 discusses physics based modeling approach and its implementation to prognostics framework. Section 3.3 and 3.4 provide implemented research methodology and capacitor degradation mechanisms as well as modeling respectively. Section 3.5 discuss the model based prognostic framework. Section 3.6 presents case study of degradation under operational conditions. Section 3.7 discusses the validation results for implemented framework. Section 3.8 concludes the chapter with a discussion, conclusions and directions for future work.

3.2 Modeling for prognostics

This section discusses approaches for PBDM and implementing of the overall prognostics framework using models. The approach to prognostics, model failure mechanisms using a first principles physics-based model of operation, as opposed to traditional approaches that employ data-driven empirical models. These are dynamic models which evolve over time as the component degrades due to usage conditions. PBDM techniques are effective for estimating the impact on lifetime due to specific failure mechanisms [9]. Failure predictions, i.e., the time to complete failure from the current time and also the evolution of degradation through this time interval, depend upon the current health state of the device/system and the conditions under which it is operated. Therefore, to apply PBDM techniques accurately, one must take into account environmental variables and operating conditions along with a device's physical properties.

In safety-critical complex systems, fault initiation and propagation needs to be determined for accurate and effective fault prognosis. The approach is to derive temporal models of fault propagation and failure using first principles physics-based methods. Recent studies in materials research have focused on microstructural characterization of materials and modeling frameworks that robustly and reliably describe anomaly initiation and propagation as a function of time [9]. These models, which address the constitutive material properties down to the molecular level, generally are not suitable for real-time treatment due to their computational complexity. Typically, more abstract models, also called lumped models, are usually derived from the microstructural characterizations [10]. They can be used to estimate and predict degradations in a component almost in real time under a given operating environment, such as temperature cycling, vibration, humidity, and electrical stress.

Typically, under normal operating conditions, where parameters are limited to manufacturer-specified threshold limits, a component does not degrade rapidly. In such cases, an experiment may have to be conducted over long periods of time (e.g., months and even years) to detect and study the underlying failure modes. Accelerated stress tests (ASTs) have been recognized to be valuable in getting much quicker estimates of the reliability and quality of electronics components [9]. ASTs are often carried out under enhanced adverse environmental conditions to accelerate the damage accumulation rate due to physical or environmental phenomena [11]. In traditional accelerated testing techniques, root-cause identification and analysis is not adequately emphasized. In the PBDM approach, it is very important to identify, understand, and model the underlying failure mechanisms to derive general (as opposed to situation-specific) models of the aging phenomena.

The main activities in implementing a PHM methodology include:

- using system structure and physical laws to derive a first principles operational model of the device/system;

- identifying failure modes based on stressors (electrical, thermal, mechanical, and chemical), which then helps to establish the underlying root cause and derive temporal models of component degradation [12];
- deriving relevant degradation/failure mechanism models;
- modifying and enhancing the physics-based models using experimental data and stochastic modeling methods; and
- using the derived physics-based models and prognostic algorithms for predicting the RUL of the system.

3.3 Research methodology

Prognostics deals with determining the health state of components and projecting this state into the future to make end of life (EOL) and RUL estimations. The approach is essential for improving system safety, reliability, and availability. Model-based prognostics approaches perform these tasks employing first principles physics models that capture knowledge about the system, its components, and their degradation mechanisms [13,14]. Faults and degradations appear as parameter value changes in the model, and this provides the mechanism for tracking system behavior under degraded conditions [15,16].

We implement the prognostics modeling process illustrated in Figure 3.1. The core work focuses on developing physics-based degradation models of components,

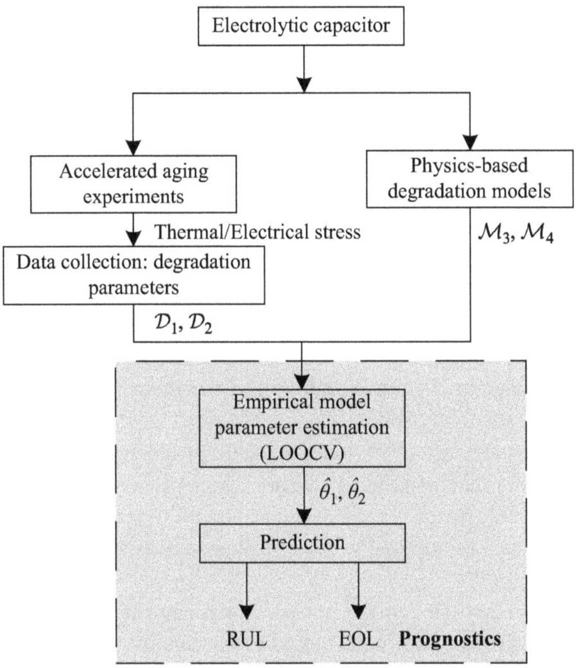

Figure 3.1 Research methodology

i.e., electrolytic capacitors that include descriptions of how fault parameters evolve in time, governed by their operating conditions. As described in Figure 3.1, we combine experimental studies on the device and the electrical and mechanical configuration information with physics-based modeling of behavior phenomena. Identifying the failure precursors and developing accurate models of degradation/failure from an understanding of the physics-based model is an important challenge that we address in this research work. Early detection and analysis may lead to more accurate estimation of parameter changes and, therefore, better prediction and EOL estimates of the capacitor. Unknown and time-varying parameters in the degradation model are estimated online. Static parameters for the model are estimated using the leave-one-out cross-validation algorithm. The derived state space models are then implemented in a Bayesian framework for prognostics.

Before discussing the modeling methodology implemented, we first discuss in the next section about the structure and type of electrolytic capacitor devices used in this work and degradation mechanisms observed under stress conditions. The study of degradation mechanisms helps in developing and refining the physics-based models.

3.4 Degradation in electrolytic capacitors

The health of electrolytic capacitors is strongly affected by their operating conditions that include parameters, such as voltage, current, frequency, and ambient temperatures among others. Degradation of capacitors used in DC–DC power converters and signal filters increases the resistive component and decreases the capacitive component of the impedance in the AC path, which results in an increase in the ripple voltage magnitude that are super imposed on the desired DC voltage. Continued degradation of the capacitor leads the converter output voltage to drop below its specified output value, and this can also affect the health and performance of downstream components. In some cases, this may cause damage to the converter and other downstream components resulting in cascading failures in the systems. We first discuss the internal structure of an electrolytic capacitor and link the different components which lead to degradation in the device.

Figure 3.2 shows a schematic of the internal components of an capacitor: cathode aluminum foil, electrolytic paper, electrolyte, and an aluminum oxide layer on the anode foil surface, which acts as the dielectric. The electrolyte paper strip is impregnated with liquid electrolyte. When in contact with the electrolyte, the oxide layer provides an excellent forward direction insulation property [17]. A magnified effective surface area is attained by etching the foil, which attains a high capacitance value in a small volume [18]. Since the oxide layer has rectifying properties, a capacitor has polarity. If both the anode and cathode foils are coated with an oxide layer, the capacitors are bipolar [19].

Figure 3.3 illustrates the unfolded separator paper into which the electrolyte is embedded. The paper height, h_c, is the height of the capacitor capsule, and l_c indicates the length of the combined roll of paper separator, anode, and cathode. The paper separator is sandwiched between the anode and the cathode. Each has a thickness

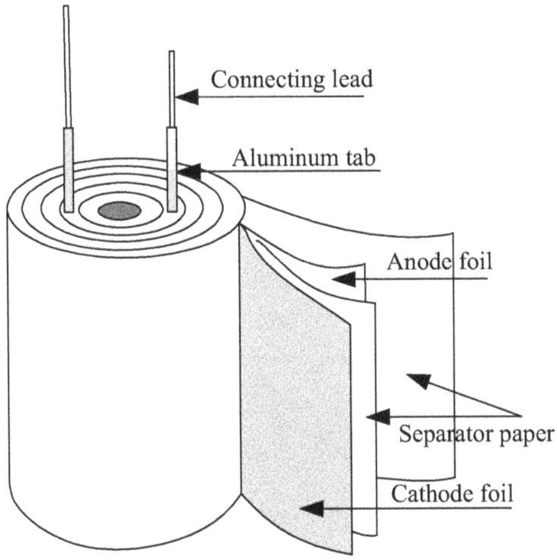

Figure 3.2 Physical model of electrolytic capacitor

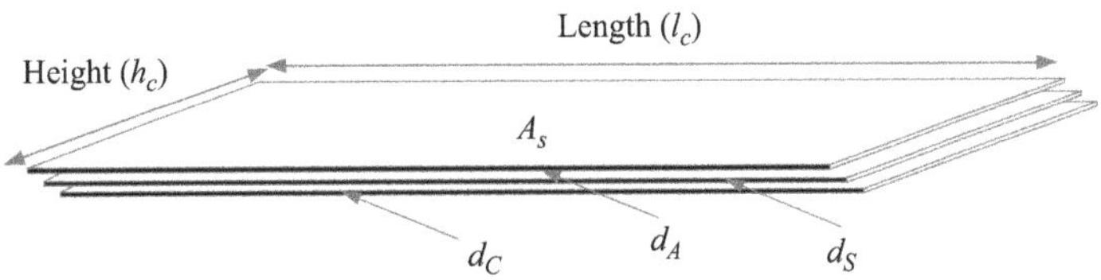

Figure 3.3 Capacitor open structure

d_S, d_A, and d_C, respectively, and can be approximated as $d_S \approx d_A \approx d_C$ for this specific type of capacitor since we are working with low-voltage devices.

Figure 3.4 presents a detailed view of the cross section of the electrolytic capacitor structure. The etched anode and cathode plates are anodized by coating them with a thin aluminum oxide layer on the surface of the foil. This layer of aluminum oxide acts as the dielectric (insulator) and serves to block the flow of direct current between anode and cathode foil surface [18]. When a DC voltage is applied to a discharged capacitor, there is an initial surge of current observed during charging. As the voltage across the terminals of the capacitor increases due to capacitor charging, the current flow drops exponentially and trends to zero as the capacitor terminal voltage equals the applied voltage [20].

3.4.1 Degradation mechanisms

Failures can be one of two types: (1) catastrophic failures, where there is complete loss of functionality, primarily short and open circuits, and (2) degradation failures, where there is gradual deterioration of capacitor parameters due to accumulated internal damage. These degradation phenomena cause a drift in the two main electrical

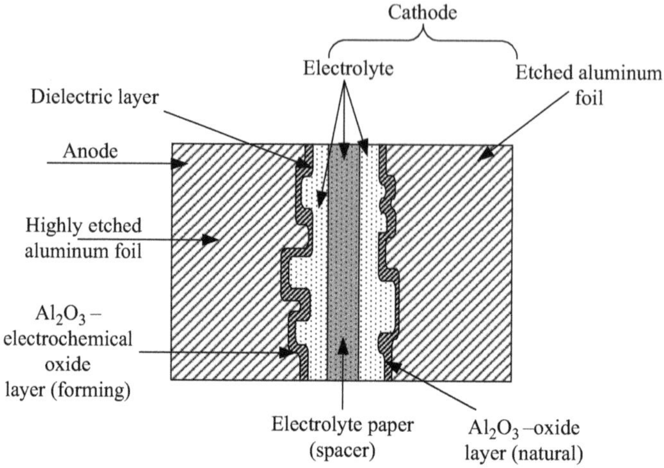

Figure 3.4 Cross-sectional details of internal capacitor structure

parameters of the capacitor: (1) the equivalent series resistance (*ESR*) and (2) the capacitance (*C*). The health of a capacitor is typically measured by the values of these two parameters. Our approach tracks the behavior of the device till complete failure, which manifests as short or open circuits, or opening of the vents. The rest of the section describes the degradation mechanisms in terms of the internal structure of the capacitors and their impact on the electrical parameters of the capacitors.

A study of the literature determines that the primary reason for degradation in electrolytic capacitors is due to deterioration of electrolyte quality, decrease in electrolyte volume due to evaporation, and weakening of the oxide layer, due to increased temperatures in the capacitor core [21,22]. The fishbone diagram in Figure 3.5 summarizes the comprehensive set of failure modes that occur in electrolytic capacitors [20,23–27]. This diagram identifies the relationship between root causes and failure modes observed in electrolytic capacitors. The major degradation phenomena, represented by the fish bones, are as follows:

- electrolyte evaporation,
- degradation of anode foil,
- degradation of cathode foil, and
- degradation/aging of oxide film.

Elaboration of the main causes, represented as minor bones linking to the degradation/failure causes, are as follows:

- prolonged use,
- increase in internal temperature,
- over voltage stress, and
- charging/discharging cycles.

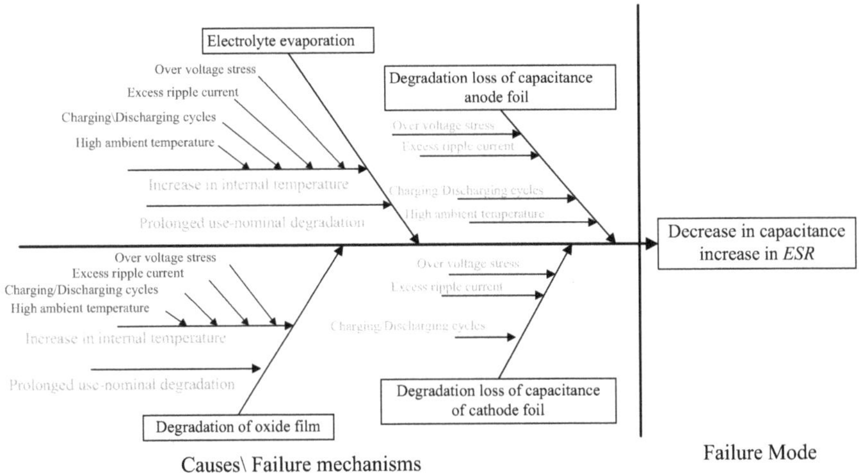

Figure 3.5 Fishbone diagram of failure mechanisms in aluminum electrolytic capacitor

These are then linked to the finer bones, which explain the root causes for the different failure mechanisms in the capacitor. These root causes can occur individually or in groups depending upon the conditions of operation.

Our primary focus in this work is to study and model degradation under thermal overstress conditions. Exposure of the capacitors to temperatures that are greater than the rated operating temperatures (T_{rated}) results in accelerated aging of the devices [15,28]. Higher ambient storage temperature accelerates the rate of electrolyte evaporation leading to degradation of the capacitance parameter [29] prominently than the *ESR*. Hence, we specifically study degradation in capacitance parameter. Referring to the fishbone diagram in Figure 3.5, thermal stress conditions have a prominent effect on degradation in capacitors due to increase in internal temperature. Models are derived based on the root causes leading to degradation which are studied by conducting experiments and underlying physics.

3.4.2 Capacitor degradation models

In the last section, we discussed degradation mechanisms and their link to the physical structure of electrolytic capacitors. Before we take the next step to deriving the physics-based models of the degradation phenomena, we represent the degraded capacitor model as an equivalent lumped parameter electrical circuit. In this section, we first discuss the derived equivalent electrical circuit model of an electrolytic capacitor. We then discuss the link between the equivalent circuit parameters to the physical variables of the capacitor, especially those that are impacted by the degradation process.

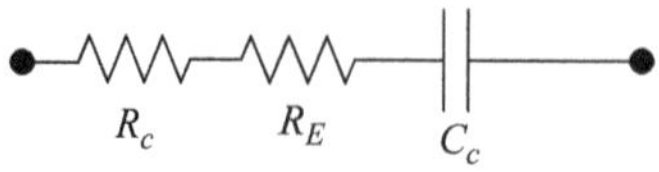

Figure 3.6 Simplified lumped parameter electrical model ($\mathcal{M}_1$)

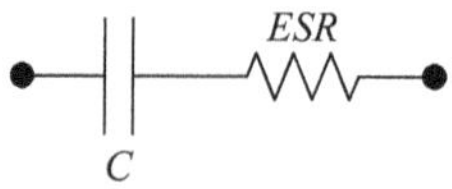

Figure 3.7 Lumped parameter electrical model ($\mathcal{M}_2$)

As explained from lumped parameter model ($\mathcal{M}_1$) in Figure 3.6, *ESR* is approximated as the sum of two component resistances

$$ESR = R_c + R_E \tag{3.1}$$

R_c, as discussed, is the combined series resistance of the lugs, cathode and anode foils, respectively, which remains almost constant since it is a combination of the resistances attributed to mechanical aspects of the capacitor that do not change much over the life of the capacitor. However, electrolyte resistance, R_E, is orders of magnitude higher than R_c [30]; therefore, changes in *ESR* are primarily influenced by changes in R_E. Reference [31] discusses that R_c remains relatively constant and R_E increases as the capacitor degrades, which can be attributed to the evaporation (i.e., decrease) of the electrolyte. R_c and R_E are combined to define the overall degradation resistance, *ESR*, as summarized in Figure 3.7.

3.4.3 Physics-based models for C and ESR

The fishbone diagram in Figure 3.5 discussed a fairly complete set of underlying phenomena that cause capacitor degradation and eventual breakdown. Given our approximate internal lumped parameter electrical circuit model of the capacitor, we have determined that the capacitance value, *C*, and the resistance value, *ESR*, are the primary determinants of capacitor health. We implement a physics modeling approach that links degradation of *C* and *ESR* parameters of the capacitor to physical phenomena associated with electrolyte evaporation and oxide layer deterioration.

3.4.3.1 Capacitance model

The input impedance of the capacitor network is defined in terms of the total lumped series and parallel impedance of the simplified circuit model [32], shown in Figure 3.7. Capacitance represents the ability of a body to store electrical charge. It can be calculated from knowledge of the configuration and the geometry of the conductors and dielectric properties of the insulator between the conductors. Based on the cylindrical geometry, the total lumped capacitance for a foil type electrolytic capacitor [32] is given by

$$C = \frac{2\epsilon_R \epsilon_0 A_s}{d_C} \tag{3.2}$$

where ϵ_R is the relative dielectric constant, ϵ_0 is the permittivity of free space, d_C is the oxide layer thickness, and A_s is the effective oxide surface area.

We hypothesize that the capacitor degrades as the effective oxide area, A_s, decreases due to evaporation of the electrolyte. The rest of the parameters in (3.2) remain relatively unchanged and assumed to be constant over the capacitor lifetime since constant thermal stress is applied and no other stress conditions are considered.

3.4.3.2 *ESR* model

The *ESR* dissipates some of the stored energy in the capacitor. An ideal capacitor would offer no resistance to the flow of current at its leads. Based on the first principles, R_E is computed as a function of the effective oxide surface area, A_s, [32] as

$$R_E = \frac{1}{2}\left(\frac{\rho_E d_C P_E}{A_s}\right) \tag{3.3}$$

where ρ_E is the electrolyte resistivity, P_E is the correlation factor related to electrolyte spacer porosity and average electrolyte pathway.

Since $R_E \approx ESR$, increase in R_E due to evaporation of electrolyte, which decreases A_s, will lead to increase in *ESR*, given by

$$ESR = \frac{1}{2}\left(\frac{\rho_E d_C P_E}{A_s}\right) \tag{3.4}$$

Decrease in the electrolyte volume due to thermal overstress also reduces the average electrolyte path length, which increases P_E. Under normal circumstances when the capacitors are stored at room temperature or below rated temperatures, no damage or decrease in the life expectancy is observed. But in cases where the capacitors are stored under high temperature conditions, the capacitors show irreversible degradation over time. In the next section, we compute the initial electrolyte volume, V_{e0}, using capacitor geometry and link the oxide surface area, A_s, parameter to change in electrolyte volume under thermal stress conditions.

3.4.3.3 Computing electrolyte volume from capacitor geometry

In Section 3.4.1, we discussed that under high thermal stress, the capacitor electrolyte evaporates at faster rates due to the higher internal temperature. The exact amount of electrolyte in the capacitor is unknown, but we can compute the initial volume of the electrolyte from the dimensions and type of the capacitor.

The approximate pristine condition volume of electrolyte, V_{e0}, based on geometry of the capacitor is expressed in terms of following equation:

$$V_{e0} = \pi r_c^2 h_c - A_s(d_A + d_C) \tag{3.5}$$

Note that, (3.5) gives us the initial electrolyte volume, V_{e0}, for a specific type of capacitor based on its geometry. The equations need to be updated for a different geometry of capacitors depending upon the number of oxide layers, thickness of oxide layer, type of separator paper used, etc. Equation (3.5) is derived in [30].

3.4.3.4 Electrolyte evaporation model

The evaporation rate, j_{eo}, is directly linked with change in the core temperature. The relationship between electrolyte evaporation rate and temperature for a particular electrolyte can be found in [31]. The specific capacitor electrolyte under study in this work consists prominently of ethyl glycol as solvent along with trialkylamines and carboxylic acid. The calculations for evaporation rate are done based on the evaporation of ethyl glycol since it forms the major percentage of the total electrolyte volume. These calculations are to be updated if a different type of electrolyte is used in the capacitor.

When the surrounding temperature of the capacitor capsule is high, heat transfer from the capsule to the capacitor core causes the internal temperature of the capacitor to increase, and as a result, the electrolyte evaporates at a faster rate. Under prolonged high temperate exposure the electrolyte evaporation rate accelerates, which in turn decreases, A_s directly affecting change in C [29,31,33], which is given by

$$V_e(t) = V_{e0} - (A_s j_{eo} w_e) \times t \tag{3.6}$$

where V_e is the dispersion volume at time t, V_{e0} is the initial electrolyte volume, w_e is the volume of ethyl glycol molecules, j_{eo} is the evaporation rate (h^{-1} cm^{-2}), and t is the aging time in hours.

From (3.6), we derive

$$A_s = \frac{V_{e0} - V_e(t)}{j_{eo}\, w_e\, t} \tag{3.7}$$

3.4.4 Time-dependent degradation models

Using the physics-based model of electrolyte evaporation, we derive the time-dependent degradation models for capacitance (C) and *ESR*. Using (3.2) and (3.7), we derive the *capacitance degradation model* as

$$\mathcal{D}_1 : C(t) = \left(\frac{2\epsilon_R \epsilon_0}{d_C}\right)\left(\frac{V_{e0} - V_e(t)}{j_{eo}\, t\, w_e}\right), \tag{3.8}$$

where V_{e0} is computed from (3.5), and the other parameters have been defined specifically for each capacitor depending upon its structure and manufacturing specifications. Evaporation rate, j_{eo}, is a temperature-dependent parameter, while electrolyte volume changes with time and evaporation rate. Similarly increase in *ESR* directly leads to increase in electrolyte resistance (R_E), (3.3) and (3.6) as

$$\mathcal{D}_2 : ESR(t) = \frac{1}{2}(\rho_E\, d_C\, P_E)\left(\frac{j_{eo}\, t\, w_e}{V_{e0} - V_e(t)}\right). \tag{3.9}$$

In model $\mathcal{D}_2$, the parameters which are temperature dependent are the rate of evaporation j_{eo} and the correlation factor, P_E related to electrolyte spacer porosity and average electrolyte pathway. As the electrolyte evaporates due to high temperature, the correlation factor P_E will increase as the average pathway of the liquid decreases. Electrolyte evaporation under thermal stress storage conditions can be attributed to increase in the temperature of the capacitor capsule. Heat travels from the surface of the body to the core, and this phenomenon is described by a thermal model in [34].

3.5 Model-based prognostics framework

A model-based prognostics algorithm based on Kalman filter and a derived physics-based degradation model for capacitors is presented. The framework implementation requires tracking changes both in C and ESR based on derived models. The prognostic methodology implemented as shown in Figure 3.8 consists two parts. The first part consists of a tracking framework where the algorithm tracks the state based on the current conditions of the device and inputs. In the second part, the algorithm is able to predict RUL of the capacitor till the health threshold of the device is crossed. This prediction is based on the accelerated degradation data from the experiments described in previous sections.

1. State tracking: The capacitance loss C_l is defined as the state variable to be estimated, and the degradation model is expressed as a discrete time dynamic model in order to estimate capacitance loss as new measurements become available. Direct measurements of the capacitance are assumed for the filter.
2. Health state forecasting: It is necessary to forecast the state variable once there are no more measurements available at time or RUL prediction t_p. This is done by evaluating the degradation model through time using the state estimate at time t_p as initial value.
3. Remaining life computation: RUL is computed as the time between time of prediction t_p and the time at which the forecasted state crosses the failure threshold value.

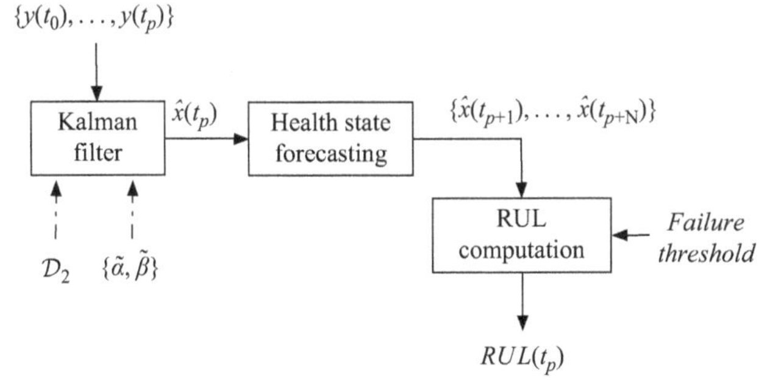

Figure 3.8 Model-based prognostics methodology

This procesł is repeated for different values of t_p through the life of the component under consideration.

3.5.1 Kalman filter for state estimation

A state-space dynamic model is needed for the filtering. The state variable $\mathbf{x}_k$ at time $\mathbf{t}_k$ is defined as capacitance loss $C_l(k)$. Since the system measurements are percentage loss in capacitance as well, the output equation is given by $\mathbf{y}_k = h\mathbf{x}_k$, where the value of h is equal to one. The following system structure is used in the implementation of the filtering and the prediction using the Kalman filter.

$$\begin{aligned} \mathbf{x}_k &= A_k \mathbf{x}_{k-1} + B_k u + \mathbf{v}, \\ \mathbf{y}_k &= h\mathbf{x}_k + \mathbf{w}, \end{aligned} \tag{3.10}$$

where

$$\begin{aligned} A_k &= (1 + \Delta_k), \\ B_k &= -\alpha\beta\Delta_k, \\ h &= 1, \\ u &= 1. \end{aligned} \tag{3.11}$$

The time increment between measurements Δ_k is not constant since measurements were taken at nonuniform sampling rate. This implies that some of the parameters of the model in (3.26) will change through time. Furthermore, v and w are normal random variables with zero mean and Q and R variance, respectively. The description of the Kalman filtering algorithm is omitted from this article. A thorough description of the algorithm can be found in [35], a description of how the algorithm is used for forecasting can be found in [36], and an example of its usage for prognostics can be found in [37].

3.5.2 Future state forecasting

The use of the Kalman filter as an RUL forecasting algorithm requires the evolution of the state without updating the error covariance matrix and the posterior of the state vector. The n step ahead forecasting equation for the Kalman filter is given below. The last update is done at the time of the last measurement t_l.

$$\hat{x}_{l+n} = A^n x_l + \sum_{i=0}^{n-1} A^i B \tag{3.12}$$

The subscripts from parameters A and B are omitted since a constant Δ_t is used in the forecasting mode (one prediction every hour).

3.5.3 Noise models

The model noise variance Q was estimated from the model regression residuals. This variance was used for the model noise in the Kalman filter implementation. The measurement noise variance R is also required in the filter implementation.

This variance was computed from the direct measurements of the capacitance with the electro-impedance spectroscopy (EIS), the observed variance is 4.99×10^{-7}.

3.5.4 Prognostics problem formulation

In prognostics studies, it is important to know when the performance of the device or system is going to lie outside an acceptable region of operation. When the capacitor parameters exceed specified thresholds, we consider the degrading device or system to have failed. For the device/system to be within the bounds of acceptable performance, we express a set of constraints, c_n, $C = \{C_i\}_{i=1}^{Cn}$, where C_i is a function

$$C_i : \mathbb{R}^{n_x} \times \mathbb{R}^{n_\theta} \to \mathbb{B}$$

that maps a given point in the joint state-parameter space,$(x(t), \theta(t))$, to the Boolean domain $\mathbb{B} = [0, 1]$, where $C_i(x(t), \theta(t)) \triangleq 1$ indicates the constraint is satisfied or 0 (failed) if the constrained is not satisfied. Each individual constraint can be combined to form a single constraint output threshold function T_{EOL}, where

$$T_{EOL} : \mathbb{R}^{n_x} \times \mathbb{R}^{n_\theta} \to \mathbb{B}$$

which is defined as

$$T_{EOL}(x(t), \theta(t)) = \begin{cases} 1, & 0 \in \{C_i(x(t), \theta(t))\}_{i=1}^{c} \\ 0, & \text{otherwise.} \end{cases} \tag{3.13}$$

Indicating the healthy state of a device/system T_{EOL} evaluates to 1, if any of the constraints are violated. At some instance in the aging cycle at time, t_p, the system is at $(x(t_P); \theta(t_P))$, and we are interested in predicting the time point t at which this state will evolve to $(x(t); \theta(t))$ such that $T_{EOL}(x(t), \theta(t)) = 1$. Using T_{EOL}, we formally define EOL of a device/system as

$$EOL_{t_p} \triangleq \inf \left\{ t \in \mathbb{R} : t \geq t_p \wedge T_{EOL}(x(t), \theta(t)) = 1 \right\}, \tag{3.14}$$

i.e., EOL is the earliest instance in the aging cycle at which at which T_{EOL} is valid for a healthy system. RUL is expressed using EOL as

$$RUL(t_p) \triangleq EOL(t_p) - t_p.$$

3.5.5 Physics-based modeling framework using unscented Kalman filter

A model-based prognostics methodology based on unscented Kalman filter (UKF) is presented. The prognostic methodology implemented in this work consists two parts. The first part consists of an estimation step where the state is based on the current conditions of the device and inputs. In the second part, the algorithm is able to predict RUL of the capacitor till the health threshold of the device is crossed. In this section, we will discuss the dynamic models for both capacitance and ESR degradation models. We first discuss the theory of UKF, and then each of the degradation models with results will be summarized.

3.5.5.1 Capacitance degradation dynamic model

From (3.8), we have the time-dependent degradation models, $\mathcal{D}_1$ for capacitance (C) given by

$$\mathcal{D}_1 : C(t) = \left(\frac{2\,\epsilon_R\,\epsilon_0\,c_b}{d_C}\right)\left(\frac{V_{e0} - V_e(t)}{j_{eo}\,t\,w_e}\right),$$

The degradation in capacitance is directly proportional to the damage variables V and c_b, respectively. As discussed earlier, increase in the core temperature evaporates the electrolyte, thus decreasing the electrolyte volume leading to a gradual decrease in the capacitance. The resultant decrease in the capacitance can be computed using (3.8). In addition to the decrease in electrolyte volume, the oxide layer also degrades causing a breakdown in the oxide layer, which further leads to degradation in capacitance (C) and *ESR*.

The dynamic capacitor degradation model can be enhanced to include the effects of the oxide layer degradation, which is given by

$$C_k = \left(\frac{2\epsilon_R\epsilon_0 c_b}{d_C}\right)\left(\frac{V_{e(k)}}{d_C}\right) \tag{3.15}$$

From the electrolyte evaporation discussion and (3.6), the first-order discrete approximation for change in electrolyte volume can be expressed as

$$\begin{aligned} \frac{dV_e}{dt} &= -(w_e A_s j_{eo}), \\ V_{e(k+1)} &= V_{e(k)} + \frac{dV_e}{dt}\Delta t, \\ V_{e(k+1)} &= V_{e(k)} - (w_e A_s j_{eo})\Delta t. \end{aligned} \tag{3.16}$$

From (3.15), we have

$$\begin{aligned} V_{e(k)} &= \frac{C_k}{2\epsilon_R\epsilon_0} d_C^2, \\ V_{e(k)} &= (C_k)\eta. \end{aligned} \tag{3.17}$$

where

$$\eta = \frac{d_C^2}{2\epsilon_R\epsilon_0\,c_b}$$

We can express (3.17) as

$$\begin{aligned} C_{k+1}\eta &= C_k\eta + \frac{dV_e}{dt}\Delta t, \\ C_{k+1}\eta &= C_k\eta - (w_e A_s j_{eo})\Delta t, \text{ hence} \\ C_{k+1} &= C_k - \frac{(w_e A_s j_{eo})}{\eta}\Delta t. \end{aligned} \tag{3.18}$$

In model $\mathcal{D}_1$, the parameters which are temperature dependent are the rate of evaporation j_{eo} and oxide breakdown factor, c_b. Electrolyte evaporation and oxide breakdown under electrical stress is described by a thermal and oxide breakdown model [30]. Rate of evaporation, j_{eo}, and resistance dependence oxide breakdown factor, c_b, are the two parameters in the model which are estimated online based on the state variable value, C.

Though the applied stress conditions are similar, there may be variation in each capacitor, and hence each capacitor may have slightly different evaporation rate. Hence the evaporation rate parameter, j_{eo}, is estimated based on the first two reading in the experimental data for each capacitor when predictions are done. The exact formulation of the parameter is incorporated as modeling error/process noise in the form of a stochastic process. Since there could be a variation in the temperature, the corresponding evaporation rate value estimated had a mean and variance of the variation values.

While the capacitance dependence breakdown factor, c_b, is a state parameter and estimated at each time step and the updated value is used in estimating the state, which is related to the electrolyte evaporation and is expressed in the state space form as

$$
\begin{aligned}
&c_b = -e^{\xi}, \\
&\text{where} \\
&\xi = (V_{e0} - V_{e(t)}), \\
&c_{b(k+1)} - c_{b(k)} = -\xi e^{\xi}.
\end{aligned}
\tag{3.19}
$$

The complete discrete time dynamic model for capacitance degradation can be summarized as

$$
\begin{aligned}
\mathcal{D}_4 : C_{k+1} &= C_k - \frac{(2\epsilon_R \epsilon_0 w_e A_s j_{eo} c_b)}{d_C^2}\Delta t, \\
c_{b(k+1)} &= c_{b(k)} - \xi e^{\xi}.
\end{aligned}
\tag{3.20}
$$

Model $\mathcal{D}_4$ is implemented in a Bayesian estimation framework. In this work, we implement the dynamic tracking function as a UKF since the degradation in capacitance (state) due to decrease in electrolyte is considered to be a dynamic nonlinear model since evaporation rate (j_{eo}) and capacitance oxide breakdown factor (c_b) are estimated online based on the state values.

The following steps are implemented for model $\mathcal{D}_4$:

1. State estimation: The current measured capacitance (C) is defined as the state variable to be estimated respectively for each of the model, and the degradation model is expressed as a discrete time dynamic model in order to estimate current capacitance due to decrease in electrolyte volume at the next available measurement. Direct measurements of the capacitance is assumed for the filter.
2. Health state forecasting: It is necessary to forecast the state variable once there are no more measurements available at the end of step 1. This is done by evaluating the degradation model through time using the state estimate at time t_p as initial value.
3. RUL computation: RUL is computed as the time between time of prediction t_p and the time at which the forecasted state crosses a certain failure threshold value or EOL.

These steps are repeated for different aging time (t_p) through the life of the capacitor device under test.

There is an uncertainty both in the estimation step as well as in the prediction step. To include uncertainty, presence of following factors with a probability distribution is required to be considered (i) sensor noise, (ii) unknown initial state of the system, (iii) derived system model, and (iv) the process noise. In addition to this even if the system state is known, uncertainty is introduced in the prediction steps as (1) the derived model may not be exactly depicting the degradation process, (2) exact value of the process noise not known, and (3) exact trajectory of the future inputs for the state space may not be known. Due to these additional sources of uncertainty inherent to the prognostics algorithm, the uncertainty in the predicted EOL/RUL will nominally be larger than the true variability in EOL/RUL [38]. In this work, we only consider the model noise and the noise variance as discussed earlier. But for the prediction step, we present the means for all the EOL/RUL predictions in our results since the focus of this work is more on deriving the physics-based degradation, implementing it in a Bayesian framework and for prognostics predictions. Next we discuss the implementation of the Bayesian framework methodology for prognostics [39–41].

3.5.5.2 UKF for capacitance state estimation

The state variable $\mathbf{x}_k$ (capacitance) at time t_k is defined as the current measured capacitance C_k. Since the system measurements are capacitance (C) as well, the output equation is given by $\mathbf{y}_k = H_k\,\mathbf{x}_k$, where H is the identity matrix. The following system structure is implemented for filtering and prediction using a UKF.

$$\begin{aligned} \mathbf{x}_k &= A_k\mathbf{x}_{k-1} + B_k u + v, \\ \mathbf{y}_k &= H_k\mathbf{x}_k + \mathbf{w}. \end{aligned} \tag{3.21}$$

where

$$\begin{aligned} A &= 1, \\ B &= -\frac{(2\epsilon_R\epsilon_0 w_e A_s j_{eo} c_b)}{d_C^2}\Delta t, \\ H &= 1, \\ u &= j_{eo}, c_b. \end{aligned} \tag{3.22}$$

In this application of UKF, the time increment between measurements Δt is not constant since measurements were taken at nonuniform time intervals, i.e., the capacitors were characterized at different time intervals. This implies that some of the parameters of the model in (3.27) will change as time progresses. Furthermore, v and w are normal random variables with zero mean and Q and R variance, respectively. The model noise (process noise) variance Q was estimated from the model regression residuals and was used for the model noise in the Kalman filter implementation. The measurement noise variance, R, was computed from the direct measurements of the capacitance with the EIS equipment, the observed variance is 4.99×10^{-7}. A detailed description of the algorithm implemented in this work can be found in [35], a description of how the algorithm is used for forecasting can be found in [42], and an example of its usage for prognostics can be found in [39,40].

3.5.5.3 *ESR* degradation dynamic model

The dynamic *ESR* degradation model can be represented as:

$$
\begin{aligned}
ESR_k &= \frac{1}{2}\left(\rho_E\, d_C\, P_E\, e_b\right)\left(\frac{d_C}{V_{e(k)}}\right),\\
V_{e(k)} &= \frac{1}{ESR_k}\left(\frac{1}{2}\right)\left(\rho_E\, {d_C}^2\, P_E\, e_b\right),\\
V_{e(k)} &= \frac{1}{ESR_k}(\gamma),
\end{aligned}
\tag{3.23}
$$

where

$$\gamma = \frac{1}{2}\left(\rho_E\, {d_C}^2\, P_E e_b\right)$$

From (3.16) and (3.23), we have

$$
\begin{aligned}
\frac{\gamma}{ESR_{k+1}} &= \frac{\gamma}{ESR_k} + \frac{dV_e}{dt}\Delta t,\\
\frac{\gamma}{ESR_{k+1}} &= \frac{\gamma}{ESR_k} - (w_e A_s j_{eo})\Delta t, \text{ hence}\\
\frac{1}{ESR_{k+1}} &= \frac{1}{ESR_k} - \frac{(w_e A_s j_{eo})}{\gamma}\Delta t.
\end{aligned}
\tag{3.24}
$$

The complete discrete time dynamic model for *ESR* degradation can be summarized as

$$
\begin{aligned}
\mathcal{D}_5 : \frac{1}{ESR_{k+1}} &= \frac{1}{ESR_k} - \frac{2 w_e A_s j_{eo}}{\rho_E\, P_E\, d_C^2\, e_b}\Delta t,\\
e_{b(k+1)} &= e_{b(k)} - \xi.
\end{aligned}
\tag{3.25}
$$

In model $\mathcal{D}_5$, the rate of evaporation, j_{eo}, and the correlation factor, P_E, related to electrolyte spacer porosity and average liquid pathway are the two time-varying parameters. Rate of evaporation, j_{eo}, and resistance dependence oxide breakdown factor, e_b, are the two parameters in the model which are estimated online based on the state variable value, *ESR*.

The following steps are implemented for model $\mathcal{D}_5$:

1. State estimation: The current measured *ESR* is defined as the state variable to be estimated respectively for each of the model and the degradation model is expressed as a discrete time-dynamic model in order to estimate current *ESR* due to decrease in electrolyte volume at the next available measurement. Direct measurements of the *ESR* is assumed for the filter.
2. Health state forecasting: It is necessary to forecast the state variable once there are no more measurements available at the end of step 1. This is done by evaluating the degradation model through time using the state estimate at time t_p as initial value.
3. RUL computation: RUL is computed as the time between time of prediction t_p and the time at which the forecasted state crosses a certain failure threshold value or EOL.

Table 3.1 Parameter values used for degradation models capacitance ($\mathcal{D}_4$) and ESR ($\mathcal{D}_5$) at 12 V EOS

Parameter	Description	Value
ϵ_R	Relative dielectric constant	10
ϵ_R	Permittivity of free space	8.8542×10^{-9} F cm^{-1}
h_c	Height of capacitor capsule	2 cm
d	Diameter of capacitor capsule	0.7 cm
l_C	Length of the spacer paper roll	118 cm
T_{rated}	Rated temperature	85 °C
j_{eo}	Evaporation rate	0.1054×10^{-3} mg h^{-1} cm^2
d_c	Thickness of oxide layer	2.22×10^{-5} cm
w_e	Volume of ethyl glycol molecule	5.66×10^{-9} cm^3
ρ_E	Electrolyte resistivity value	8×10^2 Ω cm
P_E	Correlation factor of capacitor	≈2–3.5

The uncertainty conditions for the model are the same as discussed for the capacitance degradation model.

3.5.5.4 UKF for *ESR* state estimation

The state variable $\mathbf{x}_k$ (*ESR*) at time t_k is defined as the current measured capacitance ESR_k. Since the system measurements are *ESR*-measured values as well, the output equation is given by $\mathbf{y}_k = P_k\,\mathbf{x}_k$, where P is the identity matrix. The following system structure is implemented for filtering and prediction using a UKF.

$$\begin{aligned} \mathbf{x}_k &= A_k \mathbf{x}_{k-1} + B_k u + \mathbf{v}, \\ \mathbf{y}_k &= P_k \mathbf{x}_k + \mathbf{w}, \end{aligned} \tag{3.26}$$

where

$$\begin{aligned} A &= 1, \\ B &= -\frac{2 w_e A_s j_{eo}}{\rho_E\, P_E\, d_C^2\, e_b} \Delta t, \\ P &= 1, \\ u &= j_{eo}, e_b. \end{aligned} \tag{3.27}$$

Table 3.1 summarizes the details of the parameters that are used in the degradation models. Most of the parameters remain constant since similar type of capacitors are used. The initial value of evaporation rate, j_{eo}, is referred to from [31].

3.6 Accelerated aging experiments

Accelerated life test methods are often used in prognostics research as a way to assess the effects of the degradation process through time. It also allows for the identification and study of different failure mechanisms and their relationships with different observable signals and parameters. In the following section, we present the accelerated aging methodology and an analysis of the degradation pattern induced by the aging. The work presented here is based on an accelerated electrical overstress (EOS). In the following subsections, we first present a brief description of the aging setup followed by an analysis of the observed degradation. The precursor to failure is also identified along with the physical processes that contribute to the degradation.

3.6.1 Experimental setup

Since the objective of this experiment is studying the effects of high voltage on degradation of the capacitors, the capacitors were subjected to high-voltage stress through an external supply source using a specially developed hardware. The capacitors are stressed under high-voltage conditions and specially developed hardware. The voltage overstress is applied to the capacitors as a square waveform in order to subject the capacitor to continuous charge and discharge cycles.

At the beginning of the accelerated aging, the capacitors charge and discharge simultaneously; as time progresses and the capacitors degrade, the charge and discharge times vary for each capacitor. Even though all the capacitors under test are subjected to similar operating conditions, their *ESR* and capacitance values change differently. We therefore monitor charging and discharging of each capacitor under test and measure the input and output voltages of the capacitor. Figure 3.9 shows the block diagram for the EOS experiment. Additional details on the accelerated aging system are presented in [43].

For this experiment, six capacitors in a set were considered for the EOS experimental setup. Electrolytic capacitors of 2,200 μF capacitance, with a maximum rated voltage of 10 V, maximum current rating of 1 A, and maximum operating temperature of 105 °C, were used for the study. These were the recommended capacitors by the manufacturer for DC–DC converters. The electrolytic capacitors under test were characterized in detail before the start of the experiment at room temperature.

The measurements were recorded every 8–10 h of the total 180 plus hours of accelerated aging time to capture the rapid degradation phenomenon in the *ESR* and capacitance values. The ambient temperature for the experiment was controlled and kept at 25 °C. During each measurement, the voltage source was shutdown, capacitors were discharged completely, and then the characterization procedure was carried out. This was done for all the six capacitors under test. For further details regarding the aging experiment results and analysis of the measured data, refer to [43,44].

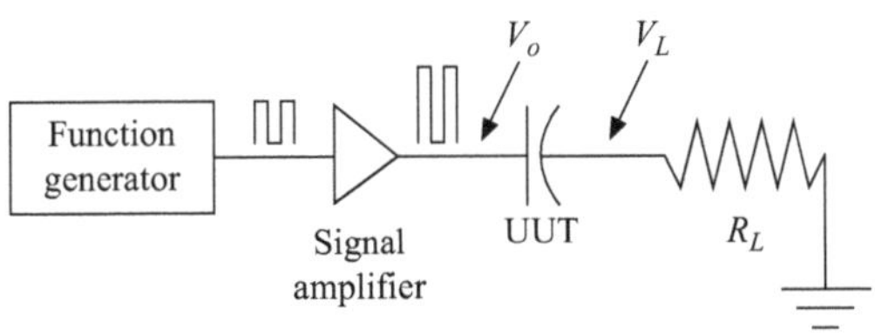

Figure 3.9 Block diagram of the experimental setup

3.6.2 Electrical overstress

In this section, a case study for capacitor degradation under EOS conditions is presented. The section discusses implementation of the developed models and prognostics framework. In EOS operation, the applied voltage on the capacitor exceeds its rated voltage, i.e., $V_{applied} > V_{rated}$. Degradation in electrolytic capacitors under EOS [45,46] can be primarily attributed to the three key phenomena: electrolyte evaporation, oxide breakdown, and increase in internal pressure.

The heat generated internally, when current passes through the capacitor during its charge and discharge cycle, leads to evaporation of the electrolyte [47]. In addition to this, as the capacitor continues to degrade overtime, other phenomena, such as oxide layer breakdown, lead to further deterioration causing the capacitance to decrease significantly. The physics-based degradation models associated with electrolyte evaporation and oxide layer breakdown were discussed in Section V.

In this section, we first discuss the electrical equivalent model and setup for the EOS experiments. Our first approach uses the experimental degradation data to implement an empirical model for capacitance degradation. We derive the empirical capacitance degradation model, validate it using a leave-one-out methodology, and then use the model to make RUL calculations. Next we present the methodology for deriving generalized physics-based degradation models for capacitance and *ESR*, respectively. We validate the dynamic physics-based model by comparing its predictions against additional data generated from the experimental study. We again use the validated model to make RUL and EOL predictions.

3.6.3 EOS experiment

For this experiment, six capacitors of 2,200 μF capacitance, with a maximum rated voltage of 10 V, maximum current rating of 1 A, and maximum operating temperature of 85 °C, were used for the study. The average pristine condition *ESR* value was measured to be 0.121 Ω and average capacitance of 2,029 μF for the set of capacitors under test. The *ESR* and capacitance values were estimated from the capacitor impedance frequency response measured using the EIS instrument. The ambient temperature for the experiment was controlled and kept at 25 °C. During each measurement, the voltage source was shut down, the capacitors were discharged completely, and then the characterization procedure was carried out. Details of the experimental setup are discussed in [43].

Electrochemical impedance spectroscopy measurements are available to characterize the electrical performance of the capacitor. Figure 3.10 shows Nyquist plots of the impedance measurements for Cap #1 at pristine condition, and after accelerated EOS aging at intervals of 71, 161, and 194 h. The degradation can be observed as the Nyquist plot shifts to the right as a function of aging time due to increase in R_E. These measurements are then used to estimate the parameters of the impedance model $\mathcal{M}_1$. The parameter estimation for computing the values of capacitance and *ESR* are performed using the EIS instrument software (EC lab). This is basically an optimization problem using an aggregate of mean squared error as an objective function. The error is aggregated at different frequencies for which measurements are available. The optimization is set up to minimize the objective function by finding

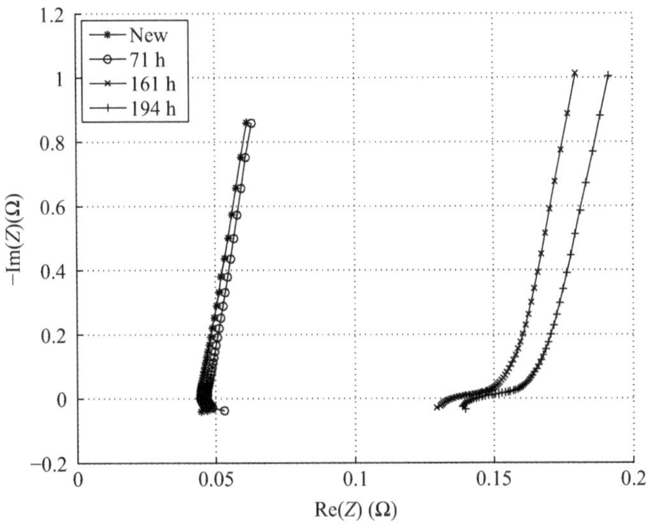

Figure 3.10 Electro-impedance measurements at different aging times

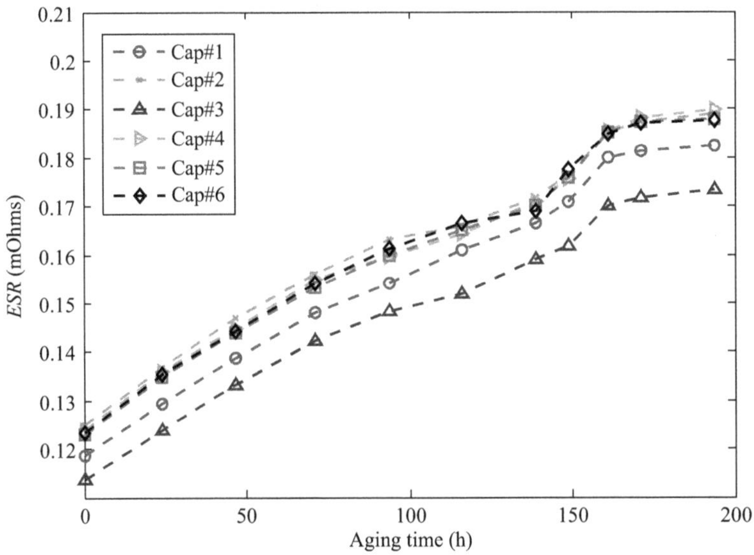

Figure 3.11 Degradation of capacitor performance, ESR *increase as a function of aging time*

optimal values for C_R^* and R_E^*. This parameter estimation is performed every time an EIS measurement is taken resulting in values of C_R and R_E at different points in time through the aging of the components.

Figure 3.11 shows increase in the *ESR* value for all the six capacitors under test over the period of time. Similarly, Figure 3.12 shows the decrease in the value of the

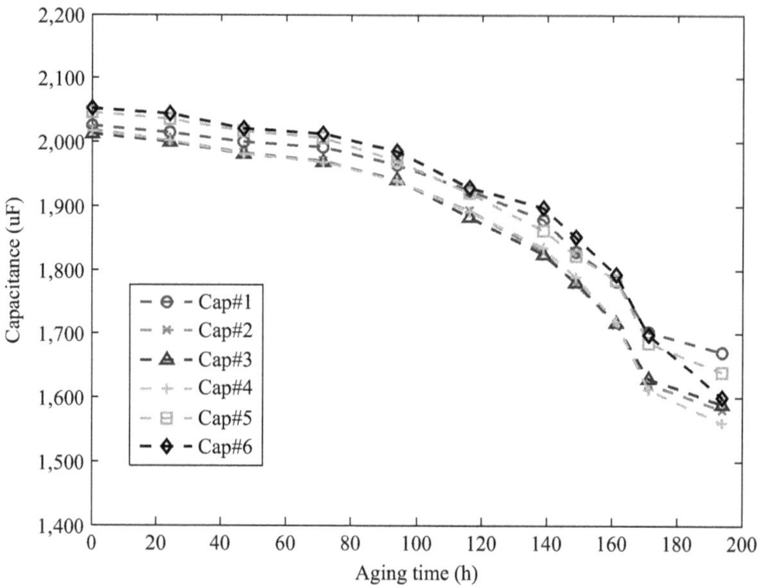

Figure 3.12 Degradation of capacitor performance, capacitance loss as a function of aging time

capacitance as the capacitor degrades over the aging period. From the plots in Figure 3.11, we observe that for the time for which the experiments were conducted, the average *ESR* value increased by 54%–55%, while over the same period of time, the average capacitance decreased by more than 20% (threshold mark for a healthy capacitor). We used the collected data from the experiments to build dynamic degradation models of capacitors.

3.7 Prediction of remaining useful life results and validation tests

This section discusses the RUL predictions and validation results using the Alpha–Lambda prognostics metric for the derived degradation models $\mathcal{D}_4$, capacitance and $\mathcal{D}_5$, *ESR*, respectively. These models were used for making RUL predictions for two different experimental data sets using similar type of capacitors. The discussions in the results and later complete one of our major goals to derive and validate generalized degradation models for electrolytic capacitors.

RUL can be estimated based on the derived physics-based degradation model till the EOL threshold of the device has reached. In the experiments conducted, all the capacitors under test did not reach EOL. The latest characterization reading for degradation in capacitance and *ESR* parameters was considered as the EOL for calculating the relative accuracy.

3.7.1 Results for capacitor degradation model ($\mathcal{D}_4$)

In this section, we discuss the RUL prediction and validation tests for the capacitance degradation model in the 12 V EOS experiment. State estimation and RUL prediction results are discussed for capacitor Cap 2 out of a batch of six available capacitors under test.

Figure 3.13 shows the result of the filter tracking for completed degradation in capacitance computed up to 200 h of aging time. The tolerance for the type of electrolytic capacitor under test is approximately 15%, and hence from the output errors it can be observed that the tracking of the model with respect to the data is acceptable.

3.7.1.1 Remaining useful life

Figure 3.14 presents results from the RUL prediction algorithm at different aging times $t_p = 24, 47, 94, 149, 171$ h, at which the capacitors are characterized and their *ESR* value is calculated. The experiments were run till almost 200 h, and hence the predictions are done till the end of experiments. EOL is defined as the time at which the forecasted capacitance value trajectory crosses the EOL threshold. Therefore, RUL is EOL minus aging times calculated at $t_p = 24, 47, 47, 94, 149, 171$ h.

3.7.1.2 Validation tests

An Alpha–Lambda prognostics performance metric [48,49] is presented in Figure 3.15 for test case of Cap 2. Performance metric identifies whether the algorithm

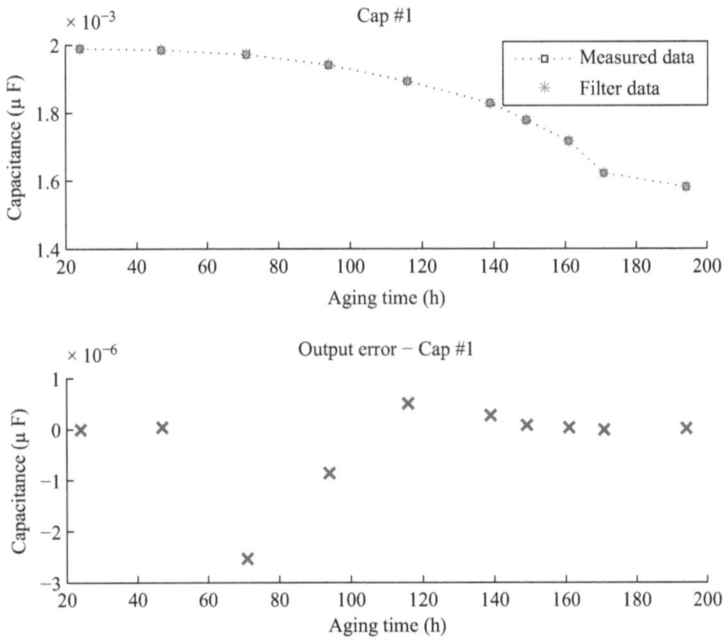

Figure 3.13 Tracking filter output against measurement data for Cap 2

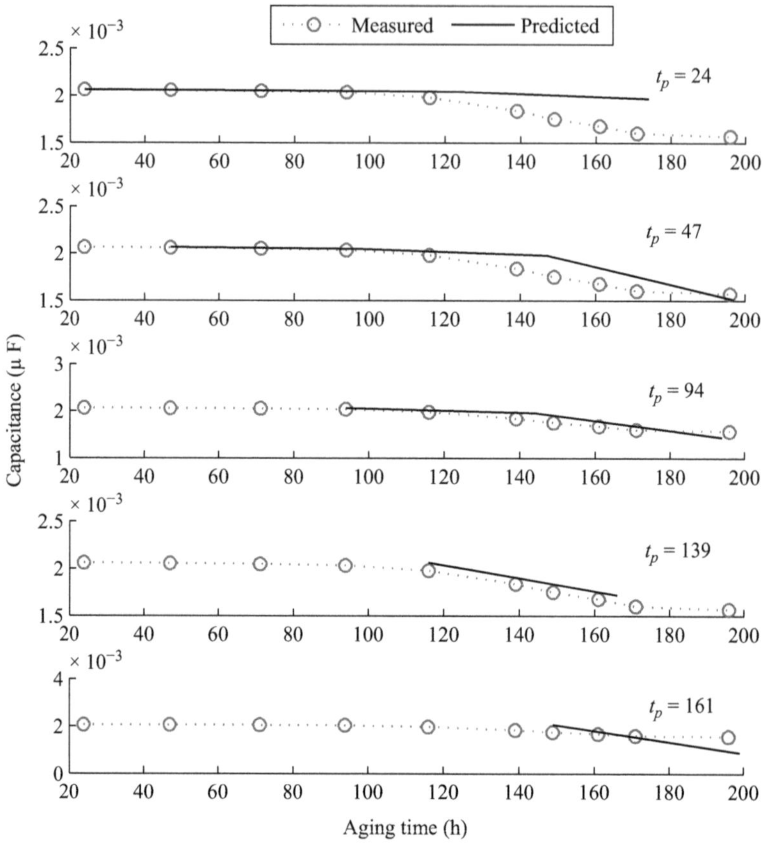

Figure 3.14 Capacitance decrease prediction at different aging time for Cap 2

performs within desired error margins (specified by the parameter α) of the actual RUL at any given time instant (specified by the parameter λ) [48]. The central dashed line represents ground truth and the shaded region corresponds to a 30% ($\alpha = 0.3$) error bound in the RUL prediction. From the metric plot in Figure 3.15, we note that the relative accuracy remains high till the end of the RUL prediction period. As mentioned earlier, this can be attributed to the accuracy of the model, and the state-based tracking approach, where the state is updated as new data points become available.

3.7.1.3 Discussion

When comparing the results generated with the empirical model discussed by Celaya et al. [50] and the developed physics-based degradation model, it is noted that the accuracy of the two models is closely comparable initially, but the empirical model results degrade toward later stages of capacitor life. This is primarily because a simple linear first-order model with static parameter values is employed for tracking the capacitor degradation.

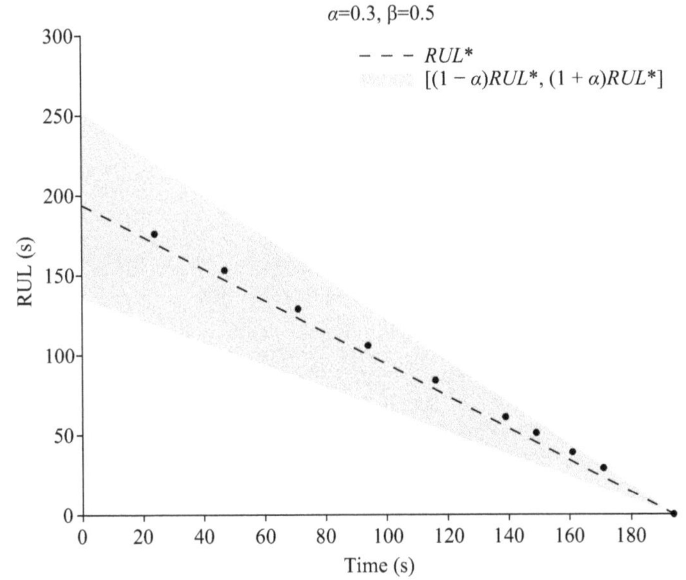

Figure 3.15 Performance based on α–λ metric for Cap 2

Table 3.2 Validation results based on RA for capacitance degradation model $\mathcal{D}_4$

Aging time	C1	C2	C3	C4	C5	C6	$\overline{RA}_a$
24	98.85	98.27	98.85	97.67	98.27	96.47	98.06
47	98.68	98.00	98.68	97.32	98.00	95.92	97.76
71	98.43	97.62	98.43	96.80	97.62	95.12	97.34
94	98.08	97.09	98.08	96.08	97.09	94.00	96.73
116	97.56	96.30	97.56	95.00	96.30	92.31	95.84
139	96.61	94.83	96.61	92.98	94.83	89.09	94.16
149	95.92	93.75	95.92	91.49	93.75	86.67	92.92
161	94.59	91.67	94.59	88.57	91.67	81.82	90.49
171	92.59	88.46	92.59	84.00	88.46	73.91	86.67
$\overline{RA}_b$	96.81	95.11	96.81	93.32	95.11	89.48	

$\overline{RA}_a$ is the mean relative accuracy of all capacitors at each prediction time (t_p).
$\overline{RA}_b$ is the mean relative accuracy of each capacitor at all prediction times.

Physics-based model more accurately represents the actual degradation phenomena, and the use of the UKF allows for more accurate tracking, especially when the behaviors are nonlinear in the later stages of the aging cycle as can be seen in the results of Table 3.2. The tuning and estimation of parameters in the empirical model using Kalman Filter is relatively simpler and also computationally less expensive. On

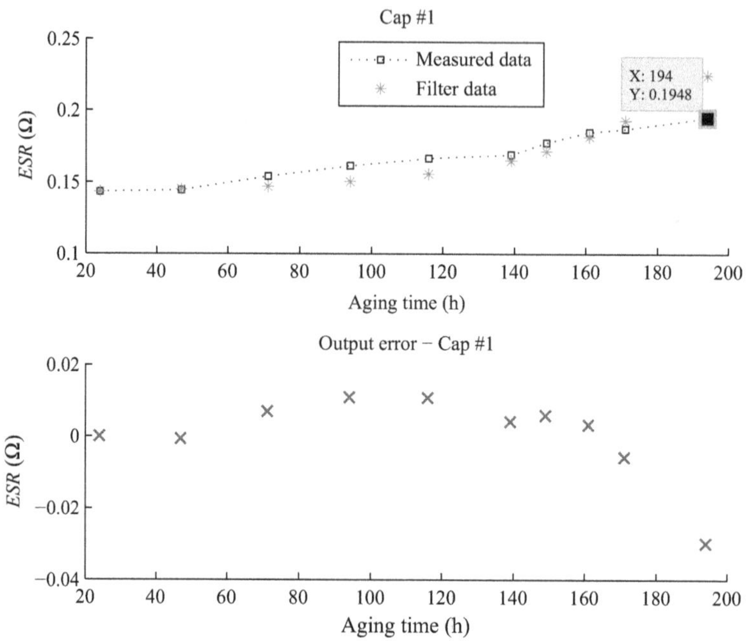

Figure 3.16 Tracking filter output against measurement data for Cap 2

the other hand, in UKF, the parameters are tuned online as updated data is received and hence computationally expensive.

Overall, both the models have their own pros and cons, the comparison of the two models and prediction results can help us decide which models to implement based on the requirements. If the model needs to be simpler and computationally less expensive, then the empirical model can be implemented where the model prediction accuracy will be compromised, while if accuracy is more important, then a complex model and computationally expensive methodology can be implemented. We next discuss the physics-based *ESR* degradation model and the results for the same data set for the *ESR* measurements.

3.7.2 Results for ESR *degradation model* ($\mathcal{D}_5$)

State estimation and RUL prediction results are discussed for capacitor Cap 2. Figure 3.16 shows the result of the filter tracking for completed degradation in capacitance up to 200 h of aging time. As can be observed from the residuals, the tracking of the model with respect to the data is acceptable.

3.7.2.1 Remaining useful life

Figure 3.17 presents results from the RUL prediction algorithm at different aging times $t_p = 24, 47, 94, 149, 171$ h, at which the capacitors are characterized and their

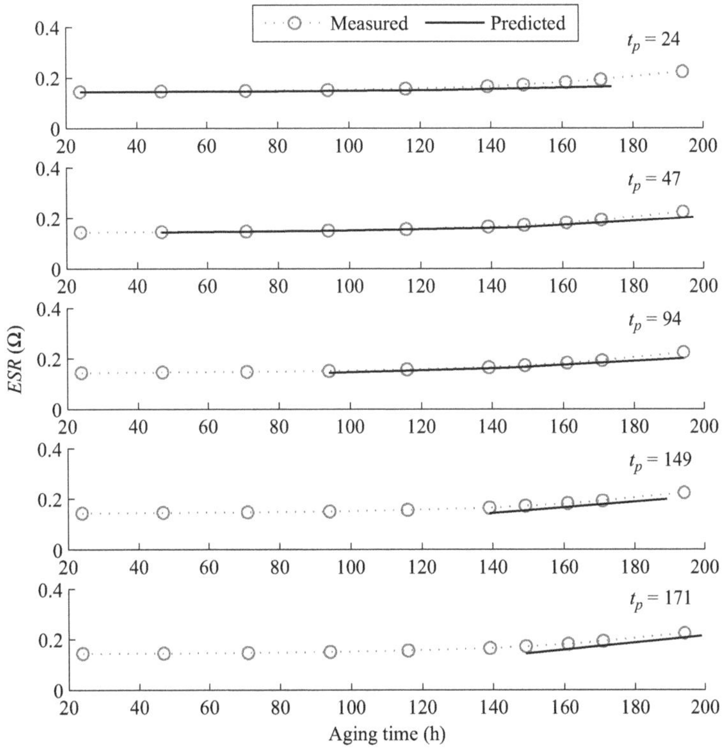

Figure 3.17 ESR *prediction at different aging time for Cap 2 by model* $\mathcal{D}_5$

ESR value is calculated. Model $\mathcal{D}_5$ is implemented for predicting the increase in *ESR*, while the measured values are the *ESR* measurements done at respective aging time intervals. The experiments were run till almost 200 h, and hence the predictions are done till the end of experiments. EOL is defined as the time at which the forecasted capacitance value trajectory crosses the EOL threshold at end of 200 h. Therefore, RUL is EOL minus aging times $t_p = 24, 47, 47, 94, 149, 171$ h.

3.7.2.2 Validation tests

An α–λ prognostics performance metric [48,49] is presented in Figure 3.18 for test case of Cap 2. The central dashed line represents ground truth and the shaded region is corresponding to a 30% ($\alpha = 0.3$) error bound in the RUL prediction. From the α–λ metric plot in Figure 3.18, it can be observed that the relative accuracy is good till the end of the experiment time and the accuracy is good enough under acceptable limits. Table 3.3 presents results for all the units under test. As mentioned earlier, this is due to inclusion of the degradation parameters, and estimating them as the *ESR* increases with degradation.

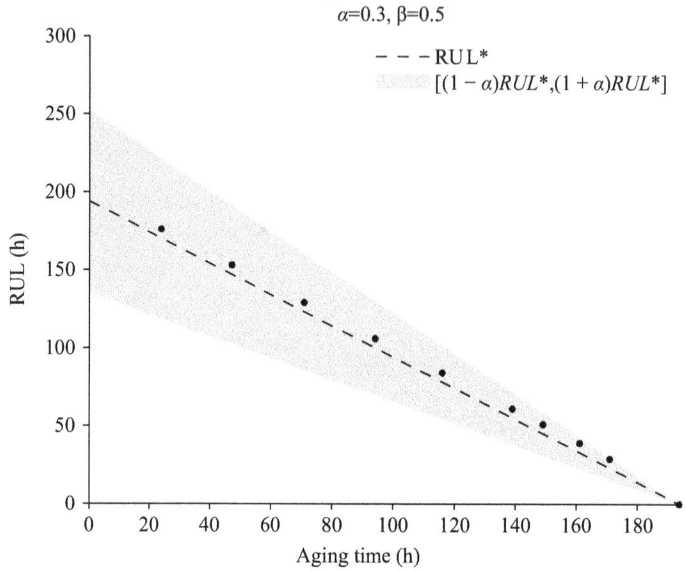

Figure 3.18 Performance based on Alpha–Lambda metric for Cap 2

Table 3.3 Summary of validation results based on RA for ESR *degradation model* $\mathcal{D}_5$*—12 V*

Aging time	**C1**	**C2**	**C3**	**C4**	**C5**	**C6**	$\overline{RA}_a$
24	96.47	96.47	96.47	96.47	96.47	96.47	96.47
47	95.92	95.92	95.92	95.92	95.92	95.92	95.92
71	95.12	95.12	95.12	95.12	95.12	95.12	95.12
94	94.00	94.00	94.00	94.00	94.00	94.00	94.00
116	92.31	92.31	92.31	92.31	92.31	92.31	92.31
139	89.09	89.09	89.09	89.09	89.09	89.09	89.09
149	86.67	86.67	86.67	86.67	86.67	86.67	86.67
161	81.82	81.82	81.82	81.82	81.82	81.82	81.82
171	73.91	73.91	73.91	73.91	73.91	73.91	73.91
$\overline{RA}_b$	89.48	89.48	89.48	89.48	89.48	89.48	

$\overline{RA}_a$ is the mean relative accuracy of all capacitors at each prediction time (t_p).
$\overline{RA}_b$ is the mean relative accuracy of each capacitor at all prediction times.

3.8 Conclusion

This chapter presents an RUL prediction algorithm based on accelerated life test data and derived physics models for electrolytic capacitors. The main elements are (1) development of the first principles based degradation models; (2) the implementation of a Bayesian-based health state tracking and RUL prediction algorithm based on the Kalman filtering framework. One major advancement reported here is the prediction of RUL for capacitors as new measurements become available. The key contribution of this work is the prediction of RUL for capacitors as new measurements become available.

The derived degradation models can be updated and developed at a finer granularity to be implemented for detailed prognostic implementation.

This capability increases the technology readiness level of prognostics applied to electrolytic capacitors. The results presented here are based on accelerated life test data and on the accelerated life timescale. Further research will focus on development of functional mappings that will translate the accelerated life timescale into real usage conditions timescale, where the degradation process dynamics will be slower and subject to several types of stresses. The performance of the proposed exponential-based degradation model is satisfactorily based on the quality of the model fit to the experimental data and the RUL prediction performance as compared to ground truth.

References

[1] Vichare, N. and Pecht, M., "Prognostics and Health Management of Electronics," *IEEE Transactions on Components and Packaging Technologies*, Vol. 29, 2006, pp. 222–229.

[2] FIDES Group, "Reliability Methodology for Electronic Systems," *FIDES Guide Issue A*, 2004. Available: https://www.fides-reliability.org/files/FIDES_guide_2004_Ed_A_EN.pdf.

[3] Celaya, J., Wysocki, P., Vashchenko, V., Saha, S., and Goebel, K., "Accelerated Aging System for Prognostics of Power Semiconductor Devices," *Proceedings of IEEE Autotestcon*, Orlando, FL, 2010, pp. 1–6.

[4] Ferrell, B. L., "JSF Prognostics and Health Management," *IEEE Aerospace Conference*, Aspen, CO, 1999.

[5] Leonard, C. and Pecht, M. G., "Improved Techniques for Cost Effective Electronics," *Proceedings of Reliability Maintainability Symposium*, Orlando, FL, 1991, pp. 174–182.

[6] Nasser, L. and Curtin, M., "Electronics Reliability Prognosis Through Material Modeling and Simulation," *Proceedings of IEEE Aerospace Conference*, Big Sky, MT, 2006.

[7] Celaya, J., Saxena, A., Wysocki, P., Saha, S., and Goebel, K., "Towards Prognostics of Power MOSFETs: Accelerated Aging and Precursors of Failure," *Proceedings of Annual Conference of the PHM Society*, Portland, OR, 2010.

[8] Balaban, E., Saxena, A., Narasimhan, S., Roychoudhury, I., Goebel, K., and Koopmans, M., "Airborne Electro-Mechanical Actuator Test Stand for Development of Prognostic Health Management Systems," *Proceedings of Annual Conference of the PHM Society 2010*, Portland, OR, 2010.

[9] Hansen, P. and McCluskey, P., "Failure Models in Power Device Interconnects," *European Conference on Power Electronics and Applications*, 2007, pp. 1–9.

[10] Kulkarni, C., Biswas, G., Koutsoukos, X., Goebel, K., and Celaya, J., "Diagnostic/Prognostic Experiments for Capacitor Degradation and Health Monitoring in DC-DC converters," *SMASIS*, 2010.

[11] Upadhyayula, K. and Dasgupta, A., "Guidelines for Physics-of-Failure Based Accelerated Stress Testing," *Annual Reliability and Maintainability Symposium*, Anaheim, CA, 1998, pp. 345–357.

[12] Orchard, M., Wu, B., and Vachtsevanos, G., "A Particle Filter Framework for Failure Prognosis," Proceedings of WTC2005 World Tribology Congress III. Washington, DC, 2005.

[13] Daigle, M. and Goebel, K., "A Model-based Prognostics Approach Applied to Pneumatic Valves," *International Journal of the PHM Society*, Vol. 2, No. 8, 2011, pp. 1–16.

[14] Saha, B. and Goebel, K., "Model Adaptation for Prognostics in a Particle Filtering Framework," *International Journal of the PHM Society*, Vol. 2, No. 6, 2011, pp. 1–12.

[15] Celaya, J., Kulkarni, C., Biswas, G., and Goebel, K., "A Model-based Prognostics Methodology for Electrolytic Capacitors Based on Electrical Overstress Accelerated Aging," *Proceedings of Annual Conference of the PHM Society*, Montreal, Canada, 2011.

[16] Kulkarni, C., Celaya, J., Goebel, K., and Biswas, G., "Physics Based Electrolytic Capacitor Degradation Models for Prognostic Studies under Thermal Overstress," *Proceedings of First European Conference of the Prognostics and Health Management Society*, 2012.

[17] Gasperi, M. L., "Life Prediction Model for Aluminum Electrolytic Capacitors," *31st Annual Meeting of the IEEE-IAS*, Vol. 4, No. 1, 1996, pp. 1347–1351.

[18] Fife, J., "Wet Electrolytic Capacitors," Patent No: 7,099 1, AVX Corporation, Myrtle Beach, SC, 2006.

[19] Chen, *Ripple Current Confusion*, KEME Electronic Corporation, 2004.

[20] Alwitt, R. S. and Parler, S. G., "Heat Generation During Pulse Operation of Prototype Aluminum Electrolytic Capacitors," *Journal of Applied Electrochemistry*, Vol. 25, No 6, 1995, pp. 533–542.

[21] Brettle, J. and Jackson, N. F., "Failure Mechanisms in Solid Electrolytic Capacitors," *Electrocomponent Science and Technology*, Vol. 3, 1977, pp. 233–246.

[22] Luo, J., Namburu, M., Pattipati, K., *et al.*, "Model-based Prognostic Techniques," *AUTOTESTCON 2003. IEEE Systems Readiness Technology Conference. Proceedings*, Anaheim, CA, 2003, pp. 330–340.

[23] Albella, J. M., Gomez-Aleixandre, C., and Martinez-Duart, J. M., "Dielectric Characteristics of Miniature Aluminium Electrolytic Capacitors under Stressed Voltage Conditions," *Journal of Applied Electrochemistry*, Vol. 14, No. 1, 1984, pp. 9–14.

[24] Wit, H. D. and Crevecoeur, C., "The Dielectric Breakdown of Anodic Aluminum Oxide," *Physics Letters A*, Vol. 50, No. 5, 1974, pp. 365–366.

[25] Ikonopisov, S., "Theory of Electrical Breakdown During Formation of Barrier Anodic Films," *Electrochimica Acta*, Vol. 22, No. 10, 1977, pp. 1077–1082.

[26] Gomez-Aleixandre, C., Albella, J. M., and Martinez-Duart, J. M., "Gas Evolution in Aluminum Electrolytic Capacitors," *Journal of the Electrochemical Society*, Vol. 131, No. 3, 1984, pp. 612–614.
[27] Roederstein, V., "Aluminum Capacitors – General Information," *Document No. 25001*, 2007. Available: https://www.vishay.com/docs/25001/alucapsintroroederstein.pdf
[28] IEC-60068-1, "Environmental Testing, Part 1: General and Guidance," *IEC Standards*, 1988.
[29] Bengt, A., *Electrolytic Capacitors Theory and Applications,* RIFA Electrolytic Capacitors, 1995.
[30] Kulkarni, C. S., *A Physics-Based Degradation Modeling Framework for Diagnostic and Prognostic Studies in Electrolytic Capacitors*, Ph.D. thesis, Vanderbilt University, 2012.
[31] Rusdi, M., Moroi, Y., Nakahara, H., and Shibata, O., "Evaporation from Water Ethylene Glycol Liquid Mixture," *Langmuir*, Vol. 21, No. 16, 2005, pp. 7308–7310.
[32] Tasca, D. M., "Pulse Power Response and Damage Characterization of Capacitors," *Electrical Overstress-Electrostatic Discharge Symposium Proceedings, EOS/ESD Association*, Vol. 3, 1981.
[33] Rodner, S. C., Wedin, P., and Bergstram, L., "Effect of Electrolyte and Evaporation Rate on the Structural Features of Dried Silica Monolayer Films," *Langmuir*, Vol. 18, No. 24, 2002, pp. 9327–9333.
[34] Kulkarni, C., Biswas, G., and Koutsoukos, X., "A Prognosis Case Study for Electrolytic Capacitor Degradation in DC–DC Converters," *Proceedings of Annual Conference of the PHM Society*, San Diego, CA, 2009.
[35] Stengel, R. F., *Optimal Control and Estimation*, Dover Books on Advanced Mathematics, 1994.
[36] Chatfield, C., *The Analysis of Time Series: An Introduction*, Chapman and Hall/CRC, 6th ed., 2003.
[37] Saha, B., Goebel, K., and Christophersen, J., "Comparison of Prognostic Algorithms for Estimating Remaining Useful Life of Batteries," *Transactions of the Institute of Measurement and Control*, Vol. 31, Nos. 3–4, 2009, pp. 293–308.
[38] Daigle, M. and Goebel, K., "Multiple Damage Progression Paths in Model-based Prognostics," *Proceedings of the IEEE Aerospace Conference*, Big Sky, MT, 2011.
[39] Saha, B. and Goebel, K., "Modeling Li-ion Battery Capacity Depletion in a Particle Filtering Framework," *Conference of the Prognostics and Health Management Society*, San Diego, CA, 2009.
[40] Daigle, M., Saha, B., and Goebel, K., "A Comparison of Filter-based Approaches for Model-based Prognostics," *Proceedings of the 2012 IEEE Aerospace Conference*, Bik Sky, MT, 2012.
[41] Orchard, M. E., *A Particle Filtering-based Framework for On-line Fault Diagnosis and Failure Prognosis*, Ph.D. thesis, Georgia Institute of Technology, 2007.

[42] Chatfield, C., *The Analysis of Time Series: An Introduction*, Chapman & Hall/CRC, 2003.
[43] Kulkarni, C., Biswas, G., Koutsoukos, X., Celaya, J., and Goebel, K., "Integrated Diagnostic/Prognostic Experimental Setup for Capacitor Degradation and Health Monitoring," *2010 IEEE AUTOTESTCON*, Orlando, FL, 2010, pp. 1–7.
[44] Celaya, J., Kulkarni, C., Biswas, G., and Goebel, K., "Towards Prognostics of Electrolytic Capacitors," *AIAA 2011 Infotech@Aerospace Conference, AIAA*, St. Louis, MO, 2011.
[45] MIL-C-62F, *General Specification For Capacitors, Fixed Electrolytic*, 2008.
[46] IEC-60384-4-1, "Fixed Capacitors for Use in Electronic Equipment," *IEC Standards*, 2007.
[47] Tye, R. P., *Thermal Conductivity*, Academic Press Inc., Burlington, MA, 1968.
[48] Saxena, A., Celaya, J., Saha, B., Saha, S., and Goebel, K., "On Applying the Prognostic Performance Metrics," *Annual Conference of the Prognostics and Health Management Society*, 2009.
[49] Saxena, A., Celaya, J., Balaban, E., *et al.*, "Metrics for Evaluating Performance of Prognostic Techniques," *International Conference on Prognostics and Health Management*, Denver, CO, 2008.
[50] Celaya, J., Kulkarni, C., Biswas, G., and Goebel, K., "Towards A Model-based Prognostics Methodology for Electrolytic Capacitors: A Case Study Based on Electrical Overstress Accelerated Aging," *International Journal of Prognostics and Health Management*, Vol. 3, No. 4, 2012, pp. 19.

Chapter 4
Embedded fault diagnosis and prognosis

Julio Viola[1], Jose Restrepo[2,3], and Ronald Harley[4]

4.1 Introduction

The increasing use of power electronics in industry applications, energy conversion, electric vehicles (EVs), aircrafts, vessels, etc. propels efforts at early detection of potential abnormalities, which can degrade system performance, in order to reduce profit losses and even risk to human lives. In the onset of the power electronics digital control era, microprocessors were used exclusively for control tasks, mainly due to the lack in computing power needed by diagnosis algorithms. Nowadays, powerful low-cost microprocessors have become standard options for the design engineer, and embedded routines that perform real-time fault diagnosis are easily added to the design. This chapter gathers the most relevant embedded techniques for real-time condition monitoring (CM) and fault diagnosis in power drives based systems that have been reported recent years, with special emphasis on those methods using electric variables (i.e., voltage and current) as health indicators of the involved devices.

4.2 Embedded systems

An embedded system involves the use of microprocessors inside of another, typically bigger and more complex, system. Control systems are the most well-known application for embedded systems and have become ubiquitous in industrial and domestic equipment. The total number of microprocessors used in embedded systems is about one hundred times the number of those used in desktop applications such as PCs, laptops, and workstations. While desktop applications are intended to be easily reconfigured by means of programming, embedded systems are designed to address more specific tasks and are not easily reconfigurable.

Another component increasingly used in embedded systems is the field programmable gate array (FPGA), which can be used as a replacement for the

[1]Electrical Engineering Department, Universidad Politécnica Salesiana – Sede Cuenca, Ecuador
[2]Facultad de Ingeniería en Ciencias Aplicadas, Universidad Técnica del Norte, Ecuador
[3]Departamento de Electrónica y Circuitos, Universidad Simón Bolívar, Venezuela
[4]School of Electrical and Computer Engineering, Georgia Institute of Technology, USA

microprocessor or as a companion device. An advantage of FPGAs is the flexibility obtained from programming directly in silicon, which can lead to embedded ad-hoc features unavailable in commercial processors, such as parallelization of data processing, embedded filter stages, customizable data bus widths, etc.

4.3 Diagnosis, prognosis, and condition monitoring

Even when there is not a universally accepted definition for these terms, it is possible to state that diagnosis deals with the detection and isolation of the fault once it has already happened. Prognosis, on the other hand, is about prediction of the device health in the future, based on a current health assessment [1,2]. CM is the real-time monitoring of one or more device indicators to determine their drift from healthy condition and thus anticipates the fault [2,3]. Some faults happen suddenly, so CM may not detect them properly, in which case fault diagnosis is required to activate, if available, fault-tolerant operation. A general architecture for diagnosis and prognosis that can be applied to general engineering systems is presented in [4]. The proposed architecture may include the following:

- Sensing and data preprocessing,
- Feature (or condition indicator) extraction,
- Sensor and feature fusion, and
- Accurate diagnostic and prognostic algorithms.

In [5], Przytula and Choi describe a general-purpose probabilistic framework for reasoning in diagnosis and prognosis. The method provides a mathematical approach to manage the uncertainty inherent to prognosis by using structured Bayesian networks as the base of the probabilistic model.

Many factors, including the monitored feature (current, voltage, temperature, etc.), the way it is sensed and converted (analogically, digitally, type of sensing circuits, etc.), how it is post-processed (online, offline), and which algorithm (probabilistic, deterministic, etc.) is used, have led to a myriad of methods reported in the literature during, at least, the last four decades. Embeddability of a given diagnosis or prognosis method will depend on required sensor bandwidth, data sampling frequency, analog-to-digital converter (ADC) resolution, and computational burden, among others considerations. The following section addresses the most relevant characteristics of current commercially available hardware that can be used to implement an embedded diagnosis and/or prognosis system.

4.4 Review of hardware used in embedded diagnosis and prognosis systems

4.4.1 Sensors

Isolated measuring of voltage and current is a critical issue in power electronics converters, considering that high- and low-voltage circuitry coexist and electrical wiring typically includes many ground references. Different methods are used to obtain the required isolation.

The oldest technique for obtaining isolated measuring of voltages is the transformer, which has advantages such as ruggedness, simplicity of use because it requires few additional components, and relatively low cost. The main disadvantages are that it can be used only for AC voltages, its limited bandwidth operation, and its bulkiness. The same is true for current sensing, where the current to be measured is passed through a current transformer.

Modern sensors use electronic devices in different ways to perform the sensing and the isolation and can be designed as entirely solid-state devices or as a combination of solid-state and other elements, such as coils, cores, etc. There are three major techniques commercially available that provide isolation: electromagnetic, optical, and capacitive. All these methods are mostly based on sigma–delta (Σ–Δ) isolation amplifiers, which take the analog input signals and convert them into digital pulses. The sensing of the variable is made in the high-voltage side of the device, typically with a resistive divider for voltage measuring or a shunt resistance for current measuring. The sensed scaled-down signal is converted with a sigma–delta technique and passed as digital pulses to the low-voltage side.

Exceptions to the previous sensing methods are Hall effect–based sensors, which are implemented as a combination of a coil wound around a ferrite core with open ends. The sensed variable generates a magnetic flux, which is directed to a Hall effect device placed in the air gap of the core. The sensing process is totally analog and can be improved with complementary circuitry to obtain better linearity and bandwidth. Figure 4.1 summarizes the most common techniques for the isolated sensing of electrical variables.

4.4.2 Microprocessors, microcontrollers, and digital signal processors

There was a time when more distinct differences existed among microprocessors, microcontrollers, and digital signal processors (DSPs). Namely, microprocessors were chips with address and data buses as almost the only interface with the outside world, whereas microcontrollers were an evolution incorporating serial and parallel digital ports, analog-to-digital and digital-to-analog converters, internal memory, etc., making them easier to include in control schemes. DSPs, meanwhile, were characterized by the inclusion of internal units aimed at fast and efficient processing of digital signals, such as digital filtering, fast Fourier transform (FFT), trigonometric operations, etc. Moreover, the use of these terms is different for every manufacturer, making differences more diffuse and making classification a subjective issue. Table. 4.1 shows a summary of the most relevant characteristics for different types of processors.

The control of power electronics systems typically requires the computation of algorithms with a moderate-to-heavy processing burden. Depending on the specific control scheme, a midrange microcontroller can meet the requirements. An example of this could be a scalar control scheme without complex mathematical operations,

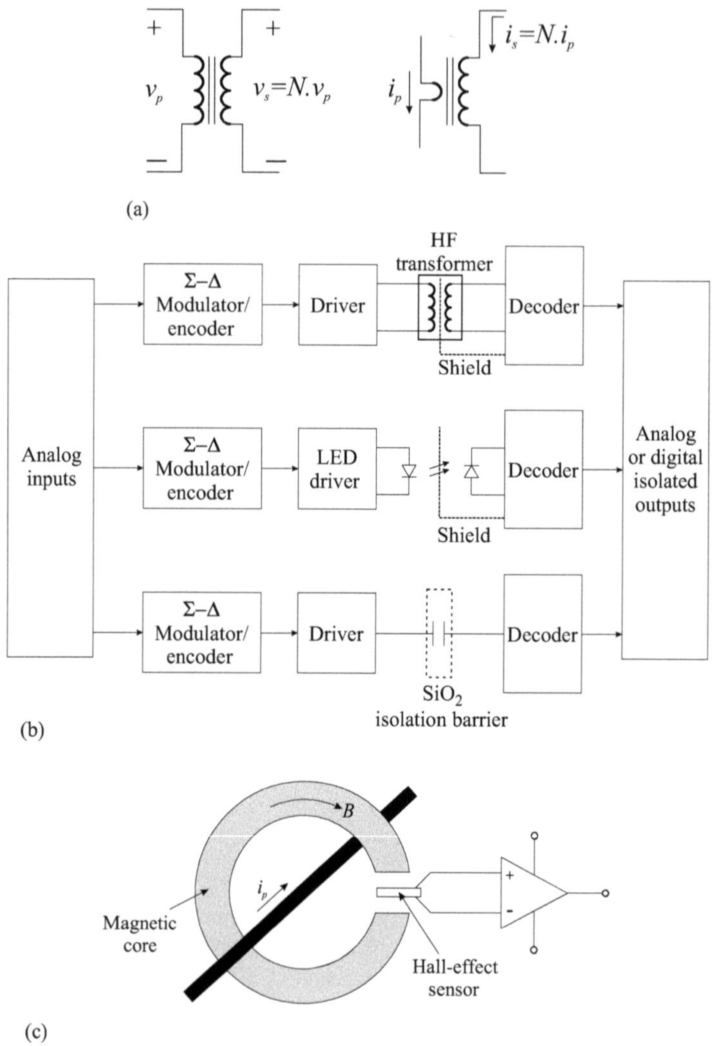

Figure 4.1 Most common methods for isolated measuring of electric variables. (a) Voltage and current transformers. (b) Digital isolation: electromagnetic, optical, and capacitive. (c) Hall effect–based sensor

requiring the generation of a few pulse width modulated (PWM) signals, and the acquisition of analog signals with low sampling frequencies. On the other hand, control schemes such as model predictive control, adaptive control, computation of neural networks (NNs), real-time computation of fast Fourier transform (FFT), and data filtering, among many other computation-intensive tasks, lie in the range of DSP, probably with floating point capabilities.

Table 4.1 Most relevant characteristics of microcontrollers and DSPs

Characteristics	Microcontrollers	DSPs
Math processing speed	Low to medium	Fastest
Power consumption	Low	High
Data format	Most operate with fixed-point	Fixed- or floating-point
Program memory	Typically have on-chip program memory	Require external program memory
Clock speed	Low to medium	Medium to high
Architecture	Mostly Von Neumann	Typically Harvard
Price	Low	Medium to high

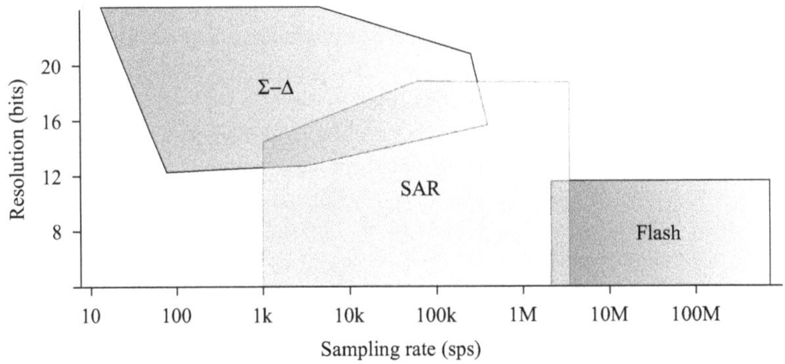

Figure 4.2 Approximated classification for current state-of-the-art ADCs

4.4.3 Analog-to-digital converters

Digital control schemes are the most commonly used in high power electronic systems. Any digital loop relies on the use of ADCs that provide a representation of analog variables in the form of digital quantities, which are processed by the microprocessor. The operation of an ADC can be divided into two processes: sampling and quantization. During the sampling process, the analog signal is discretized by holding the values in every sampling time. These retained values are fed to the quantization stage, which uses different techniques to convert every discretized value to a digital word, which is a representation in the digital domain of the original analog signal.

Three major techniques are the most used in the quantization stage: successive approximation register (SAR), sigma–delta, and flash.

The first criteria taken into account for the designer of an embedded acquisition system for the selection of a particular technology of ADC is sampling rate (samples per second) and resolution (bits). These parameters follow an approximated inverse relationship between them, that is, the higher the sampling rate, the lower the resolution for the chips commercially available. Figure 4.2 shows an approximated distribution for ADCs belonging to SAR, Σ–Δ, and flash families; some overlapping is evident.

SAR technology has been the most popular since the 1970s, but in recent decades, Σ–Δ competes in terms of cost and performance, especially in industrial signal acquisition. While the theory behind Σ–Δ operation has been well known since the 1950s, the operating principles strongly rely on highly demanding digital processing such as digital filtering, oversampling, noise shaping, and decimation. It was not until the 1990s that the required processing power could be implemented in silicon chips, leading to hundreds of commercial ADCs based on the Σ–Δ principle.

The implementation of a typical digital control loop requires sampling at the switching frequency: one sample of every involved variable is processed and used to take a control action in the next control cycle. Standard sampling rates are therefore on the order of several kilohertz, and many embedded prognosis and diagnosis techniques continue to be developed around these typical values. Nowadays, however, inexpensive ADCs with much higher sampling rates are available, and methods that previously were thought to be only for offline applications (involving, e.g., oscilloscopes for data acquisition) can be implemented in embedded systems.

After sampling rate and resolution, an embedded systems designer is concerned with the number of analog channels, signal-to-noise ratio, and, of course, ADC price. Both SAR and Σ–Δ can meet current design requirements, and options such as ≥16 bits resolution, ≥10 ksps, ≥80 dB, and ≥8 are offered by major manufacturers. The trend, however, shows that Σ–Δ ADCs fit better and have more available parts for industrial applications.

Even though most modern microcontrollers incorporate ADCs as a standard part of their internal structure, many control schemes have specific requirements for sampling rates and number of analog channels, leading to the need for using external ADCs.

4.5 Switching devices and their faults

In the last two decades, the insulated gate bipolar transistor (IGBT) has become omnipresent as the main switch from which medium- and high-voltage power converters are built. The key to its success stems from it having the best of metal oxide semiconductor field effect transistors (MOSFETs) and bipolar junction transistors (BJTs) properties, namely, high input impedance and low-voltage drop during on-state in its main terminals [6]. Nowadays, the voltage blocking capability for IGBT modules is above 1,000 V, with the 1,200-V standard being the most common, and the current management capability can be in the order of thousands of amperes [7]. Faults in IGBTs can be classified as follows:

- Intrinsic: those related to the silicon die, including electrical stress (overvoltage, overcurrent), latch-up, and electrostatic discharge [9].
- Extrinsic: those related to package and soldering degradation, including bond wire lift-off, bond wire heel cracking, and aluminum pad reconstruction [10].

Figure 4.3 shows a classification of the most common open-circuit and short-circuit faults in IGBTs.

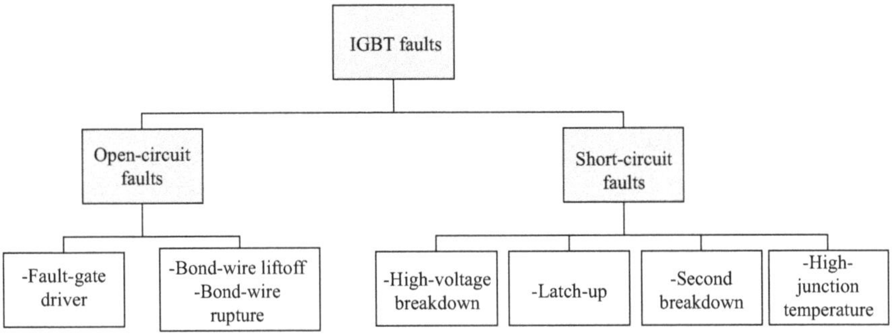

Figure 4.3 Open-circuit and short-circuit faults in IGBTs [8]

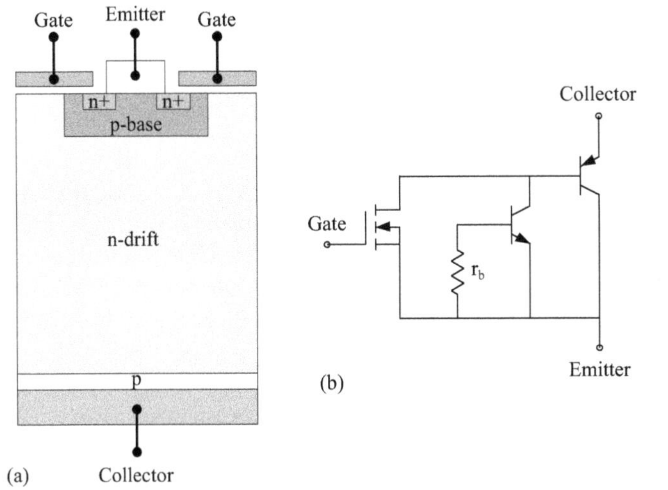

Figure 4.4 (a) Internal structure for an IGBT. (b) Equivalent circuit for the IGBT

To better understand of the intrinsic IGBT faults, a diagram of an IGBT's internal structure and equivalent circuit is shown in Figure 4.4(a) and 4.4(b). The control terminal is equivalent to a MOSFET gate, which is insulated from the rest of the chip by a layer made of silicon dioxide (SiO_2). The body of the device has a p–n–p–n structure, which under certain collector current densities can act as a parasitic thyristor, leading to the latch-up phenomenon through which the gate loses control of collector current, and can lead to a fault of the device because of hot spots formation [11]. Even though many modifications of the internal structure have been proposed to overcome this issue [12], it still is one of the main drawbacks of the IGBT [13].

The most common package for commercially available IGBTs is the power module, which has several advantages when compared to discrete devices, namely:

- typically housing more than one IGBT with an associated freewheeling diode, which facilitates the construction of power converters;

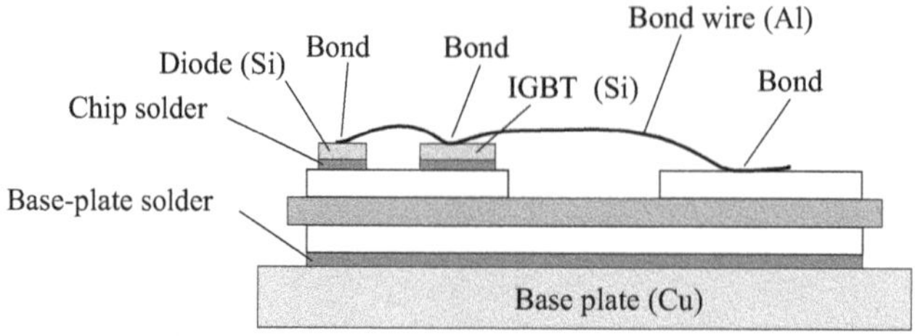

Figure 4.5 Internal structure for an IGBT module

- a ruggedness and surface finishing suitable for mounting directly on the heat sink (typically with thermal grease as an interface);
- insulation between the silicon die and the case metal plate;
- a lower thermal resistance between junction and case and therefore greater power management; and
- rugged terminals with different attaching methods such as screw, tabs, etc.

Figure 4.5 shows a cross-sectional view of a typical IGBT module. The base of the module is made of a copper alloy and has a polished surface to ensure good heat transfer to the heat sink. Inside the module, the semiconductor chips for the IGBT and the antiparallel diode are insulated by means of a direct bond copper ceramic substrate. Connection from silicon die to power terminals on the outside of the module is made with aluminum or gold bond wires soldered with Pb or Sn/Ag-based alloys.

The main operating issues in IGBT power modules are related to their internal structure, which is composed of different materials: since each material has a different thermal coefficient, a mismatch in expansions and contractions along the heat-up and cool-down cycle occurs. These thermal cycles happen every time a load is connected and disconnected, causing a temperature swing in the internal parts of the module. When these thermal cycles are repeated many times, mechanical stresses between firmly bonded materials lead to material fatigue and eventually cracks in the solder joints. The most common fault in IGBT modules is bond wire lift-off which is characterized by the disconnection of the bond wire [14]. Unlike faults derived from latch-up effect, the bond wire lift-off leads to open-state faults.

Bond wire heel cracking and aluminum pad reconstruction are other common faults in power modules produced by aging effects after repetition of thermal cycles. The study of these kinds of faults began as early as the mid-1990s [15] and were, in turn, a continuation of research developed for faults in power bipolar transistors in the 1970s [16].

It is worth noting that not all power modules use bond wires. Figure 4.6 shows the cross-section structure of a press-pack IGBT. Among their major features press-pack technology offers higher current ratings, easier stackability, and less ohmic losses. It is a preferred choice in high-power applications such as flexible AC transmission systems (FACTSs) and high voltage direct current (HVDC).

Beyond these examples the technique of bond wires is, by far, the most used for low- and midrange applications.

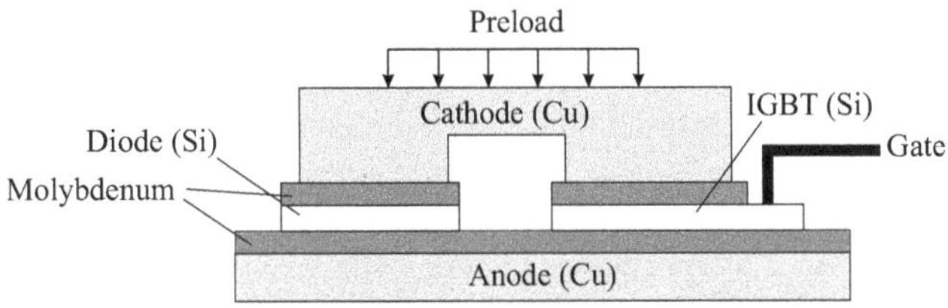

Figure 4.6 Internal structure for a press-pack IGBT

Table 4.2 Classification of monitoring techniques for power switches [2,17,18].

Methods	**Advantages or application fields**	**Limitations**
$V_{CE\,ON}, R_{ON}$	Strongly related to degradation of power switches (IGBT, MOSFET)	No additional hardware needed
Thermal resistance/impedance (R_{TH}/Z_{TH})	Can detect solder degradation	Difficulty in measuring junction temperature
Gate signal	Can detect electrical faults	Difficulty in measuring junction temperature
Switch time	Gate driver failure detection	Short switching measurement needed
Sensor based	Reliability and accuracy	Cost and modifications in layout of devices
System identification	No additional hardware modification needed	Complex correlation and complicated algorithms

4.6 Analysis of aging in IGBT power modules

The identification of features that can be used as indicators of the aging process in power modules is the basis of any successful prognosis method and was a major subject of study in the last decade. Based on [2,17,18], Table 4.2 shows a possible classification of monitoring techniques depending on the degradation indicator.

For a better understanding of the aging and its influence on the evolution of the parameters, to be used as precursors, artificial processes resembling the natural aging of switching devices have been proposed. These methods are referred as artificial or accelerated aging and may involve thermal cycling, power cycling, vibrations, etc.

An example of thermal cycling is presented in [19], where the devices under test are actively heated without attaching a heat sink and cooled down by forced air-cooling through fans. To shorten the time duration of the tests, the junction temperature is increased to levels above the maximum recommended by the device's manufacturer. This process triggered degradation mechanisms in the gate oxide, solder joints, and bond wires.

Power cycling instead uses the heat generated in the silicon die to obtain temperature swings in the device. An early work addressing a fast power cycling technique, to be used in the study of reconstruction of metallization and crack growth in bond wires, is proposed by Held *et al.* in [20]. Power cycles are reduced to less than 10 s, producing approximately the same damage as other slower methods.

Another example is proposed in [21], where the aging method is based on imposing a high collector current value and a high switching frequency. These applied conditions ensure that the IGBT temperature will exceed the maximum value specified by the manufacturer. In [22], power cycles lasting about 40 s are used to generate stress and accelerate the aging of the power modules. V_{CE} and R_{THj-c} are monitored to determine a threshold at which the device is considered degraded. A rise of 2%–3% in V_{CE} and 10% in R_{THj-c} are used as decision criteria. In [23], a software and hardware platform for accelerated aging of power switches is proposed. The test bed supports thermal cycling, dielectric overvoltage, and current overstress and allows the validation of prognostic algorithms. In [24], an offline test rig for monitoring switching characteristics of aged transistors is proposed. Turn-on and turn-off current and voltage waveforms are recorded with high-frequency sampling to study behavior evolution of the device under test. In [25], Katsis *et al.* evaluate the increase in thermal impedance in MOSFETs. Devices under test are power cycled, imaged, and thermally analyzed to generate a correlation between void percentage in the solder die-attach layer and thermal impedance. In [26], the authors use an aging process based on repetitive short-circuit operations to analyze the current distribution flowing in the power dies of MOSFETs and IGBTs, investigating the effect of the local degradation of aluminum sheet resistance and bond wire lift-off.

In some cases, to detect evolution of a given parameter, the aging can be accelerated by manually degrading the power module. For example, in [17,27,28], the aging is emulated by sequentially disconnecting the bond wires between the silicon die and the case. To this end, the protective gel layer in the IGBT module was removed. Since accelerated aging is very expensive and can take a long time, in [29], a general framework for designing aging tests is presented, so that such experiments can be optimized. Additionally, thanks to finite element simulations, thermomechanical behavior of the assembly under realistic operating conditions for long period of time can be studied in shorter times. The goal of the simulations was to determine how many thermal cycles of a given accelerated aging test protocol are necessary to lead to the same damage in the solder as would occur in a given period in real working conditions.

The methods described above are not intended to operate online, as they are thought to validate the study of parameters that are related to degradation mechanisms. Online prognosis methods can be developed based on knowledge of the behavior of those parameters.

4.7 Prognosis and condition monitoring of power switches

This section addresses the most common techniques of monitoring a relevant device's parameters in order to estimate the health of the device and/or predict the future

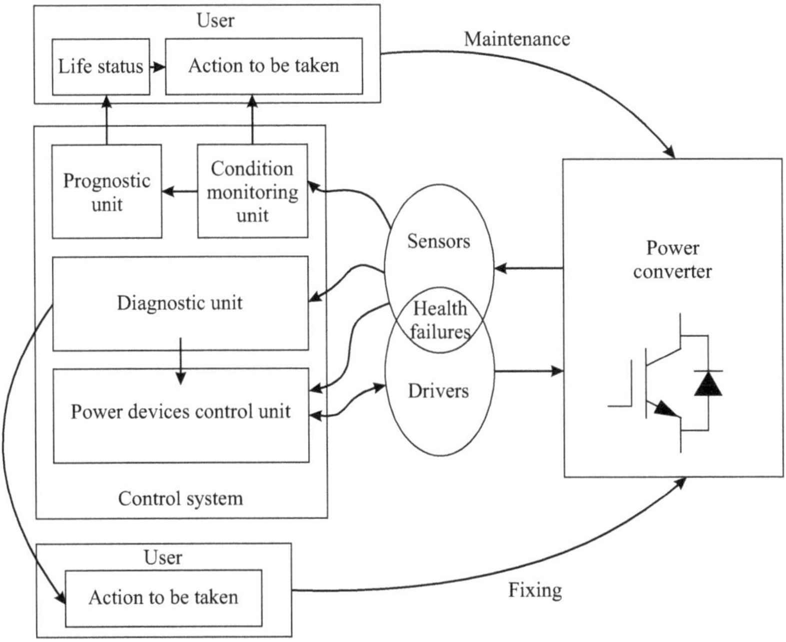

Figure 4.7 Health monitoring system for power converters [3]

appearance of faults. The emphasis is put on those methods using electrical variables as the main source of information. Figure 4.7 depicts a scheme of a CM system [3].

Different authors report various aging precursors as viable: collector-emitter voltage during on-state ($V_{CE\,ON}$) for IGBTs, drain-source resistance during on-state ($R_{DS\,ON}$) for MOSFETs, the body diode voltage drop, the parasitic capacitances, and gate threshold voltage (V_{th}) for die attach solder and gate-oxide degradation mechanisms [19]. The authors in [3] summarize the advantages and shortcomings of different damage indicators.

The following sections study the monitoring of signals intended for early detection of failures in power switches.

4.7.1 V_{CE} monitoring

A review of literature shows that the monitoring of $V_{CE\,ON}$ drift is the most used technique for the evaluation of wire bond degradation. The reported drift for this parameter is in the order of hundreds of millivolts after several million thermal cycles. This fact imposes strong restrictions on the design of the measuring circuit, since the ADC has to have enough resolution, so that the digital data can retain the information carried in the analog signal. For example, the least significant bit (LSB) of a 12-bit ADC with a standard input range for analog signals of ± 5 V has a value of 2.44 mV, which is probably the lowest resolution usable for this kind of detection. For online embedded monitoring of $V_{CE\,ON}$, another restriction is that in most applications, the

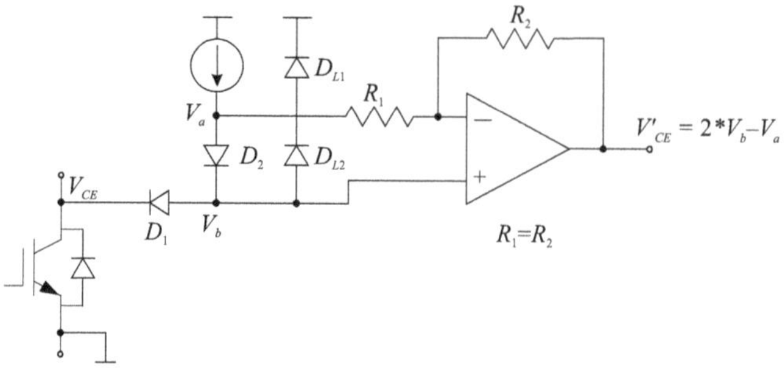

Figure 4.8 Circuit for measuring $V_{CE\,ON}$ with limiting function

same V_{CE} voltage reaches values of several hundreds of volts when the IGBT is in off-state, requiring the use of additional circuitry to protect the ADC. A voltage divider can be used, but it has a major drawback that for low values of V_{CE}, the drift is also divided making its detection even more difficult. A clamping circuit can help to fix this problem, allowing the low values of $V_{CE\,ON}$ to reach the ADC without any attenuation and limiting the voltages above a predefined value. This limiting value must be below the operating input range of the ADC. Figure 4.8 shows a possible scheme for this circuit [30,31]. During the on-state, the current source forward biases D_1 with the result that in node *Vb* appears $V_{CEON} + V_\gamma$, where V_γ is the forward voltage drop of diode D_1. This diode is selected so that it can withstand the reverse voltage appearing in its cathode when the IGBT goes to the off-state, thus effectively protecting the acquisition circuitry. Since the diode forward bias is obtained with a known current source, the value of V_γ can be previously obtained and subtracted to obtain real value of $V_{CE\,ON}$.

The sampling of the V_{CE} voltage must be done when the IGBT is completely in on-state, so that a stable measurement is obtained. The most common technique for synchronization of ADC sampling uses the peaks and/or valleys of the carrier signal from the PWM generator. This method is shown in Figure 4.9 and ensures that the sampling is done in the center of the on-state time window. For different pulse widths, however, the sampling time will be closer to or further from of the falling edge, depending on the modulation index value. This can be an issue if a specified amount of time is desired between the switching and the measurement. To overcome this, only those values of V_{CE} obtained for specific pulse widths can be considered for the monitoring of this parameter.

Another restriction in measuring V_{CE} is the electrical isolation required between the power device and the processing circuitry. This is especially important for measurements taken in the upper devices of each converter leg. Different sensors described in Section 4.4.1 are essential to this purpose of providing the required isolation.

Ghimire *et al.* [30] present a review in wear-out prediction methods of IGBT power modules and freewheeling diodes based on the real-time V_{CE} measurement. Monitoring of $V_{CE\,ON}$ is also used in [22] by Smet *et al.* as an aging indicator in IGBTs power modules. A circuit using an auxiliary MOSFET is proposed for

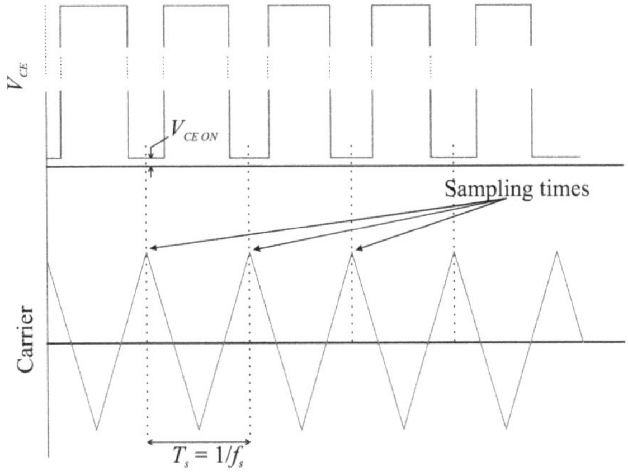

Figure 4.9 Synchronization of ADC sampling with PWM carrier signal

real-time measuring of $V_{CE\,ON}$. In [21], Tlig *et al.* analyze the effects of aging in the common mode collector-emitter voltage during turn-on and turn-off. A conducted electromagnetic interference (EMI) analysis of V_{CE} voltage shows an increase of oscillations as a function of the aging time considered. Also, a change in switching times as a function of aging time is reported.

4.7.2 $R_{DS\,ON}$ monitoring

Nasrin *et al.* [32] and Das *et al.* [17] propose the use of spread spectrum time domain reflectometry for $R_{DS\,ON}$ monitoring of power MOSFETs and IGBTs. The technique is based on reflectometry, which has been used for decades to detect faults in electrical wiring. A high-frequency spread spectrum generator is used to send a pulse through a cable that is connected to the drain and source terminals of the power switch. The system measures the reflected power, which is a function of the impedance at the far end of cable, that is, $R_{DS\,ON}$. A correlation between the sent and the received signals is obtained and used for the analysis. The method, however, is not easily embeddable due to the high frequencies involved and the amount of data processing it requires.

4.7.3 Other indicators

In [28], an online estimation of thermal resistance $R_{TH\,j-a}$ is used as a CM method for IGBT packages. Changes in the transient thermal impedance can be related to the progressive degradation of the thermal path that results from solder fatigue at the die-attach. Given the difficulty of directly measuring this parameter during the normal operation of power converters, an estimation is performed with a Kalman filter.

Monitoring of $R_{TH\,j-c}$ is also reported in [33], where Sarkany *et al.* use power cycling over an IGBT module to produce thermal variations inside the module.

Thermal cycling is continued until the failure of all IGBTs in the module occurs. After that, the IGBTs are disassembled and the cracking of bond wires as well as an accelerated degradation in gate oxide is observed. At the same time, a continuous increase of $R_{TH\,j-c}$ can be observed along the aging process.

In [34], Ginart *et al.* propose a method that uses the ringing in the current produced in every IGBT switching for a three-phase motor drive. Since these oscillations are produced by the internal parasitic elements, a change is expected with component aging. Characterization of the ringing shows that the main frequency components are in the order of tens of megahertz, which is a drawback for embedded implementation, since a very high sampling frequency would be required.

In [27], Ji *et al.* use an indirect method for measuring virtual junction temperatures with temperature sensitive electrical parameters (TSEPs), which is based on the method proposed in [35]. Also, in [36], Chen *et al.* develop a thermal modeling for a MOSFET using Cauer and Foster networks to represent thermal impedances between junction and air. In this case, the TSEP threshold voltage V_{th} is selected due to its high dependence on the junction temperature. A circuitry for threshold voltage measurement is proposed that also involves measuring the drain current and the temporal disconnection of the power switch from the DC link.

4.8 Fault diagnosis techniques of power switches

Independent of whether a real-time health monitoring technique for power switches was implemented or not, there is always a chance for a switch to fail; the detection of these anomalies is nonetheless an advantage in helping to make decisions that can greatly improve the safety of human operators and consumers, minimize the costs associated with stop times of industrial processes, and limit the damages to remaining healthy power devices. Fault tolerant operation is an entire area of research on its own, and it is based on the detection of faults in order to start alternative control schemes using the healthy power switches. In Section 4.5, the most common faults in power switches were classified as intrinsic and extrinsic. While intrinsic faults typically lead to short-circuit conditions between main terminals, extrinsic faults mostly produce open-circuit conditions.

As stated before in this review, emphasis is given on detection and identification of faults based on electrical signals already used in the standard operation of the power converter, namely, output phase currents and voltages. While output voltages can immediately show the origin and type of a fault, their content of high-frequency harmonics makes them difficult to use in embedded systems. Sampling at lower frequencies and at specific times can be useful in showing high or low levels of voltage, which can be used for diagnosis, but sensors at the output of converters are not commonly used. Instead, output currents are typically low-pass signals, which are easier to acquire and process, even when their load dependence makes the design of the detection algorithm challenging. The vast majority of diagnosis methods, however, is based on their use. In the following sections, current- and voltage-based fault-diagnosis methods are covered.

4.8.1 Open-circuit fault detection

The most common reported methods for open-circuit faults diagnosis can be classified as follows [37,38]:

1. Methods based on current Park's vector,
2. Others methods based on current properties: waveform, frequency, spectrum analysis, mean values, deviation from a reference, etc.,
3. Comparison between actual AC voltage and references, and
4. Converter behavior rules diagnosis in time domain.

Methods 1 and 2 use the current as the monitored parameter, method 3 uses the comparison of reference with measured voltages, and method 4 uses a combination of measured currents and voltages. The following sections provide details of different reported methods.

4.8.1.1 Current-based techniques

The modulus of current Park's vector is the most widely used technique in reported methods for open-circuit fault detection. During normal operation, the current space vector for a three-phase inverter driving a balanced three-phase load describes a circular trajectory that can be tracked by the embedded detection algorithm with relatively low computational effort. When an open-circuit fault occurs, the trajectory becomes distorted, and the form and position of the distortion can be used to detect and identify the faulty power switch. In Figure 4.10(a) and 4.10(b), the current Park's vector trajectories for a healthy and a faulty inverter are depicted. It is noteworthy that this method is useful only for three-phase inverters and for some types of two-phase or n-phase inverters in which the concept of a rotating vector in a d–q plane can be applied. An example of fault detection in single-phase inverters can be found in [39,40], where the first harmonic of the only current is used.

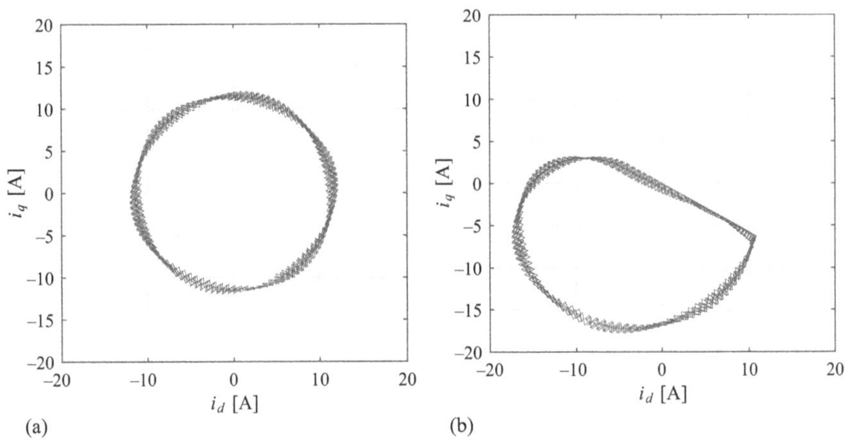

Figure 4.10 (a) Current Park's vector trajectory for a healthy inverter. (b) Current Park's vector trajectory for an inverter with one faulty switch

From the point of view of a detection algorithm, the tracking of circular trajectory is performed with the modulus of Park's vector, which can be obtained from measured currents in *a*, *b*, and *c* phases. Park's vector *d* (direct) and *q* (quadrature) components are calculated as follows:

$$i_d = \sqrt{\frac{2}{3}} i_a - \frac{1}{\sqrt{6}} i_b - \frac{1}{\sqrt{6}} i_c \tag{4.1}$$

$$i_q = \frac{1}{\sqrt{2}} i_b - \frac{1}{\sqrt{2}} i_c \tag{4.2}$$

from which the modulus is obtained as

$$|\bar{i}| = \sqrt{i_d^2 + i_q^2} \tag{4.3}$$

The value $|\bar{i}|$ is used to normalize measured currents, and the average values of the normalized currents are used for detection. Normalization makes the method independent of the load connected to the inverter. Included among the numerous works using this method or slight variations of it are [41] and [42], where Estima *et al.* propose a method for real-time detection and identification of single and multiple open-circuit faults in voltage source inverters feeding AC machines. The method uses the already existent current sensors in each phase of the AC machine. As previously explained, a Park transformation and averaging are used to obtain the values used as the table entries, allowing for the identification of single and double open-circuit faults. This method can detect a faulty switch within a time equivalent to one-ninth of the motor phase currents' fundamental period. Because it only requires the measuring of standard phase currents, it is easily embeddable.

In [43], Bae *et al.* propose a fault detection method based on monitoring the time instant that the phase current Park's vector is in each sector of the *d–q* plane when compared with the current reference. In normal operating conditions, the current vector spends the same amount of time over each sector when the frequency of phase currents is constant. After detection, the specific faulty IGBT is identified by observing the polarity of each phase current. The method is experimentally tested in an embedded system and can identify 18 combinations of IGBT open-circuit faults in a three-phase two-level inverter. Detection times are, on average, 50.8% of a fundamental current period.

A method using just one current sensor is proposed by Yan *et al.* in [44] Three-phase currents are reconstructed through a single-current Hall effect sensor strategically placed and knowledge of the switching states applied. The zero-voltage vector sampling method [45] is used, which consists in sampling the current during the two zero-voltage vectors applied in every switching period. The proposed method can detect and locate both single and double open-circuit faults in about one cycle of output current.

In [46], Abadi *et al.* present an extension of the method for the case of neutral point clamping (NPC) three-level inverters achieving a detection time for open-circuit faults that is less than one cycle of the output current.

A major drawback for some of these methods is the need for setting comparison values as thresholds that may be dependent on the connected load, which means the diagnosis methods are not totally independent from the specific applications of the VSI. Others methods using different characteristics of the current signal can be found. For example, in [47], Wu *et al.* propose a real-time scheme for single and double switch open-circuit fault diagnosis in pulse-width modulated voltage source inverters for induction motor drives. The method is based on three-phase output currents, which are used to compute allelic points, which are functions based on present and past values of measured currents. Asymmetries in values and signs between the functions allows for the classification of faults in four classes: a single transistor open-circuit fault, a double transistor fault involving two transistors in the same inverter leg, a double transistor fault involving two upper transistors or two lower transistors, and a double transistor fault involving an upper transistor in one leg and a lower transistor in another leg. Detection times obtained by simulations and by experimental tests are in the order of 1/4 of the grid voltage cycle. In [48], Sim *et al.* propose a method for the detection of open-switch faults in cascaded H-bridge (CHB) multilevel inverters. The method is based on the distortion of output current when an open-switch fault occurs in an H-bridge cell. By using two redundant zero-voltage switching states and by calculating current slope for both, the faulty cell and the specific faulty switch can be accurately identified. A probabilistic approach is proposed by Zhou *et al.* in [49], where a fault-diagnosis scheme based on probability density analysis of the sampling currents is proposed. The method is based on the fact that variance and skewness of the current are very different in healthy and faulty conditions. The current has an almost homogeneous distribution from $-Imax$ to $+Imax$ in healthy condition, but when an open-circuit fault occurs, the distribution changes to a range from 0 to $+Imax$ for faults in the upper switch and from $-Imax$ to 0 for faults in the lower switch. A sliding window with a length of one current period is used to calculate variance and skewness of the measured current in each phase. The method requires the experimental adjustment of thresholds to detect the change in the skewness. Detection times will depend on the period of the current. Correlation coefficients between acquired output current data in a three-phase inverter are proposed as indicators of open-faults by Wu *et al.* in [50]. During healthy condition, correlation coefficients are close to one, but after a fault occurs, a deviation can be observed and used for the detection.

4.8.1.2 Voltage-based techniques

A paper from Cai *et al.* [51] proposes a Bayesian network-based data-driven fault-diagnosis methodology of three-phase inverters driving permanent magnet synchronous machine (PMSM) in the oil industry. Data-driven fault diagnosis does not require a known model or signal patterns (as model- and signal-based methods do), but it does require a large amount of historical data. Single and double open-circuit faults are analyzed for IGBTs, and for this, two output line-to-line voltages at the inverter outputs are measured for different fault modes. An FFT analysis is performed on the voltages, and harmonics contents are used to determine the existence of a fault.

Sensing of the switched voltage in any leg of a power converter is a highly demanding task because of the great swing in voltage levels, which go from several hundreds or even thousands of volts when the upper switch is in the on-state to almost zero volts when the lower switch is in the on-state. Additionally, frequencies of switched voltages are typically in the order of several kilohertz, therefore requiring sampling frequencies of hundreds of kilohertz. A common approach is to process the signals in the analog domain and then convert them to digital samples so they can be stored and processed. An example of this is found in [52], where a diagnosis method for the detection of open switch faults of inverter-fed PMSM drives is introduced. Voltage in the middle-point of each inverter leg is measured and processed with a low-pass filter to remove the high-frequency components. The obtained signal is then compared with the reference value sent to the PWM controller. The method is embeddable if an analog low-pass filter is used. Another example of open-circuit fault detection using a measurement circuit for V_{CE} voltage is proposed by Sun *et al.* in [53], where a resistor divider and a comparator are used to compare the measured voltage with its estimation, which is based on the switching function of a three-phase two-level inverter. In [54], Baocheng *et al.* propose a simple method also based on a resistor divider and an op-amp applied to an NPC single-phase inverter.

In [55], Sobanski and Orlowska-Kowalska propose a method for the detection of single open-circuit faults in power converters driving three-phase induction machines (IMs) with direct torque control (DTC) schemes. The proposed technique is voltage based since it detects the voltage vector presence time in the sectors of the complex *d–q* plane. Depending on fault location, in steady state, the reference voltage vector is forced to be in one specific sector for a longer time period than in others. Detection time is shorter than one period of the stator current.

In [56], Sen *et al.* propose an open-circuit fault-diagnosis method for a modular multilevel converter (MMC). Detection is based on measured voltages for each submodule and circulating currents.

In [57], Hao *et al.* use the output voltage signal of a CHB inverter as the fault characteristic signal of this fault diagnosis strategy. A sampling frequency of 50 kHz is used for output voltage signals, which are processed with FFT to obtain the first 34 harmonics, and then the fault is classified by using a support vector machine. Considering the computation involved, the method can be implemented with standard ADCs and a midrange microprocessor.

Pires *et al.* take a probabilistic approach in [58], where an open-circuit fault detection method for three-phase three-level T-type grid connected inverter (GCI) is presented. Output voltages are used as the analyzed variable, and the asymmetry in the histogram of voltages is used for detection of fault. Specific patterns in the histogram allow the identification of the faulty switch. The high-frequency sampling used, however, makes it impractical for embedding. Another example using high-frequency sampling of the output voltage in an NPC three-level inverter is proposed in [59]. After acquisition the voltage signal is decomposed by a wavelet analysis, and a Bayesian classifier is used to diagnose the inverter's fault modes. The processing burden and high sampling frequency makes the method not appropriate for embedding.

4.8.1.3 Hybrid techniques

Some techniques mixing current and voltage-based detection of open-circuit faults have been reported, such as [60], where Lamb *et al.* present an open-circuit fault detection method for CHB multilevel converters based on a voltage sensor and a current sensor to measure the output voltage in each leg and the direction of current flowing through the leg. Another example is [61], where Li *et al.* propose an open-circuit fault detection method for MMC. The method is based on the already available voltage measurements and on the resulting error between the measured currents and estimated currents obtained from a state observer. When the absolute value of the calculated error surpasses a given threshold, a fault is detected. Detection times are in the order of a few milliseconds. Another hybrid voltage/current-based technique for open-circuit faults detection in an NPC N-level converter is proposed by Caseiro *et al.* and Mendes *et al.* in [62,63], respectively. The error between expected voltages and estimated voltages in the midpoint of each converter leg is used as the basis of comparison for fault detection. Expected voltages are calculated with known switching states and DC-link voltage, while estimated voltages are calculated departing from the AC side line-to-line voltages and the current circulating through the coupling inductors.

4.8.2 Short-circuit fault detection

For any application involving a voltage source inverter, short-circuit faults are always more serious than open-circuit faults, and minimal detection times are the main concern. Detection methods based on mathematical transformations, even when they successfully detect the damaged device, are not fast enough to activate any protective action and must be used accompanying another, faster method. In [38], Lu *et al.* present a review of protection and current-limiting methods:

1. Desaturation detection,
2. Current mirror method,
3. Gate voltage sensing,
4. dv/dt Feedback control,
5. Protection by gate voltage limiting,
6. Protection by snubbers and clamp circuits,
7. Protection by slow turn-off of IGBT using additional passive components,
8. Protection by two-step gate pulse,
9. Average current Park's vector approach, and
10. Vector composition of the inverter output voltage.

Most commercial drivers' circuits include the desaturation detection feature, which is designed to detect and protect potentially destructive situations in the IGBTs or MOSFETs. The method is based on the measurement of the voltage between the main terminals of the IGBT or MOSFET after the turn-on signal is set in the gate terminal. In normal operation, a low-voltage for on-state is expected, and a threshold level is set to compare if $V_{CE\,ON}$ or $V_{DS\,ON}$ is below the threshold. In Figure 4.11, the connection for desaturation detection is shown. Diode D_{desat} is used to protect the driver circuit when the switch is in off-state. Current source I_{BIAS} forward biases

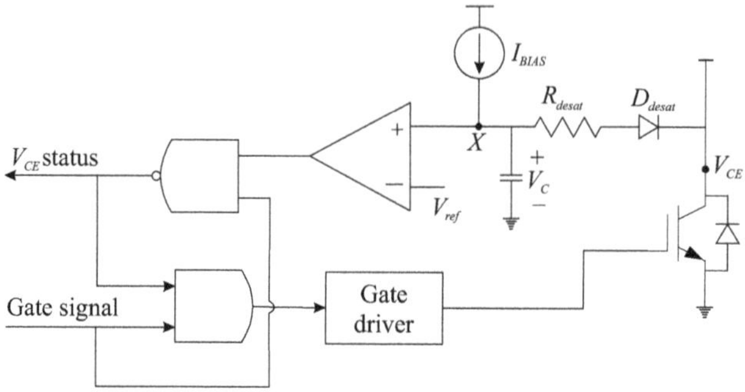

Figure 4.11 Desaturation detection circuit

the diode when the switch is in on-state, with the result that approximately $V_{CE\,ON}$ appears in node **X**, and this voltage is internally compared with a reference value. This value depends on the specific model of the driver circuit but is around 6–7 V. If $V_{CE\,ON}$ is greater than this value, the switch is turned off for the remainder of the current switching period and a fault signal is activated. This signal can be used for the microprocessor to activate a remedial action such as controlled shut down, alternative control algorithm, etc.

For this method to work, it is necessary to wait until the collapse of V_{CE}, which takes some time after the turn-on signal, is set in the gate. All commercial driver circuits allow for the setting of what is called a blanking time, which is a time between when the turn-on signal is applied and when the resulting V_{CE} is compared. The setting of this time will depend on the switching speed of the power switch and typically is in the order of a few microseconds. This restriction is in fact the main weakness of this method, since this required waiting time can be excessively long in order to avoid a damage in the switch. This is particularly true when a DC-link pole-to-pole short-circuit occurs.

In [64], Amini *et al.* propose a fault diagnosis scheme capable of detection of open-circuit and short-circuit faults. The authors advise, however, that short-circuit fault detection time can be longer than accepted in order to prevent permanent damage to the IGBT under fault.

In [65], de Mello *et al.* present a proposal of fault diagnosis for short-circuit switches in an H-bridge inverter. By using values of control voltages (V_{GE}) and current in the load, a circuit composed by logic blocks is designed for the detection. A protection resistance is added in the DC bus to activate (when needed) a limitation of short-circuit current. Detection times are not reported.

In [66], Khomfoi *et al.* propose a method for detection of fault types in CHB multilevel inverters based on their output voltage waveform. Two neural networks are trained: one for open-circuit fault detection and the other for short-circuit fault detection. The phase voltages at the output of the inverter are used as inputs to the NN. Simulations and experimental results show that fault detection and clearing times are in the order of 6–9 line cycles.

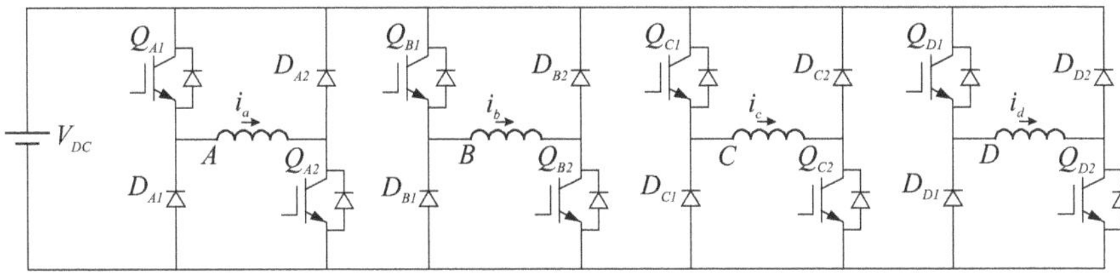

Figure 4.12 Four-phase asymmetric converter driving a switched reluctance machine

Gate–drain depletion capacitance (C_{GDJ}) is one of the electrical parameters most affected during a malfunction of the IGBT such as short-circuit faults. This fact is used by Rodriguez *et al.* in [67,68] for a detection technique based on gate-signal monitoring and can detect failures of short-circuit and open-circuit in the IGBTs. A comparison of gate-emitter voltage with specific voltage thresholds is proposed for detecting open- and short-circuit faults based on the effects that the variation of C_{GDJ} will have on the evolution of V_{GE} during IGBT turn-on. Analog instrumentation and differential amplifiers are used to obtain the comparisons. The delay time obtained for the fault-detection circuit is less than 3 μs.

For some specific converter topologies, short-circuit faults are not more dangerous than open-circuit faults. That is the case for an asymmetric converter controlling a four-phase switched reluctance machine (SRM), which is reported by Marques *et al.* in [69]. In this topology, the SRM windings are connected in series between IGBTs as shown in Figure 4.12. The position of winding inductance prevents a DC-link pole-to-pole short-circuit if an IGBT presents a short-circuit fault. Also, the rise rate of the current is limited, allowing the use of techniques such as those reported for open-circuit faults. In [69], phase currents are normalized with its reference, and averaged taking an averaging period that depends on the machine mechanical speed. Error coefficients are then calculated between the four averaged currents, and their values are used as entries in two tables: one for detecting open-circuit faults and the other for detecting short-circuit faults. Detection times are around 1 ms for both types of faults, and the method can be easily embedded. Another example is the Z-source inverter and Z-source derived inverters such as the quasi-Switched Boost Inverter (qSBI), shown in Figure 4.13, where inductance is connected in series with the DC-link, allowing more time for short-circuit fault detection as it is reported in [70].

4.9 Fault prognosis and diagnosis in sources and loads

An additional benefit of sensing and processing electrical signals is the possibility of using them for monitoring the sources and the loads connected to the power converter. Typically, the same signals carry information that can be analyzed to detect potential abnormalities in the load and/or the source. The most common industrial loads are

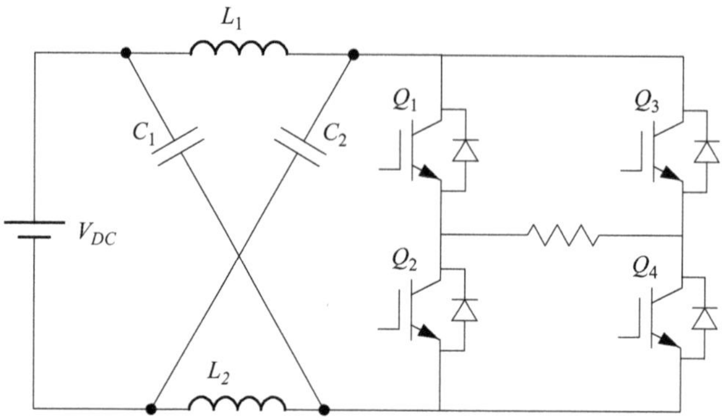

Figure 4.13 Single phase Z-source boost inverter

electric machines (IM, PMSM, DC machine, etc.), while the most common source is the electrical grid. Some sources can behave as loads under certain conditions and vice versa, for instance, batteries, the main grid, and some electric machines can absorb or inject power from or to the converter. Other sources such as photovoltaic (PV) arrays are very common and behave only as sources. In the same way, resistive loads (electrodes in metal welding and melting, loads in inductive heating systems, etc.) can act only as loads every time.

4.9.1 PV arrays

With the massive adoption of solar energy systems in both domestic and utility installations, safe and optimal operation are essential to ensure that the PV array will last the expected useful life and obtain the return on investment in the shortest time. Partial shading is a common issue affecting PV arrays, and it occurs when only some portion of a panel is shaded, while the remaining part is under solar light. Two major concerns are associated with this situation:

- Occurrence of multiple maximum power points: this can lead to the algorithm of maximum power point tracking (MPPT) becoming trapped in a nonoptimal power point.
- Formation of hot spots inside the panel due to reverse biasing of shadowed cells, which can lead to permanent destruction of the silicon in the affected area.

Figure 4.14 shows a scheme of a PV array with objects (e.g., tree leaves) blocking a small area of its surface and provoking a partial shading. For this example, the characteristic I–V curve at the panel's output is shown in a continuous line in Figure 4.15; the altered results are represented in the dashed line shown in the same figure.

MPPT algorithms are already embedded in any microprocessor controlling an inverter in a PV system and typically use the "perturb and observe" technique, which involves continuously trying to catch the operating point where the product $V * I$ is

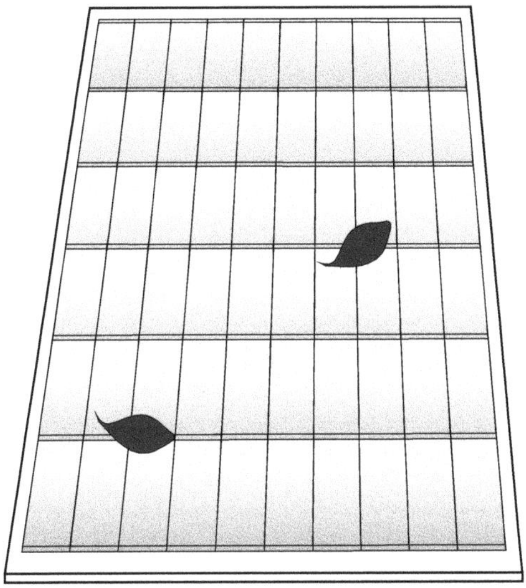

Figure 4.14 Example of a PV array with the effect of partial shading

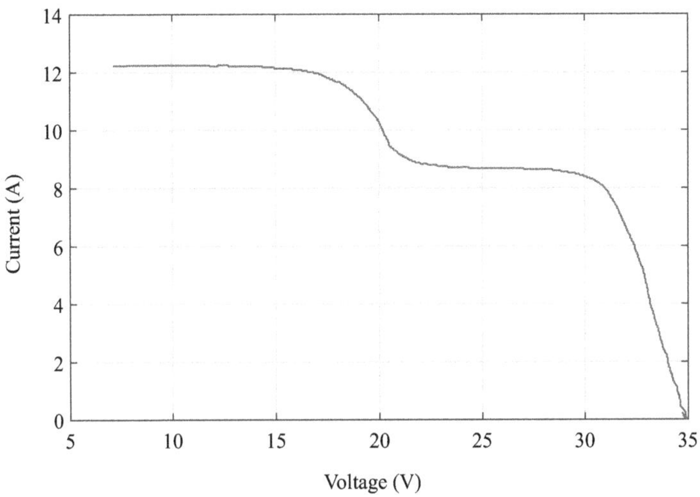

Figure 4.15 Characteristic curve V vs. I for a PV array under the effect of partial shading

at its maximum. An equivalent technique can be used to detect the inflection in the I–V curve by calculating first and second derivatives and using these results as an indication of a partial shading event happening. In Figure 4.16, an example of derivatives is shown for the curve shown in Figure 4.15.

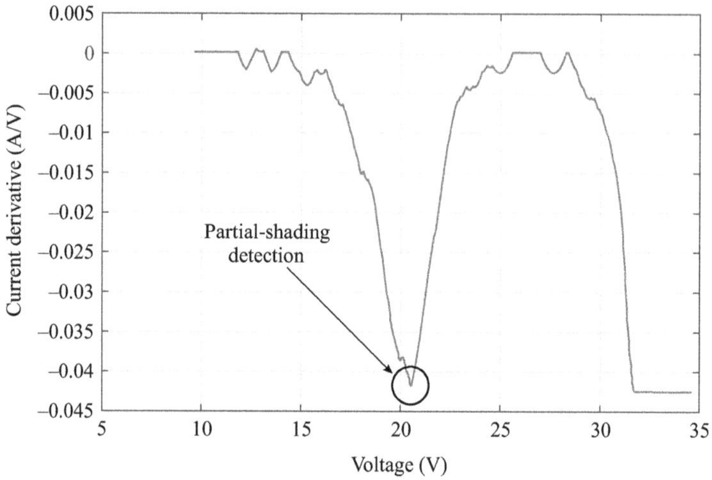

Figure 4.16 Detection of partial shading by using numerical derivatives in the V vs. I curve

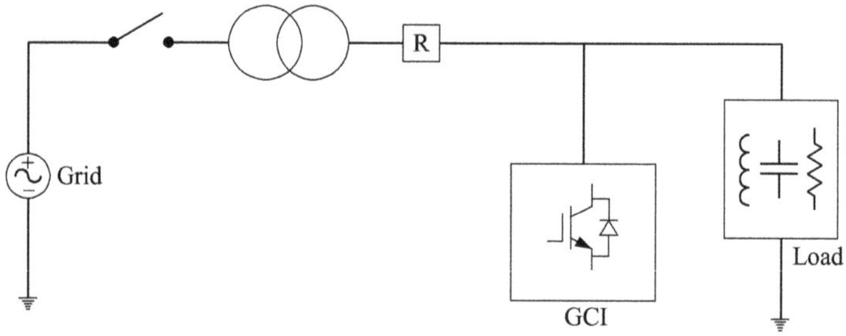

Figure 4.17 Grid connected inverter operating in islanding mode

4.9.2 Detection of islanding in grid connected inverters

The massive installation of domiciliary GCIs that use PV arrays makes possible the injection of surplus power to the grid. In most countries where this kind of operation is allowed, it is required that the GCI not operate when a blackout in the mains occurs, which is known as islanding mode. Figure 4.17 shows a GCI operating in islanding mode when an interruption of electrical service from the utility occurs. In those situations, it is required that relay *R* disconnects the GCI from the grid.

The most common standards for islanding detection methods are IEEE 1547 [71] and IEC 62116 [72]; however, many countries have specific regulations that may differ from these standards. Detection methods have to be designed to accomplish with the disconnection times stated in the corresponding norms. Figure 4.18 shows a classification of islanding detection methods, where in turn, those resident in the inverter can be classified as either passive or active. While passive methods are based

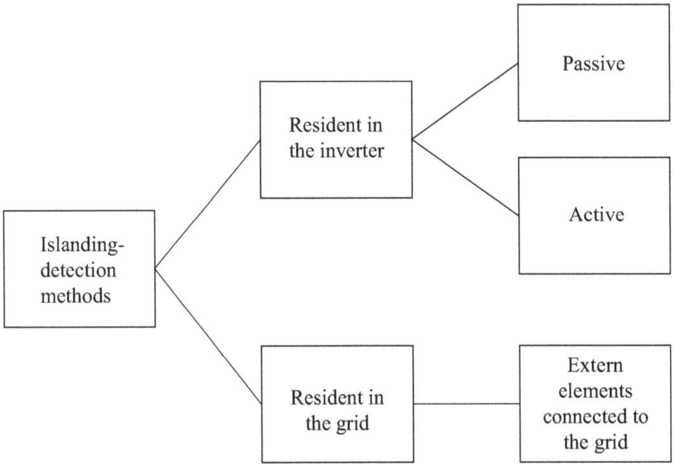

Figure 4.18 Classification of anti-islanding methods

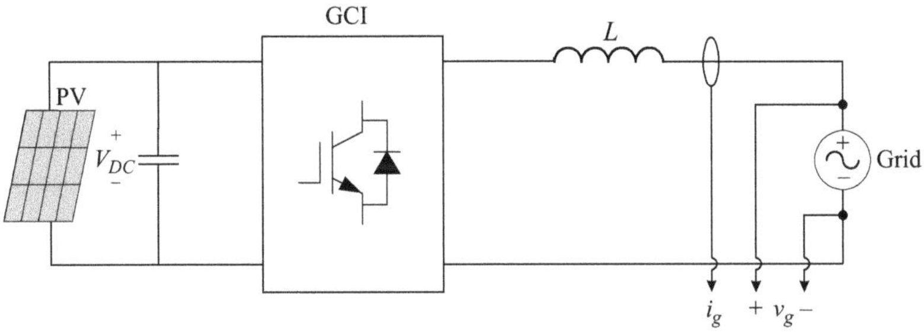

Figure 4.19 Electrical scheme for a grid connected inverter

on the monitoring of some grid parameters such as frequency, voltage amplitude, voltage phase jump detection, and voltage harmonic contents, active methods exert some kind of electrical perturbation and observe the grid behavior. The only goal of any islanding detection scheme present in the inverter is the disconnection from the grid, so they have to be embeddable by definition.

Normal operation of GCI involves the measuring of mains voltage and current through the coupling inductor as is depicted in Figure 4.19. Embedded algorithms typically do not require additional sensors, and their operation is based on these two variables.

Quoc-Tuan presents in [73] the following classification of active methods for islanding detection:

- Active methods by impedance monitoring;
- Active methods by frequency: frequency shift, active frequency drift, and frequency jump; and
- Active methods by voltage: voltage shift and voltage feedback.

All active methods alter the inverter's ongoing control signal to generate the desired perturbation on the grid and use the sensors to capture the grid response. The main advantage of active methods over passive ones is a better detection performance, but at the cost of introducing perturbation in the grid and having longer detection times [74]. In [75], Deshbhratar *et al.* present a comparison of passive and active detection methods, concluding that the selection of a specific method will be strongly dependent on the country's electrical code.

4.9.3 Condition monitoring in electric machines

The use of measured currents using the same sensors included in the inverter operating as the drive is the most common approach for detection of faults in electrical machines [76]. Major faults in electric machines can be classified as electrical or mechanical. Among the electrical anomalies can be mentioned: degradation of insulation or short-circuit in stator and rotor windings (phase-to-phase, phase-to-ground, or turn-to-turn), and broken rotor bar and end ring faults. The most common mechanical faults are rotor eccentricity, and degradation of gearboxes and bearings [77]. The major challenge for a CM or a fault detection method lies in identifying the motor current signature under healthy operation, which is mixed with electrical noise due to switching effects. Another drawback of this technique is that it cannot distinguish among faults that have the same signature frequencies. Figure 4.20 shows a stator current spectrogram for an IM where the ridges corresponding to fundamental and third harmonic are compared with those corresponding to broken bars and eccentricity fault [78].

Depending on the control strategy applied to the inverter, the switching frequency can be fixed (carrier-based modulation schemes) or variable (direct torque control and other noncarrier-based controls). Having a fixed switching frequency simplifies the filtering of undesired spectrum components. Additionally, some phenomena generate a more pronounced effect on the measured current than others. For example, an easier detection is expected from a short-circuit between turns in the stator winding than from a damaged ball inside a bearing. While many reported methods use analysis of temperature or mechanical signals such as acoustic and/or torque pulsations [79], in this section, a review of methods based exclusively on currents measured at the output of the inverter are considered.

Over the years, many methods originally proposed as offline have been revised and implemented as online ones, because of the required computation power and ADC resolution. In [80], a method and an experimental setup for stator winding turn faults detection in induction motors is proposed by Lee *et al.* The method relies on an off-diagonal term of the sequence component impedance matrix, which is immune to supply voltage unbalance variation and measurement errors. An online embedded implementation is shown that uses existing current and voltage sensors. The method presented by Mihet-Popa *et al.* in [81] uses readings from a digital oscilloscope to compare with the measurements obtained from the drive's sensors when they are used for CM and fault detection in an IM. In [82] Akin *et al.* propose a real-time fault diagnosis system where the reference frame theory used in control of electric machines is applied to find signatures associated with different machine faults such as broken

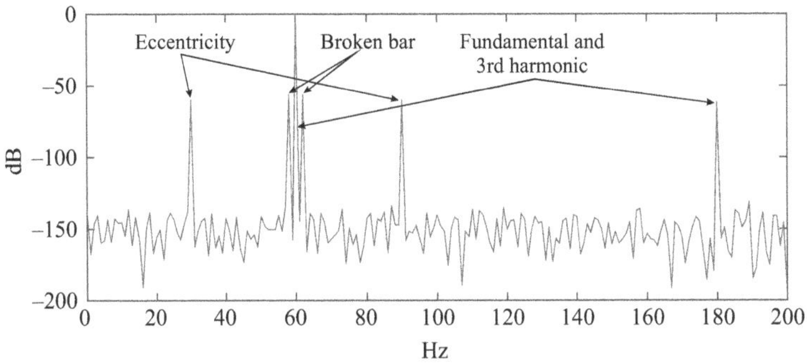

Figure 4.20 Spectrogram for the stator current in an induction machine

rotor bar and eccentricity. The system requires a DSP board and the same sensors used for the control loop of the machine. By using the reference frame theory approach, the analysis of the spectrum of the current is simplified, and few fault-related harmonics are checked by FFT computation. Sampling frequencies of 20 kHz and ADC resolution of 12 bits prove to be enough for reliable detection of faults. In [83,84], Rajagopalan *et al.* present two embedded methods for detection of rotor faults in brushless DC machines. In the first method, the stator current is processed by using a windowed Fourier transform to determine the ridges of the spectrogram. The second method is based on Wigner–Ville distributions and allows a better frequency resolution at the cost of being computationally more intensive. Both methods are implemented in a DSP and tested in the detection of mechanically unbalanced rotor and rotor dynamic eccentricity.

Electrical machines used as generators can also be monitored for detection of anomalies. That is the case, for example, with wind turbines (WTs), which mostly use doubly fed induction generators. The signals used for CM and fault diagnosis in WT mainly include vibration, acoustic emission, strain, torque, temperature, and electrical variables. In the same way as in electrical motors, due to electromechanical coupling between rotatory and static parts of an electrical generator in WT, electrical signals contain information of possible faults in mechanical parts, for example, surface condition of WT blades, generator rotor imbalance, and gearbox and bearing failures. An example is presented by Duviella *et al.* in [85] where a classification algorithm for WTs in a wind farm is proposed. The method allows the detection, isolation, and forecast of sudden and gradual faults in the wind farm. A typical gradual fault would be debris build-up on the turbine blades, which modifies the aerodynamics of the WT and the generated electric power for a given wind speed. The method is based on differences in generated electric power, it is data-driven and does not require prior physical knowledge.

4.9.4 State of health in batteries

Batteries are a fundamental part of any power electronics system requiring energy storage. The use of batteries is widespread in off-the-grid PV systems, EV and

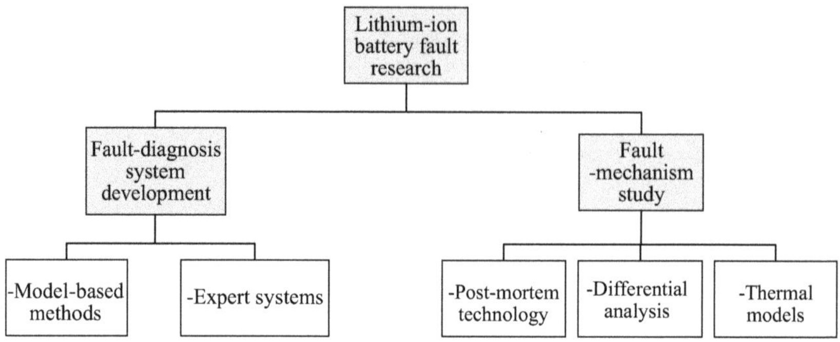

Figure 4.21 Current fault research for Li-ion batteries

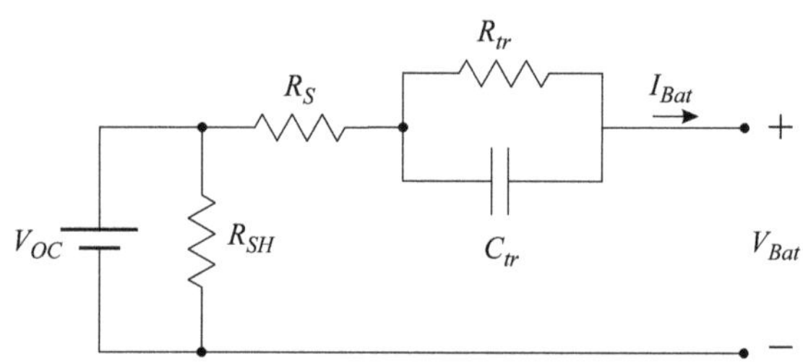

Figure 4.22 Basic Thevenin electrical battery model

hybrid-electric vehicle (HEV), and in uninterruptible power supplies. At present lead acid, nickel–metal hydride (NiMH), and lithium ion (Li-ion) are the most used technologies; the latter has the greatest use increase because of its massive incorporation in EVs and HEVs [86]. In [87], Wu *et al.* present an overview of the fault research for lithium-ion batteries which is shown in Figure 4.21. More specifically, the fault diagnosis techniques applied to this technology can be classified as model-based methods or expert systems. Many battery equivalent circuit models have been proposed that try to precisely estimate the battery state of charge and state of health (SoH). While some models require information about the battery's inside temperature and/or electrolyte characteristics, those methods based on the measuring of electric signals are the best candidates for an embedded implementation intended for online monitoring. For those models, there are just two electrical variables to consider: voltage in the battery main terminals and input/output current. Figure 4.22 shows one the simplest electrical models for a battery, based on a constant voltage source V_{OC} representing the open-circuit voltage, a resistance R_{SH} representing the internal self-discharge, a resistance R_S representing ohmic effects in steady state, and a resistive-capacitive (RC) network (R_{tr} and C_{tr}) representing the dynamics of the battery during transient states [88,89].

In a similar work by Wu *et al.* [87], a fault diagnosis and prognosis method for Li-ion batteries based on fuzzy logic is proposed. The authors use the technique of incremental capacity (IC) analysis, which is based on low rate charge/discharge voltage curves to derive information from the inner electrodes. The method evaluates the degradation of the battery. For real applications, the reliability and real-time issue should be considered, that is, the computation power required by the parameter estimation algorithm and IC analysis should fit the target CPU.

In [90], Pastor-Fernández *et al.* use IC and differential voltage (DV) techniques to identify and quantify degradation modes as well as to estimate and to predict the SoH of Li-ion batteries based on capacity. These techniques involve the measuring of voltage and charge/discharge current. Integration of current is used to obtain battery charge, and derivation is used to obtain gradients of charge with respect to battery voltage. While that work mainly covers an offline analysis based on the study of peaks in IC/DV curves, which allows the identification of degradation modes, it is potentially embeddable.

A real-time model-based monitoring system is proposed by Kim *et al.* in [91]. It uses a second-order RC electrical model, where the parameters are tuned by means of an extended Kalman filter. The entire algorithm is tested using an Arduino 2560, which is a low-cost microcontroller, demonstrating its embeddability.

4.9.5 Fault diagnosis in sensors

In recent years, some works have been reported addressing the issue of fault detection in the current and voltage sensors included in the inverters. In general, the proposed methods use different approaches based on state observers and knowledge of the switching state of the inverter. The authors in [92] present a case of fault detection in sensors inside a power converter. In that study, a failure in the DC-link voltage sensor is detected and the measured voltage is replaced by an estimation. In [93], a method for fault detection in current sensors in an electric railway traction system is proposed. The method is based on the fact that the sum of stator currents in a three-phase star connected induction motor is zero whether the IGBTs are faulty or not. This method can identify three different fault modes in the current sensors within one fundamental period. The method presented in [94] can detect faults in sensors caused by gain, offset, or open-circuit faults. The method is applied to a three-phase four-leg inverter operated with a predictive control algorithm.

References

[1] George Vachtsevanos, George Georgoulas, and George Nikolakopoulos. Fault diagnosis, failure prognosis and fault tolerant control of aerospace/unmanned aerial systems. In *2016 24th Mediterranean Conference on Control and Automation (MED)*, pages 366–371. IEEE, Jun 2016.

[2] Shaoyong Yang, Dawei Xiang, Angus Bryant, Philip Mawby, Li Ran, and Peter Tavner. Condition Monitoring for Device Reliability in Power Electronic Converters: A Review. *IEEE Transactions on Power Electronics*, 25(11): 2734–2752, Nov 2010.

[3] Yvan Avenas, Laurent Dupont, Nick Baker, Henri Zara, and Franck Barruel. Condition Monitoring: A Decade of Proposed Techniques. *IEEE Industrial Electronics Magazine*, 9(4):22–36, Dec 2015.

[4] Canh Ly, Kwok Tom, Carl S. Byington, Romano Patrick, and George J. Vachtsevanos. Fault diagnosis and failure prognosis for engineering systems: A global perspective. In *2009 IEEE International Conference on Automation Science and Engineering*, pages 108–115. IEEE, Aug 2009.

[5] K. Wojtek Przytula and Arthur Choi. Reasoning Framework for Diagnosis and Prognosis. In *2007 IEEE Aerospace Conference*, pages 1–10. IEEE, Mar 2007.

[6] Noriyuki Iwamuro and Thomas Laska. IGBT History, State-of-the-Art, and Future Prospects. *IEEE Transactions on Electron Devices*, 64(3):741–752, Mar 2017.

[7] Vinod K. Khanna. IGBT Fundamentals and Status Review. In *Insulated Gate Bipolar Transistor IGBT Theory and Design*, pages 35–100. John Wiley & Sons, Inc., Hoboken, NJ, USA, Jan 2005.

[8] Ui-Min Choi, Frede Blaabjerg, and Kyo-Beum Lee. Study and Handling Methods of Power IGBT Module Failures in Power Electronic Converter Systems. *IEEE Transactions on Power Electronics*, 30(5):2517–2533, May 2015.

[9] Rahul S. Chokhawala, Jamie Catt, and Laszlo Kiraly. A Discussion on IGBT Short-Circuit Behavior and Fault Protection Schemes. *IEEE Transactions on Industry Applications*, 31(2):256–263, Mar/Apr 1995.

[10] Rui Wu, Frede Blaabjerg, Huai Wang, Marco Liserre, and Francesco Iannuzzo. Catastrophic failure and fault-tolerant design of IGBT power electronic converters—an overview. In *IECON 2013—39th Annual Conference of the IEEE Industrial Electronics Society*, pages 507–513. IEEE, Nov 2013.

[11] Ly. Benbahouche, A. Merabet, and A. Zegadi. A comprehensive analysis of failure mechanisms: Latch up and second breakdown in IGBT(IXYS) and improvement. In *2012 19th International Conference on Microwaves, Radar & Wireless Communications*, pages 190–192. IEEE, May 2012.

[12] Bantval J. Baliga. Insulated gate bipolar transistor with reduced susceptibility to parasitic latch-up. U.S. Patent US5396087, 1995.

[13] Fei Yang, Ji Tan, Guangyin Zhang, *et al.* A Novel IGBT Structure With Floating N-Doped Buried Layer in P-Base to Suppress Latch-Up. *IEEE Electron Device Letters*, 37(9):1174–1177, Sep 2016.

[14] Kristian Bonderup Pedersen and Kjeld Pedersen. Bond wire lift-off in IGBT modules due to thermomechanical induced stress. In *2012 3rd IEEE International Symposium on Power Electronics for Distributed Generation Systems (PEDG)*, pages 519–526. IEEE, Jun 2012.

[15] Wuchen Wu, Guo Gao, Limin Dong, *et al.* Thermal reliability of power insulated gate bipolar transistor (IGBT) modules. In *Twelfth Annual IEEE Semiconductor Thermal Measurement and Management Symposium. Proceedings*, pages 136–141. IEEE, Mar 1997.

[16] G. A. Lang, B. J. Fehder, and W. D. Williams. Thermal Fatigue in Silicon Power Transistors. *IEEE Transactions on Electron Devices*, 17(9):787–793, Sep 1970.

[17] Swagat Das, Faisal Khan, Mohammed K. Alam, and Preetham Goli. Detection of aging related IGBT bond-wire lift-off using spread spectrum time domain reflectometry (SSTDR). In *2017 IEEE Applied Power Electronics Conference and Exposition (APEC)*, pages 789–794. IEEE, Mar 2017.

[18] Michael G. Pecht. *Prognostics and Health Management of Electronics*. John Wiley & Sons, Inc., Hoboken, NJ, USA, Aug 2008.

[19] Serkan Dusmez, Syed H. Ali, Mehrdad Heydarzadeh, Anant S. Kamath, Hamit Duran, and Bilal Akin. Aging Precursor Identification and Lifetime Estimation for Thermally Aged Discrete Package Silicon Power Switches. *IEEE Transactions on Industry Applications*, 53(1):251–260, Jan 2017.

[20] M. Held, P. Jacob, G. Nicoletti, P. Scacco, and M.-H. Poech. Fast power cycling test of IGBT modules in traction application. In *Proceedings of Second International Conference on Power Electronics and Drive Systems*, volume 1, pages 425–430. IEEE, May 1997.

[21] Mohamed Tlig, Jaleleddine B. H. Slama, and M. A Belaid. Experimental study on the EMI and switching time evolution of IGBT devices after operating aging tests. In *2014 International Conference on Electrical Sciences and Technologies in Maghreb (CISTEM)*, pages 1–5. IEEE, Nov 2014.

[22] Vanessa Smet, François Forest, Jean-Jacques Huselstein, Amgad Rashed, and Frédéric Richardeau. Evaluation of Vce Monitoring as a Real-Time Method to Estimate Aging of Bond Wire-IGBT Modules Stressed by Power Cycling. *IEEE Transactions on Industrial Electronics*, 60(7):2760–2770, Jul 2013.

[23] Greg Sonnenfeld, Kai Goebel, and Jose R. Celaya. An agile accelerated aging, characterization and scenario simulation system for gate controlled power transistors. In *2008 IEEE AUTOTESTCON*, pages 208–215. IEEE, Sep 2008.

[24] Jan Knobloch, Jan Martis, Radoslav Cipin, and Ziad Nouman. Test stand for obtaining power transistors switching characteristics during aging. In *2017 IEEE International Conference on Environment and Electrical Engineering and 2017 IEEE Industrial and Commercial Power Systems Europe (EEEIC/I&CPS Europe)*, pages 1–5. IEEE, Jun 2017.

[25] Dimosthenis C. Katsis and Jacobus Daniel van Wyk. Void-Induced Thermal Impedance in Power Semiconductor Modules: Some Transient Temperature Effects. *IEEE Transactions on Industry Applications*, 39(5):1239–1246, Sep 2003.

[26] Tien A. Nguyen, Stephane Lefebvre, Pierre-Yves Joubert, Denis Labrousse, and Serge Bontemps. Estimating Current Distributions in Power Semiconductor Dies Under Aging Conditions: Bond Wire Liftoff and Aluminum Reconstruction. *IEEE Transactions on Components, Packaging and Manufacturing Technology*, 5(4):483–495, Apr 2015.

[27] Bing Ji, Volker Pickert, and Bashar Zahawi. In-situ bond wire and solder layer health monitoring circuit for IGBT power modules. In *Integrated Power Electronics Systems (CIPS), 2012 7th International Conference on*, volume 9, pages 1–6. Mar 2012.

[28] Mohd. A. Eleffendi and C. Mark Johnson. In-Service Diagnostics for Wire-Bond Lift-off and Solder Fatigue of Power Semiconductor Packages. *IEEE Transactions on Power Electronics*, 32(9):7187–7198, Sep 2017.

[29] Jean-Marc Thébaud, Eric Woirgard, Christian Zardini, et al. Strategy for Designing Accelerated Aging Tests to Evaluate IGBT Power Modules Lifetime in Real Operation Mode. *IEEE Transactions on Components and Packaging Technologies*, 26(2):429–438, Jun 2003.

[30] Pramod Ghimire, Szymon Beczkowski, Stig Munk-Nielsen, Bjorn Rannestad, and Paul B. Thogersen. A review on real time physical measurement techniques and their attempt to predict wear-out status of IGBT. In *2013 15th European Conference on Power Electronics and Applications (EPE)*, pages 1–10. IEEE, Sep 2013.

[31] Szymon Bęczkowski, Pramod Ghimre, Angel Ruiz de Vega, et al. Online Vce measurement method for wear-out monitoring of high power IGBT modules. In *2013 15th European Conference on Power Electronics and Applications (EPE)*, pages 1–7. Sep 2013.

[32] M. Sultana Nasrim and Faisal H. Khan. Characterization of aging process in power converters using spread spectrum time domain reflectometry. In *2012 IEEE Energy Conversion Congress and Exposition (ECCE)*, pages 2142–2148. IEEE, Sep 2012.

[33] Zoltan Sarkany, Andras Vass-Varnai, Gusztav Hantos, and Marta Rencz. Failure prediction of IGBT modules based on power cycling tests. In *19th International Workshop on Thermal Investigations of ICs and Systems (THERMINIC)*, pages 270–273. IEEE, Sep 2013.

[34] Antonio E. Ginart, Douglas W. Brown, Patrick W. Kalgren, and Michael J. Roemer. Online Ringing Characterization as a Diagnostic Technique for IGBTs in Power Drives. *IEEE Transactions on Instrumentation and Measurement*, 58(7):2290–2299, Jul 2009.

[35] Ralf Schmidt and Uwe Scheuermann. Using the chip as a temperature sensor -the influence of steep lateral temperature gradients on the Vce(t)-measurement. In *2009 13th European Conference on Power Electronics and Applications*, pages 1–9. Sep 2009.

[36] Huifeng Chen, Bing Ji, Volker Pickert, and Wenping Cao. Real-Time Temperature Estimation for Power MOSFETs Considering Thermal Aging Effects. *IEEE Transactions on Device and Materials Reliability*, 14(1):220–228, Mar 2014.

[37] Kai Rothenhagen and Friedrich W. Fuchs. Performance of diagnosis methods for IGBT open circuit faults in three phase voltage source inverters for AC variable speed drives. In *2005 European Conference on Power Electronics and Applications*, pages P.1–P.10. IEEE, Sep 2005.

[38] Bin Lu and Santosh Sharma. A literature review of IGBT fault diagnostic and protection methods for power inverters. In *2008 IEEE Industry Applications Society Annual Meeting*, pages 1–8. IEEE, Oct 2008.

[39] Tamer Kamel, Yevgen Biletskiy, Christopher P. Diduh, and Liuchen Chang. Open circuit fault diagnoses for power electronic converters. In *2015 IEEE 24th International Symposium on Industrial Electronics (ISIE)*, pages 361–366. IEEE, Jun 2015.

[40] Tamer Kamel, Yevgen Biletskiy, and Liuchen Chang. Fault Diagnoses for Industrial Grid-Connected Converters in the Power Distribution Systems. *IEEE Transactions on Industrial Electronics*, 62(10):6496–6507, Oct 2015.

[41] Jorge O. Estima and Antonio J. M. Cardoso. Single power switch open-circuit fault diagnosis in voltage-fed PWM motor drives by the reference current errors. In *8th IEEE Symposium on Diagnostics for Electrical Machines, Power Electronics & Drives*, pages 364–371. IEEE, Sep 2011.

[42] Jorge O. Estima and Antonio J. M. Cardoso. A New Approach for Real-Time Multiple Open-Circuit Fault Diagnosis in Voltage-Source Inverters. *IEEE Transactions on Industry Applications*, 47(6):2487–2494, Nov 2011.

[43] Cheol-Ju Bae, Sang-Min Lee, and Dong-Choon Lee. Diagnosis of multiple IGBT open-circuit faults for three-phase PWM inverters. In *2016 IEEE 8th International Power Electronics and Motion Control Conference (IPEMC-ECCE Asia)*, pages 2504–2509. IEEE, May 2016.

[44] Hao Yan, Yongxiang Xu, Jibin Zou, Yuan Fang, and Feiyang Cai. A Novel Open-Circuit Fault Diagnosis Method for Voltage Source Inverters with Single Current Sensor. *IEEE Transactions on Power Electronics*, 1–12, Nov 2017.

[45] Yongxiang Xu, Hao Yan, Jibin Zou, Baochao Wang, and Yunhui Li. Zero Voltage Vector Sampling Method for PMSM Three-Phase Current Reconstruction Using Single Current Sensor. *IEEE Transactions on Power Electronics*, 32(5):3797–3807, May 2017.

[46] Mohsen B. Abadi, A. M. S. Mendes, and S. M. A. Cruz. Three-level NPC inverter fault diagnosis by the Average Current Park's Vector approach. In *2012 XXth International Conference on Electrical Machines*, pages 1893–1898. IEEE, Sep 2012.

[47] Feng Wu and Jin Zhao. A Real-Time Multiple Open-Circuit Fault Diagnosis Method in Voltage-Source-Inverter Fed Vector Controlled Drives. *IEEE Transactions on Power Electronics*, 31(2):1425–1437, Feb 2016.

[48] Hyun-Woo Sim, June-Seok Lee, and Kyo-Beum Lee. Detecting Open-Switch Faults: Using Asymmetric Zero-Voltage Switching States. *IEEE Industry Applications Magazine*, 22(2):27–37, Mar 2016.

[49] Dehong Zhou, Yunhua Li, Jin Zhao, Feng Wu, and Hui Luo. An Embedded Closed-Loop Fault-Tolerant Control Scheme for Nonredundant VSI-Fed Induction Motor Drives. *IEEE Transactions on Power Electronics*, 32(5):3731–3740, May 2017.

[50] Yang Liu, Feng Wu, and Jin Zhao. Symmetry-Analysis-based Diagnosis Method with Correlation Coefficients for Open-Circuit Fault in Inverter. *Electronics Letters*, 51(21):1688–1690, Oct 2015.

[51] Baoping Cai, Yubin Zhao, Hanlin Liu, and Min Xie. A Data-Driven Fault Diagnosis Methodology in Three-Phase Inverters for PMSM Drive Systems. *IEEE Transactions on Power Electronics*, 32:5590–5600, Sep 2016.

[52] C. Choi and W. Lee. Design and Evaluation of Voltage Measurement-Based Sectoral Diagnosis Method for Inverter Open Switch Faults of

Permanent Magnet Synchronous Motor Drives. *IET Electric Power Applications*, 6(8):526–532, Aug 2012.

[53] L. Z. Sun, Q. T. An, and L. Sun. Hardware-Circuit-based Diagnosis Method for Open-Switch Faults in Inverters. *Electronics Letters*, 49(17):1089–1091, Aug 2013.

[54] Wang Baocheng, Wang Jie, Sun Xiaofeng, Wu Junjuan, and Wu Weiyang. Phase multilevel inverter fault diagnosis and tolerant control technique. In *2006 CES/IEEE 5th International Power Electronics and Motion Control Conference*, pages 1–5. IEEE, Aug 2006.

[55] Piotr Sobanski and Teresa Orlowska-Kowalska. Experimental validation of the simple voltage-vector-location-based method of open-circuit IGBTs faults in DTC-SVM induction motor drive. In *2015 International Conference on Electrical Drives and Power Electronics (EDPE)*, pages 19–24. IEEE, Sep 2015.

[56] Mustafa Sen, Muhannad Alaraj, and Jae-Do Park. Open circuit fault detection and localization in modular multilevel converter. In *2016 North American Power Symposium (NAPS)*, pages 1–6. IEEE, Sep 2016.

[57] Xu Hao, Zhang Jian, Qi Jie, Wang Tianzhen, and Han Jingang. RPCA-SVM fault diagnosis strategy of cascaded H-bridge multilevel inverters. In *2014 First International Conference on Green Energy ICGE 2014*, pages 164–169. IEEE, Mar 2014.

[58] V. Fernao Pires, D. Foito, and Tito G. Amaral. Fault detection and diagnosis in a PV grid-connected T-type three level inverter. In *2015 International Conference on Renewable Energy Research and Applications (ICRERA)*, pages 933–937. IEEE, Nov 2015.

[59] Danjiang Chen, Yinzhong Ye, and Rong Hua. Fault diagnosis of three-level inverter based on wavelet analysis and Bayesian classifier. In *2013 25th Chinese Control and Decision Conference (CCDC)*, pages 4777–4780. IEEE, May 2013.

[60] Jacob Lamb and Behrooz Mirafzal. Open-Circuit IGBT Fault Detection and Location Isolation for Cascaded Multilevel Converters. *IEEE Transactions on Industrial Electronics*, 64(6):4846–4856, Jun 2017.

[61] Binbin Li, Shaolei Shi, Bo Wang, Gaolin Wang, Wei Wang, and Dianguo Xu. Fault Diagnosis and Tolerant Control of Single IGBT Open-Circuit Failure in Modular Multilevel Converters. *IEEE Transactions on Power Electronics*, 31(4):3165–3176, Apr 2016.

[62] Luís M. A. Caseiro and André M. S. Mendes. Generalized IGBT open-circuit fault diagnosis algorithm with no additional sensors for grid-connected N-level NPC converters. In *IECON 2015—41st Annual Conference of the IEEE Industrial Electronics Society*, pages 2421–2426. IEEE, Nov 2015.

[63] André M. S. Mendes, Mohsen B. Abadi, and Sérgio M. Á. Cruz. Method to Diagnose Open-Circuit Faults in Active Power Switches and Clamp-Diodes of Three-Level Neutral-Point Clamped Inverters. *IET Electric Power Applications*, 10(7):623–632, Aug 2016.

[64] Jalal Amini and Mehrdad Moallem. A Fault-Diagnosis and Fault-Tolerant Control Scheme for Flying Capacitor Multilevel Inverters. *IEEE Transactions on Industrial Electronics*, 64(3):1818–1826, Mar 2017.

[65] Andre B. de Mello Oliveira, Robson L. Moreno, and Enio R. Ribeiro. Short circuit fault diagnosis in switches of a single-phase full-bridge inverter. In *2013 Brazilian Power Electronics Conference*, pages 1107–1113. IEEE, Oct 2013.

[66] Surin Khomfoi, Leon M. Tolbertt, and Burak Ozpineci. Cascaded H-bridge multilevel inverter drives operating under faulty condition with AI-based fault diagnosis and reconfiguration. In *2007 IEEE International Electric Machines & Drives Conference*, pages 1649–1656. IEEE, May 2007.

[67] Marco Rodríguez, Abraham Claudio, Didier Theilliol, and Luis Vela. A new fault detection technique for IGBT based on gate voltage monitoring. In *2007 IEEE Power Electronics Specialists Conference*, pages 1001–1005. IEEE, Oct 2007.

[68] Marco A. Rodriguez-Blanco, Abraham Claudio-Sanchez, Didier Theilliol, *et al.* A Failure-Detection Strategy for IGBT Based on Gate-Voltage Behavior Applied to a Motor Drive System. *IEEE Transactions on Industrial Electronics*, 58(5):1625–1633, May 2011.

[69] Jorge F. Marques, Jorge O. Estima, Natalia S. Gameiro, and António J. Marques Cardoso. A new diagnostic technique for real-time diagnosis of power converter faults in switched reluctance motor drives. In *2013 Twenty-Eighth Annual IEEE Applied Power Electronics Conference and Exposition (APEC)*, pages 1192–1198. IEEE, Mar 2013.

[70] Vivek Sharma, Nikita Rawat, and Pulkit Garg. Fault diagnosis of single phase Z-source inverter. In *2015 Second International Conference on Advances in Computing and Communication Engineering*, pages 108–111. IEEE, May 2015.

[71] IEEE standard for interconnection and interoperability of distributed energy resources with associated electric power systems interfaces. *IEEE Std 1547-2018 (Revision of IEEE Std 1547-2003)*, pages 1–138, Apr 2018.

[72] Test procedure of islanding prevention measures for utility-interconnected photovoltaic inverters. *IEC Std. 62116*, 2008.

[73] Tran Quoc-Tuan. New methods of islanding detection for photovoltaic inverters. In *2016 IEEE PES Innovative Smart Grid Technologies Conference Europe (ISGT-Europe)*, pages 1–5, Oct 2016.

[74] S. C. Paiva, H. S. Sanca, F. B. Costa, and B. A. Souza. Reviewing of anti-islanding protection. In *2014 11th IEEE/IAS International Conference on Industry Applications*, pages 1–8. Dec 2014.

[75] Prajakta Deshbhratar, Rahul Somalwar, and Sumant G. Kadwane. Comparative analysis of islanding detection methods for multiple dg based system. In *2016 International Conference on Electrical, Electronics, and Optimization Techniques (ICEEOT)*, pages 1525–1530. Mar 2016.

[76] Yao Da, Xiaodong Shi, and Mahesh Krishnamurthy. Health monitoring, fault diagnosis and failure prognosis techniques for Brushless Permanent Magnet

Machines. In *2011 IEEE Vehicle Power and Propulsion Conference*, pages 1–7. IEEE, Sep 2011.

[77] Subhasis Nandi, Hamid A. Toliyat, and Xiaodong Li. Condition Monitoring and Fault Diagnosis of Electrical Motors—A Review. *IEEE Transactions on Energy Conversion*, 20(4):719–729, Dec 2005.

[78] Seungdeog Choi, Bilal Akin, Mina M. Rahimian, and Hamid A. Toliyat. Performance-Oriented Electric Motors Diagnostics in Modern Energy Conversion Systems. *IEEE Transactions on Industrial Electronics*, 59(2):1266–1277, Feb 2012.

[79] Siliang Lu, Qingbo He, Tao Yuan, and Fanrang Kong. Online Fault Diagnosis of Motor Bearing via Stochastic-Resonance-Based Adaptive Filter in an Embedded System. *IEEE Transactions on Systems, Man, and Cybernetics: Systems*, 47(7):1111–1122, Jul 2017.

[80] Sang B. Lee, Rangarajan M. Tallam, and Thomas G. Habetler. A Robust, On-line Turn-Fault Detection Technique for Induction Machines based on Monitoring the Sequence Component Impedance Matrix. *IEEE Transactions on Power Electronics*, 18(3):865–872, May 2003.

[81] Lucian Mihet-Popa and Mario Pacas. Failure detection in converter fed induction machines under different operation conditions. In *IEEE International Conference on Electric Machines and Drives, 2005*, pages 967–974. IEEE, May 2005.

[82] Bilal Akin, Seungdeog Choi, Umut Orguner, and Hamid A. Toliyat. A Simple Real-Time Fault Signature Monitoring Tool for Motor-Drive-Embedded Fault Diagnosis Systems. *IEEE Transactions on Industrial Electronics*, 58(5):1990–2001, May 2011.

[83] Satish Rajagopalan, Thomas Habetler, Ronald Harley, Jose Restrepo, and Jose Aller. Non-stationary motor fault detection using recent quadratic time-frequency representations. In *Conference Record of the 2006 IEEE Industry Applications Conference Forty-First IAS Annual Meeting*, volume 5, pages 2333–2339. Oct 2006.

[84] Satish Rajagopalan, Jose Aller, Jose Restrepo, Thomas Habetler, and Ronald Harley. Detection of Rotor Faults in Brushless dc Motors Operating Under Nonstationary Conditions. *IEEE Transactions on Industry Applications*, 42(6):1464–1477, Nov 2006.

[85] Eric Duviella, Lisa Serir, and Moamar Sayed-Mouchaweh. An evolving classification approach for fault diagnosis and prognosis of a wind farm. In *2013 Conference on Control and Fault-Tolerant Systems (SysTol)*, pages 377–382. IEEE, Oct 2013.

[86] Aoxia Chen and P. K. Sen. Advancement in battery technology: A state-of-the-art review. In *2016 IEEE Industry Applications Society Annual Meeting*, pages 1–10. Oct 2016.

[87] Chao Wu, Chunbo Zhu, and Yunwang Ge. A New Fault Diagnosis and Prognosis Technology for High-Power Lithium-Ion Battery. *IEEE Transactions on Plasma Science*, 45(7):1533–1538, Jul 2017.

[88] Min Chen and Gabriel. A. Rincon-Mora. Accurate electrical battery model capable of predicting runtime and *I–V* performance. *IEEE Transactions on Energy Conversion*, 21(2):504–511, Jun 2006.

[89] Maciej Świerczyński, Daniel Ioan Stroe, Ana-Irina Stan, Remus Teodorescu, and Dirk Uwe Sauer. Selection and Performance-Degradation Modeling of $LiMO_2/Li_4Ti_5O_{12}$ and $LiFePO_4/C$ Battery Cells as Suitable Energy Storage Systems for Grid Integration with Wind Power Plants: An Example for the Primary Frequency Regulation Service. *IEEE Transactions on Sustainable Energy*, 5(1):90–101, Jan 2014.

[90] C. Pastor-Fernández, W. Dhammika Widanage, G. H. Chouchelamane, and J. Marco. A SoH diagnosis and prognosis method to identify and quantify degradation modes in Li-ion batteries using the IC/DV technique. In *6th Hybrid and Electric Vehicles Conference (HEVC 2016)*, pages 1–6. Institution of Engineering and Technology, Nov 2016.

[91] T. Kim, A. Adhikaree, R. Pandey, et al. A real-time condition monitoring for lithium-ion batteries using a low-price microcontroller. In *2017 IEEE Energy Conversion Congress and Exposition (ECCE)*, pages 5248–5253. Oct 2017.

[92] Gwang-Seob Kim and Kyo-Beum Lee. Fault diagnosis and fault-tolerant control of a dc-link voltage sensor for PV inverters. In *Proceedings of The 7th International Power Electronics and Motion Control Conference*, pages 1408–1412. IEEE, Jun 2012.

[93] Xing-Lai Ge, Xiao-Yun Feng, Bin Gou, and Yong-Chao Liu. Load-current-based current sensor fault diagnosis and tolerant control scheme for traction inverters. *Electronics Letters*, 52(20):1717–1719, Sep 2016.

[94] Mohammadreza Eskandari, Arash Ahmadi, Mohammad F. Samadi, and Mehrdad Saif. Hardware realization of a dq based fault detection scheme for 3ϕ four-leg inverter. In *2016 World Automation Congress (WAC)*, pages 1–6. IEEE, Jul 2016.

Chapter 5
Fault-tolerance strategies for power converters

Jose A. Restrepo[1,2], *Julio C. Viola*[3], *and Ronald Harley*[4]

Nowadays, with the high availability of power switches, power converters are present in most electric equipment, from industrial down to consumer grade applications. Reliable operation of these converters is still a concern [1,2], especially for critical operation systems with costly down time, and much research has been devoted to providing a possible solution [3–5]. Voltage source inverters (VSIs) are the most common type of converter used in practical industrial applications. According to [6], failures in the switching devices amount to 38% of their faults. For [7], semiconductor and soldering joints failures in power devices take up to 34% of failures in power electronic systems. Reliability in power converters can be improved either by increasing the power rating in the components or by using configurations with a certain degree of fault-tolerance capability [5]. The typical failures in power devices are short-circuit faults, open-circuit faults, and gate misfiring faults [1,6,8]. Also, much research has been devoted to fast detection of these faults [9]. Once a device enters into a fault state, its impact on the circuit is often seen by changes in voltages and currents. These changes or signatures can be detected and identified using different methods. Lu *et al.* [6] present a very comprehensive list for open-circuit fault-diagnosis methods, which includes the methods' effectiveness, detection time, implementation complexity, and variables employed for detection, among others. They are the following:

1. Park's method [8,10]. This method uses the current space vector trajectory's midpoint over one period for fault detection and faulty device location. This method depends on the load–current magnitude.
2. Normalized DC current and modified normalized DC current methods [8,11]. This method uses a normalized current and its first order discrete Fourier Transform (DFT) coefficients in each phase to identify the faulty device. The modified normalized DC current method uses only the largest coefficient among the three phases.

[1] Facultad de Ingeniería en Ciencias Aplicadas, Universidad Técnica del Norte, Ecuador
[2] Departamento de Electrónica y Circuitos, Universidad Simón Bolívar, Venezuela
[3] Carrera de Ingeniería Eléctrica, Universidad Politécnica Salesiana – Sede Cuenca, Ecuador
[4] School of Electrical and Computer Engineering, Georgia Institute of Technology, USA

3. Current trajectory [12]. This method detects the fault either by using an analysis of the current space vector trajectory in the Concordia frame or by using the instantaneous frequency. The first case uses a parameter representing the current trajectory's slope. The second case detects the fault when the current space vector's instantaneous frequency falls below a certain threshold. This method allows for detection of inverter faults but does not pinpoint the faulty device.
4. Simple direct current method [8,11]. This method uses the largest magnitude of the three-phase DC current values, and when compared with a threshold, it identifies the faulty leg. The faulty switch is identified by the polarity of the mean value.
5. Comparison of AC voltage actual and reference quantity approach in time domain [2,13]. This detection technique employs a direct comparison of the measured voltages to their reference voltages obtained from the pulse width modulation (PWM) reference signals. Detection times are in the range of one-fourth of a period [6].
6. Method using voltage across lower switch [14]. This method uses the voltage across the lower switch. It uses specialized circuitry for fault detection. The detection time achieved in simulations is about 2.7 ms.
7. Centroid-based fault detection [15]. The fault's type and location are identified with the help of the centroid. The algorithms use pattern symmetry across the positive and negative α–β axes after the three-phase currents are transformed from *abc* to $\alpha\beta$ plane using the Concordia transformation. An open-switch fault is detected if a non-absolute value of the centroid is not at the plane's origin.
8. Current pattern recognition in time domain [16]. The inverter's faults are characterized by the current waveform's signatures in time domain. The signature is defined by parameters, such as DC value and average current's polarity during a predefined amount of time. For a semiconductor fault, 9 out of 27 flags should become active to detect the fault.
9. Converter behavior rules diagnosis in time domain [17]. This approach uses the measured DC bus current in each leg, the AC currents, and voltages. This knowledge base of behavior rules has been proven offline, and defining the complete switching behavior of the power converter needs high computational efforts and extra sensors.
10. Spectrum analysis [18]. In this method, the current spectrum is analyzed in order to detect the signatures of an open transistor. The downside of this method is its high computational demands.
11. Current deviation method [19]. This method uses the normalized inverter's output current. For inverter-connected motor drives, the method uses the machine state variables. During fault-free operation, the mean value of current deviation should be zero. A faulty semiconductor leads to a considerable deviation, which is easily detected. This method uses discrete Fourier analysis on the clustered amplitude of current deviation versus evaluation segments. The first harmonic has the largest magnitude (coefficient c_1), followed in magnitude by the mean value coefficient (c_0). The ratio c_1/c_0 serves as a fault indicator. When $c_0 > 0.1$, a fault condition happens for $|c_1/c_0| \geq 0.5$.
12. Subtractive clustering-based mean current vector method [20]. This method is based on subtractive clustering analysis of the stator current vector and a quick

mean current vector calculation method. The amplitude and angle of the different mean current vector cluster are used to determine the fault modes and fault localization.

13. Wavelet-based methods. In [21], the condition-monitoring mechanism is based on DWTs and fuzzy logic; in [22], the authors apply discrete wavelet transformations (DWT) combined with neural networks (NNs) to a typical three-phase inverter. Identification and classification use approximation and six levels of details for the currents. In [23], the DC-link current is monitored over one electrical cycle for healthy and faulty operations, the time-domain waveform is processed using the wavelet transform, and suitable indices are derived to train an adaptive neuro-fuzzy inference system.
14. Rule-based methods. In [24], the authors diagnose the conditions of an inverter by transforming the current to the *d–q* axis. Then, they extract the features consisting of *d–q* phase currents' trajectories for each fault mode. In the ideal case, a set of fault modes can be classified by using the shape of these trajectories. In [25], a model and various faulted modes of the inverter–motor combination were developed, and voltages and current signals serve to train an artificial NN for fault diagnosis. The authors in [26] use a rule-based expert system for fault diagnosis of the VSI. It requires user interaction with the expert system, so it is applied in offline diagnosis.

From [6], there are a couple of methods for open-circuit faults that have medium implementation complexity, good effectiveness, fast detection time (less than 8 ms), and low dependence on detection variables.

As for short circuit, this operating point is an unsafe state, and although insulated gate bipolar transistors (IGBTs) have an inherent capability to limit the external current during short circuits [27], there are many mechanisms that result in destruction of the power switch. The failure mechanisms during short circuits are not unique, and many studies try to throw some light on them. For example, one destruction mode is induced by high electron densities during short-circuit, resulting in a dynamic change of the vertical distribution of the electrical field. The device can be destroyed by local hot spots arising during the short-circuit mode [28]. In other instances, when the short-circuit current is rapidly turned-off, the IGBT is forced into the self-clamping mode due to the parasitic inductance and high turn-off current [29,30]. According to [27,31–33], high-voltage IGBTs destruction usually occurs during or at the end of the short-circuit pulse. On the other hand, low-voltage IGBTs (600–1,200 V) typically can sustain the short-circuit pulse and safely turn-off but fail some hundred microseconds to milliseconds after the short-circuit pulse has completed if a critical amount of energy has been dissipated [34,35]. Also, according to [6,7], an IGBT short-circuit fault may result from the following mechanisms:

1. **High-voltage breakdown** [36,37]. High values in V_{GE} or V_{CE} can lead to destruction of the device during turn-on or turn-off. In the first case, high-voltage levels induced by fast reduction in collector current and stray inductance can destroy the IGBT. In the second case, a sudden destruction and peak current can happen after several microseconds during turn-on. This can be avoided by proper clamping of $G - E$ and $C - E$ voltages.

2. **Static/dynamic latch-up** [31,38–41]. Inherent to the original IGBT structure, new devices incorporate structures to mitigate latch-up occurrence, but this is still a cause of IGBT destruction.
3. **Second breakdown** describes a sudden drop in collector-emitter voltage that occurs as collector current increases. Second breakdown occurs because of localized increase in temperature (hot spots produced by filament currents). This in turn results in current increase through a positive feedback loop, leading to the device's destruction [39,41–45].
4. **Energy shocks** [33,46]. The high power dissipation within a short time is defined as energy shock. The high short-circuit current will result in energy shock and high temperature. The IGBT may not fail immediately even when the junction temperature exceeds the rated temperature. An increase in temperature results in an exponential increase in carrier concentration and uncontrolled thermal growth. The authors in [47] propose a "critical energy (CE)" to explain the mechanisms of catastrophic failure and wear-out failure under repetitive short-circuit operations. For short circuits below CE, the IGBT could last for more than 10^4 cycles of repetitive short-circuit conditions before failure.

The short-circuit fault is an extreme working point for the IGBT, resulting from unplanned device's turn-on [48], or low load impedance [46]. In this extreme condition, high current flows through the power switch, limited mainly by its capacities. Short-circuit faults pose more danger to the converter since they can quickly lead to the power stage breakdown and sometimes produce collateral damage. The large amount of energy during an IGBT short-circuit fault, when allowed to evolve, affects the maximum current during the fault [7,49]. This also produces latch-up or self-turn-off impairment [41,50–52], which leads the power switch to its destruction during the V_{CE} desaturation process. The short-circuit high currents, if untreated, produce a temperature increase that destroys the switching element and could spread to other subsystems in the converter [53].

Many high performance commercial power devices can withstand short-circuit times around 10 μs. Proper protective action gives permanent damage protection to the device and to the system. Usually, this is possible by applying gate turn-off before losing gate control of the device. In case gate control is lost, i.e., to latch-up, the offending device can be isolated by proper fusing [54].

The aim of this chapter is to review diagnosis methods for the switching devices, as well as methods for maintaining limited operation of the converter. Several authors have proposed methods for aspects related to the switching devices' health, mainly related to health prognosis, fault diagnosis, and fault-tolerant operation topologies and control methods [3,55,56].

5.1 Fault prognosis/diagnosis and health management

A key part of sustaining or extending operation of VSIs is detecting problems in the normal operation of the switching devices. This is achieved before an actual fault occurs in the switching device (prognosis), or as soon as the fault starts (health management).

In [3], the use of prognostic and health management (PHM) techniques serves to operate the converter with diminished capabilities, while still being able to provide quasi-normal operation to some extent. Once the possibility of a fault is detected, the converter can still be operated in an alternate "normal" mode, depending on the converter's topology for fault-tolerance operation, or even in a limited "limping mode" when no additional provisions are taken in its topology for fault-tolerant operation.

The authors in [56] define diagnosis, prognosis, and condition monitoring as part of PHM as:

1. Diagnosis: Identifies the cause of a fault, once the fault has occurred.
2. Prognosis: Gives a health status at some point and predicts its status in the future.
3. Condition monitoring: Measures some parameters to establish current health according to baseline measurements.

5.1.1 Condition monitoring of IGBTs

In [56,57], the authors define condition monitoring so that proper maintenance can be applied before serious damage occurs. A first approach to PHM is using built-in tests; they consist in specialized circuitry for health monitoring. The built-in test can operate in two ways [55]: either in interruption mode (I-monitor) or in continuous (C-monitor) mode. The authors in [58] propose the use of a built-in circuit operating in I-monitor. The circuit is designed to detect bond wire defects by injecting current from two current sources. One of the sources injects 100 mA continuously, while the other is a predefined high-current pulse train. The deterioration in the bond wires can be detected by changes in $V_{ce(on)}$ or V_F against baseline measurements. The authors in [59] present a similar approach for detecting bond wire defects by monitoring $V_{ce(on)}$, this time using an external circuit for online monitoring (C-monitoring).

5.1.2 Health prognosis of IGBTs

In power converters, prognosis allows for the forecast of likely behavior of the converters' components. The components' future behavior is in general assessed by the deviation of their operation from their normal or expected conditions. Health management provides the methods for extending the useful life of the components by using prognosis information to take preemptive actions against the components' deterioration and to avoid catastrophic failures [55].

Current power switching devices have a certain capability to withstand short-circuit conditions until proper protection measures clear the fault. For example, in [60], the authors state that a 1,200-V Si IGBT (SGW15N120) can withstand a short-circuit for 33 μs under conditions of $V_{ce} = 540$ V, $V_{ge} = 15$ V, $T_{case} = 125$ °C; for the same device, the manufacturer datasheet gives a short-circuit withstand time of 10 μs @ $V_{GE} = 15$ V, 100 V $\leq V_{CC} \leq 1{,}200$ V and $T_j \leq 150$ °C. However, most failures of IGBTs are due to aging mechanisms such as thermal and thermomechanical stress [61,62]. The thermal stress in turn is caused in many cases by electrical overstress [63].

In [7], the IGBTs' failures are divided into catastrophic failures and wear-out failures. The first type of failure, catastrophic failure, is commonly the result of single-event overstresses (e.g., overvoltage, overcurrent, over-temperature, etc.). The second type, wear-out failure, is generally induced by accumulated degradation over time.

With a proper PHM in place, wear-out failures can be predicted in time to take preventive measures, to keep limited operation of the converter using fault-tolerant control strategies and to provide remedial action to the failing device [64]. Catastrophic failures, on the other hand, where prognosis and health management does not apply—as they can happen randomly and are more difficult to predict [7]—could use the same fault-tolerant control techniques, provided that proper protections prevent failure propagation to other parts of the converter.

Catastrophic failures in IGBTs can be grouped into open-circuit and short-circuit failures [64]. Open-circuit failures in general do not propagate to other parts of the converter but without proper precautions can lead to unsafe or erratic operation of the system. Short-circuit faults, on the other hand, could propagate to other parts of the converter, due to the high currents produced during the short-circuit event.

5.1.3 Diagnosis

Once the fault has occurred, diagnosis gives the means to adjust the current estimated to the fault behavior of the components, and to select provisions to sustain the faults. Postmortem analysis of the faulted device can also shed light on possible faults in other systems, or even faults in the design stage (i.e., device dimension).

Once the fault happens, or even better, during the onset of a fault, fast diagnosis of the compromised device can allow the protection elements in the system to isolate and prevent the propagation of the fault to other parts in the system. Also, fast diagnosis of faulty devices allows for effective use of fault-tolerant control techniques, or for online topology modification to keep continuous operation.

5.2 Fault-tolerant topologies

Semiconductor reliability has been an important factor in the use of power electronic converters. Even though the failure rate of power elements has been dropping—for example, the average failure rate for power modules in traction dropped from 1,000 failures in time (FITs) in 1995 to 20 FITs in 2000 [65]—still, the chance of device failure has led to several works over the years [66].

5.2.1 2L-VSI with middle-point connection

Nowadays, two-level voltage source converters such as the one represented in Figure 5.1 are the most widespread type of power converter in use by industry. The topologies for fault-tolerant operation range from redundant switching legs, shown in Figure 5.2, to the most common middle-point connection depicted in Figure 5.3.

Figure 5.4(a) shows the instantaneous space vectors generated by the standard voltage source converter (VSC) and the maximum circular trajectory the space vectors can follow in a PWM sense. Space vectors $\vec{v}_0$ and $\vec{v}_7$ are redundant states for this converter, corresponding to the null vector. For fault-tolerant operation, the

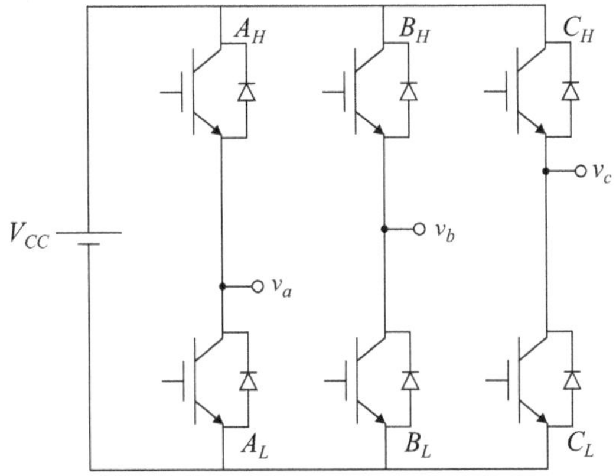

Figure 5.1 Standard two-level voltage source converter

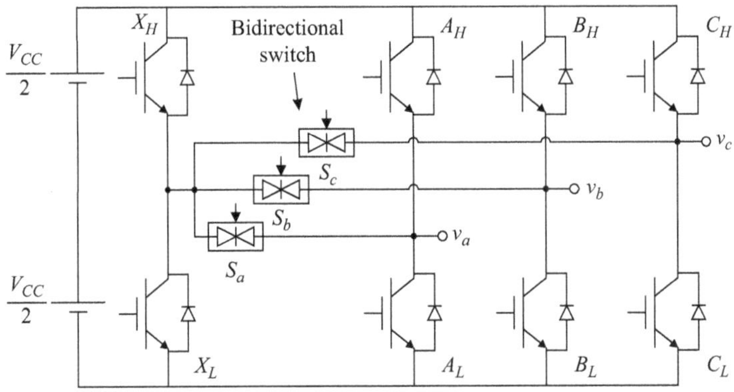

Figure 5.2 VSI topology for fault-tolerant operation with redundant leg. Adapted from Reference [5]

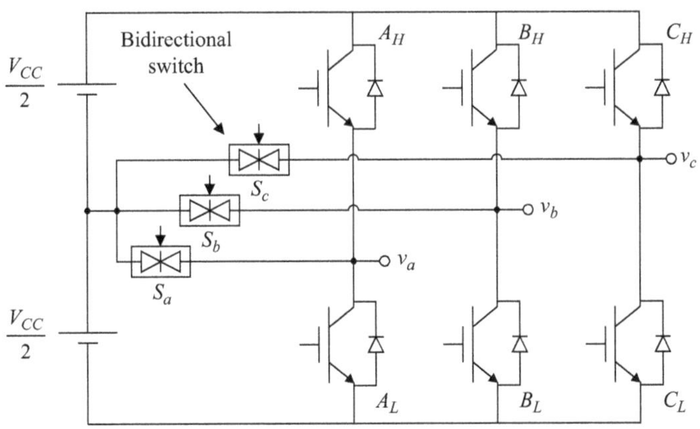

Figure 5.3 VSI topology for fault-tolerant operation. Adapted from Reference [5]

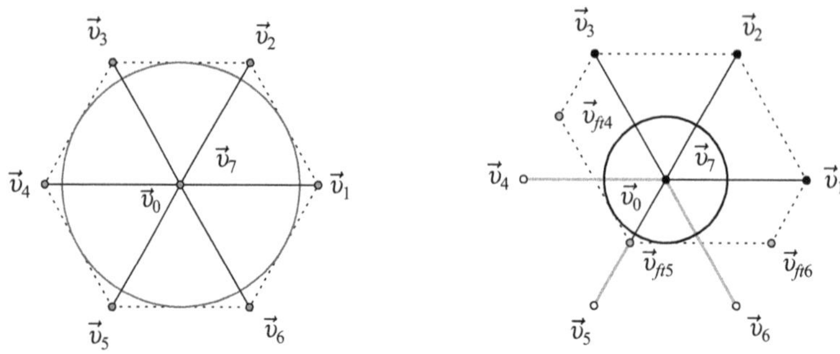

(a) Space vectors available in a healthy VSC.

(b) Space vectors available in the middle-point connected 2L-VSC.

Figure 5.4 VSI switching states: (a) healthy case and (b) fault-tolerant case

middle-point connection topology generates additional states, shown in Figure 5.4(b) for a fault in phase "*c*," so that the available space vectors can follow a circular trajectory in a PWM sense. The space vectors in the middle-point connected topology, under fault-tolerant operation, can only follow a circular trajectory with half the diameter of its healthy counterpart.

5.2.2 *Space vector generation during fault-tolerant operation*

It is assumed that for proper transition from faulty to fault-tolerant operation, a proper detection method should be in place. The sequence of operations is as follows:

1. Fault detection algorithm,
2. Clearing of the fault,
3. Connection of the faulty branch to the auxiliary circuit, and
4. Change in duty cycle generation, according to the topology used for fault-tolerant operation.

It is expected that the fault-tolerant operation would be in place for a short period of time, until a scheduled stop of the system allowed replacement of the faulty components. During a fault in one device of the standard VSC, the corresponding phase is isolated by the fault detection system implemented in the converter, and the middle-point connection is enabled. The space vector synthesis will be the contribution of the two healthy branches of the converter using PWM, plus the fixed space vector resulting from the faulty branch's connection to the supply's middle point. For example, for a fault in phase "*c*" such as the one depicted in Figure 5.4(b), the space vector trajectory will be synthesized using PWM on the base space vectors $\vec{v}_1$ and $\vec{v}_3$ plus a space vector with fixed $V_{CC}/2$ magnitude in $\vec{v}_5$ direction [67–69].

According to [67,69], the algorithm for generating the duty cycle demands in the healthy branches that are needed to follow a maximum circular trajectory during fault operation is shown in Algorithm 1.

Figure 5.5 shows a comparison for the maximum amplitude circular trajectories achieved by the available space vectors. The trajectory for healthy operation is naturally centered and sweeps a larger area. On the other hand, the trajectory for

Algorithm 1: Duty cycle demands during fault-tolerant operation

procedure FTDUTY(Faulty_branch)
    $x = x/V_{CC}$
    $y = y/V_{CC}$
    if Faulty_branch=a **then**
        $x \leftarrow x - 0.5$
        $y \leftarrow y/\sqrt{3}$
    if Faulty_branch=b **then**
        $x \leftarrow x + 0.25$
        $y \leftarrow y/\sqrt{3} - 0.5$
    if Faulty_branch=c **then**
        $x \leftarrow x + 0.25$
        $y \leftarrow y/\sqrt{3} + 0.5$
    if $y = 0$ **then**
        $N = 2.5 - ((fx > fy) + (fx > -fy) + 0.5)$
    else
        $N = 2.5 - \text{sign}(fy) * ((fx > fy) + (fx > -fy) + 0.5)$
    $\delta \leftarrow 1$
    switch N **do**
        case 0
            $Da = \delta(x + y - 1) + 1$
            $Db = Da - x + y$
            $Dc = Da - x - y$
        case 1
            $Db = \delta(2y - 1) + 1$
            $Da = Db + x - y$
            $Dc = Db - 2y$
        case 2
            $Db = \delta(-x + y - 1) + 1$
            $Da = Db + x - y$
            $Dc = Db - 2y$
        case 3
            $Dc = \delta(-x - y - 1) + 1$
            $Da = Dc + x + y$
            $Db = Dc + 2y$
        case 4
            $Dc = \delta(-2y - 1) + 1$
            $Da = Dc + x + y$
            $Db = Dc + 2y$
        case 5
            $Da = \delta(x - y - 1) + 1$
            $Db = Da - x + y$
            $Dc = Da - x - y$
                ▷ The duty cycle for the faulty branch gets dismissed

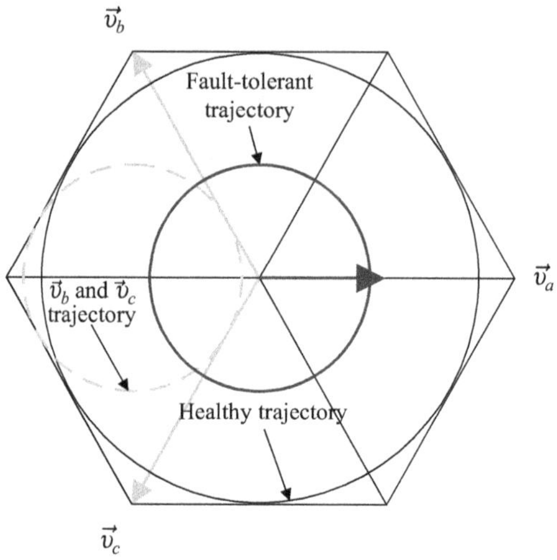

Figure 5.5 Ideal VSI trajectories under healthy and faulty (fault tolerant) trajectory

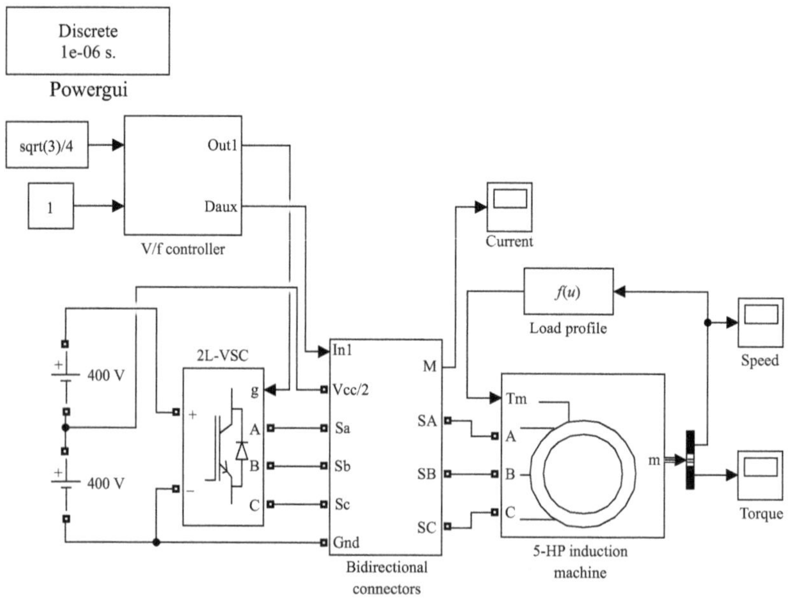

Figure 5.6 Simulink diagram of the fault-tolerant strategy for a two-level VSI with middle-point connection

fault-tolerant operation—in this case, the fault is in phase "c," shown in Figure 5.5—is generated by the remaining base space vectors and becomes centered by the contribution from the middle-point connection, fixed space vector in Figure 5.5.

Figure 5.6 shows the MATLAB®/Simulink™ diagram used for simulating the middle-point connection topology. Figure 5.7 shows the duty cycle demand, stator

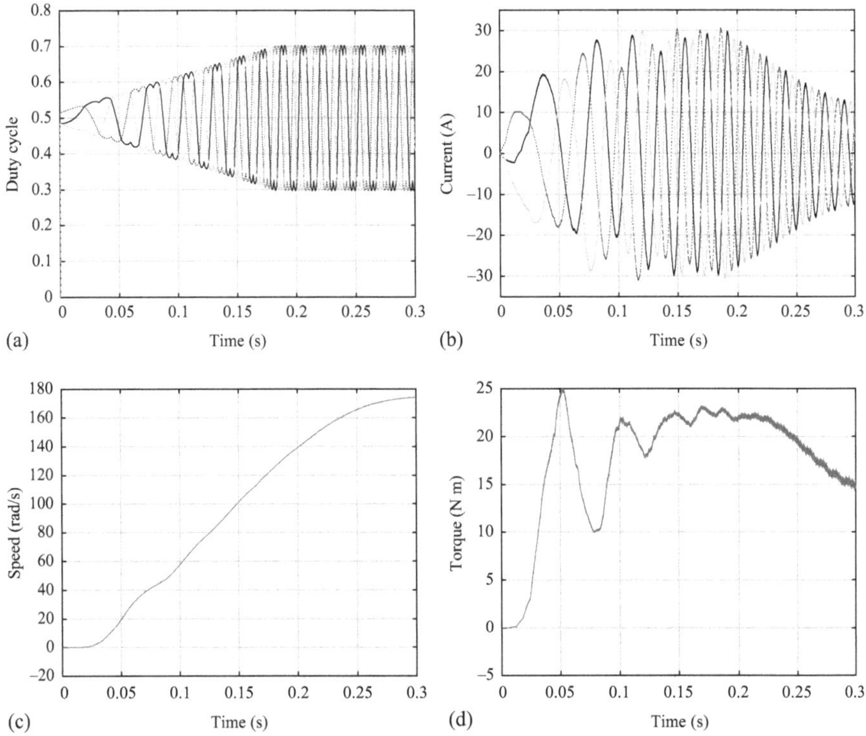

Figure 5.7 Converter's healthy operation with poor duty cycle utilization due to higher DC-link voltage: (a) duty cycles from controller, (b) load currents, (c) induction machine's speed, and (d) electric torque

currents, rotor speed, and electric torque for a V/f controlled induction machine during healthy operation.

Figure 5.8 shows the results for the V/f controlled induction machine during trigger suppression in a single device "A_L" at $t = 0.15$ s. In this case, as shown in Figure 5.8(b), the current in phase "a" becomes unipolar (positive only).

Figure 5.9 shows the result for the same V/f controller, with trigger suppression in two devices "A_L" and "A_H." In this case, the machine runs as a single-phase machine, and it keeps working because there is already a rotating field produced before trigger suppression in both devices.

Figure 5.10 shows how the stator current, rotor speed, and torque evolve in time during the transition from healthy operation to fault-tolerant operation. For the test, at $t = 0.15$ s, the inverter's phase "a" is disconnected, and the bidirectional switch associated to phase "a" is connected to $V_{CC}/2$. Figure 5.10(a) shows how the duty cycles generated by the controller change from the typical space vector form to sinusoidal references in the remaining healthy legs of the inverter. At the same time, the currents from the machine depicted in Figure 5.10(b) and the speed in Figure 5.10(c) do not seem affected by the transition from healthy-to-fault-tolerant

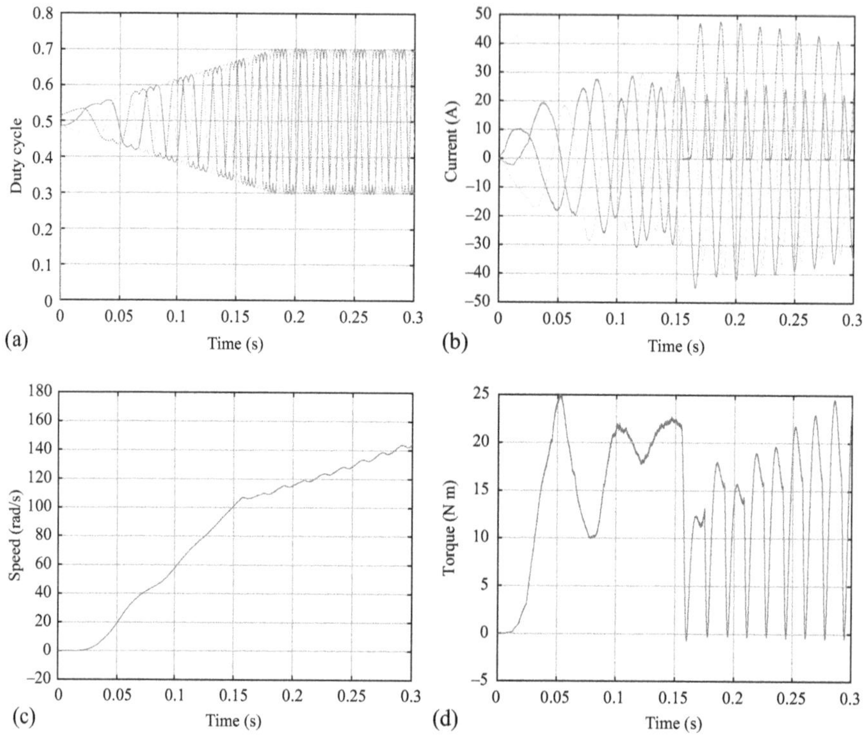

Figure 5.8 Converter's operation during transition between healthy and faulty operation (trigger suppression in IGBT A_L): (a) duty cycles from controller, (b) load currents, (c) induction machine's speed, and (d) induction machine's torque

operation. Figure 5.10(d) shows how the voltage between phase "*a*" and the negative DC link gets fixed at $V_{CC}/2$ at the transition from healthy to fault-tolerant operation.

A problem with this strategy is that the fault-tolerant trajectory reduces to half the DC-link voltage availability. Therefore, an unperturbed converter's operation requires higher DC-link voltage, and this increases voltage stress in the device and results in larger load–current ripple.

A better topology uses an auxiliary leg that rests disconnected until a fault is detected in any phase and the operation switches to the auxiliary leg. In this way, the converter does not need an increase in DC-link voltage, but the number of devices with its corresponding drive circuit increases.

5.2.3 Fault-tolerant operation of the back-to-back converter

The fault tolerance is obtained by reconfiguring the converter topology to allow continuous operation of the system even with complete loss of one leg.

Figure 5.11 shows a back-to-back configuration for fault-tolerant operation [70]. This configuration ensures system control for one leg isolation from either the input

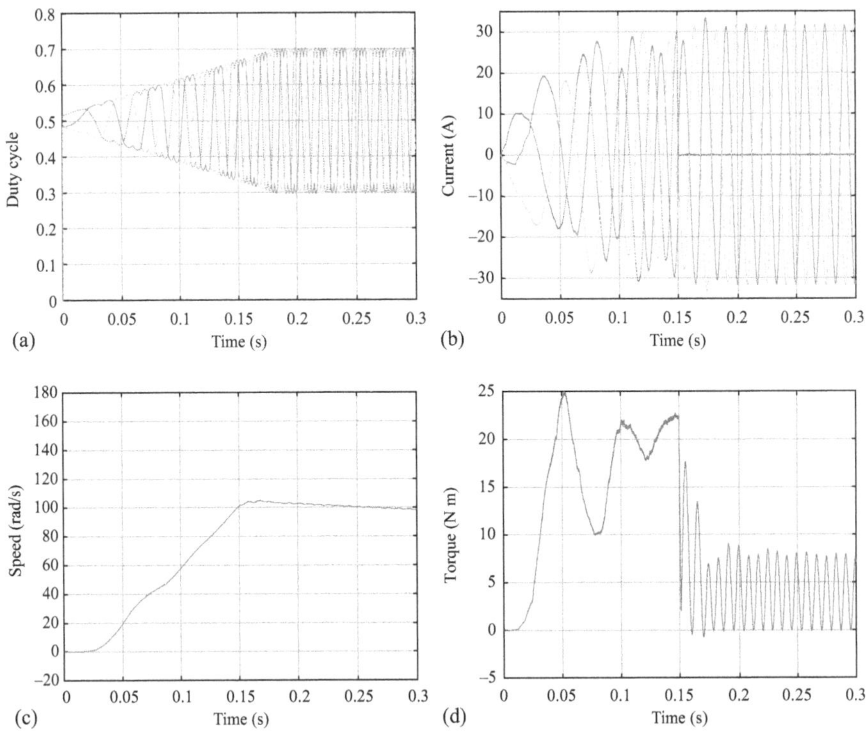

Figure 5.9 Converter's operation during transition between healthy and faulty operation (trigger suppression in IGBTs A_L and A_H): (a) duty cycles from controller, (b) load currents, (c) induction machine's speed, and (d) induction machine's torque

rectifier or the inverter output. The back-to-back converter keeps continuous operation by isolation and phase reconnection when a device or even a complete leg fails.

Proper fault-tolerant operation needs effective and timely fault detection. Existing fault detection strategies take a few milliseconds, in practice from half a cycle to a couple of cycles [6].

During fault-tolerant operation, the converters operate using the double zero-sequence injection [71]. For example, for a fault in phase "c_2," the controller isolates that branch and triggers the associated bidirectional switch to share "c_1" between both converters. The modulation demands become as shown in (5.2) [71]:

$$\begin{aligned} \widetilde{D}_{a1} &= D_{a1} + D_{c2} \\ \widetilde{D}_{b1} &= D_{b1} + D_{c2} \\ \widetilde{D}_{c1,c2} &= D_{c1} + D_{c2} \\ \widetilde{D}_{a2} &= D_{a2} + D_{c1} \\ \widetilde{D}_{b2} &= D_{b2} + D_{c1} \end{aligned} \tag{5.1}$$

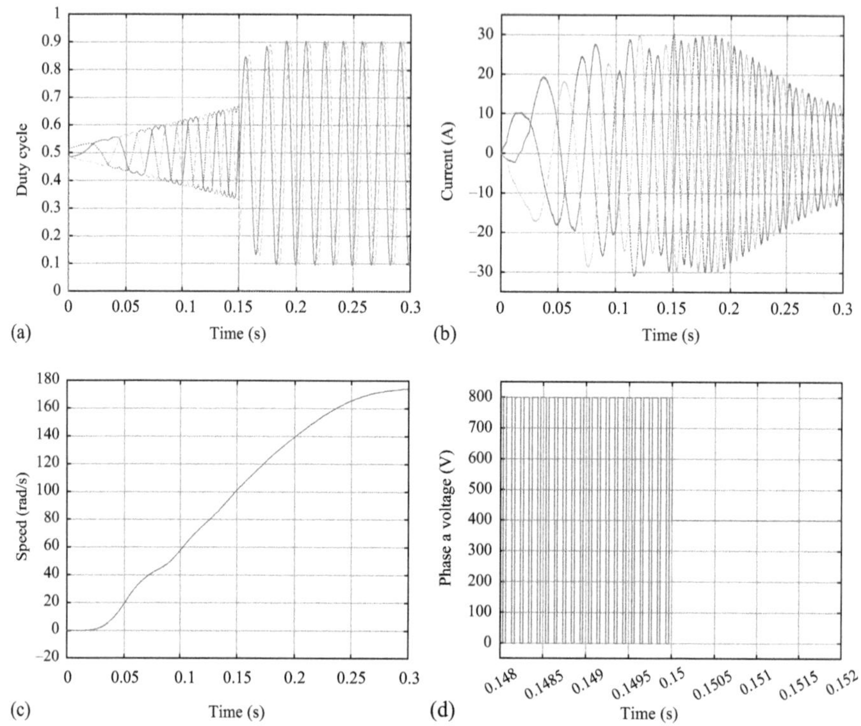

Figure 5.10 Converter's operation during transition between healthy and fault-tolerant mode for fault in phase "a" devices: (a) duty cycles from controller, (b) load currents, (c) induction machine's speed, and (d) zoom for the voltage between the faulty phase and the negative DC bus link

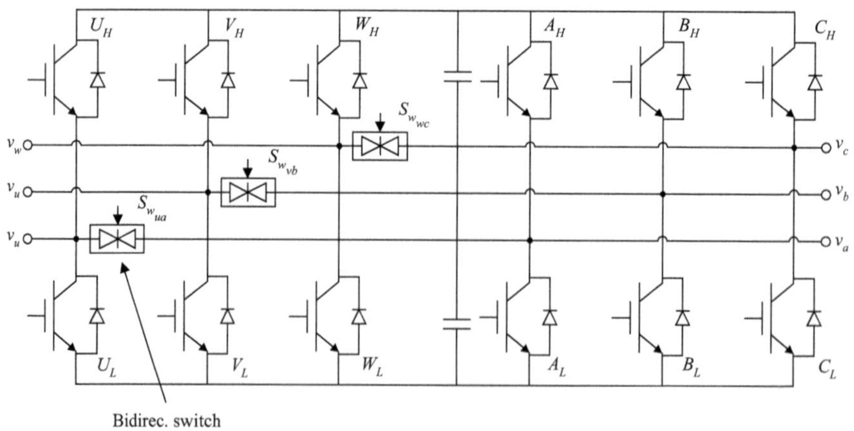

Figure 5.11 Back-to-back configuration for fault-tolerant operation. Adapted from Reference [70]

In this way, each line–line voltage remains unchanged. However, operation at the same maximum pre-fault voltage level needs twice the pre-fault DC-link voltage [70]. The modulation signals for both inverters can be unrelated. So, each converter is free to operate at different fundamental frequencies.

As long as the new modulating signals ($\widetilde{D}_{a1}, \widetilde{D}_{b1}, \ldots$) given in (5.1) do not exceed the carrier–signal magnitude, both inverters will operate without entering the overmodulation zone. According to [71], using the full DC link requires adding an additional zero-sequence component D_0, shown in (5.2), to each modulation signal.

$$D_0(t) = -\frac{1}{2}\left[D_{max}(t) + D_{min}(t)\right] \tag{5.2}$$

For a general space vector modulation, as defined in [67,69,72], the duty cycles as a function of the generalized space vector pulse width modulation (GSVPWM) parameter become

$$v_{\text{offset}} = \begin{cases} (1-\delta)(1-D_{\max}) - \delta D_{\min} & \text{for healthy operation} \\ (1-\delta)\left(\dfrac{1}{2} - D_{\max}\right) - \delta D_{\min} & \text{for fault-tolerant operation} \end{cases} \tag{5.3}$$

where δ is defined in Table 5.1 for different modulation methods [69].

Again, for optimum DC-link use, either the duty cycle demands must be limited to the range [0,1/2] to avoid overmodulation during fault-tolerant operation or the DC-link needs to double while halving the duty cycles when entering fault-tolerant operation. The latter is impractical and still requires power switches with higher voltage blocking for fault-tolerant operation than for healthy operation.

Table 5.1 Some modulation cases using the GSVPWM algorithm [68]. ©2018 IEEE. Reprinted, with permission, from Applied Power Electronics Conference and Exposition (APEC), 2010 Twenty-Fifth Annual IEEE, pp. 343–348, 25–25 Feb. 2010

Modulation	δ
DPWM_{min}	1
DPWM_{max}	0
SVPWM	$\frac{1}{2}$
DPWM_0	$(1/2)\left[1+(-1)^{n_1}\right]$
DPWM_1	$(1/2)\left[1+(-1)^{n_2}\right]$
DPWM_2	$(1/2)\left[1+(-1)^{(n_1+1)}\right]$
DPWM_3	$(1/2)\left[1+(-1)^{(n_2+1)}\right]$

$n_1 = N = 2.5 - \text{sign}(v_{bN} - v_{cN})\left[(v_{aN} > v_{bN}) + (v_{aN} > v_{cN}) + 0.5\right]$
$n_2 = 3.5 + \text{sign}(v_{cN})\left[(v_{aN} > 0) + (v_{aN} > -v_{cN}) + 0.5\right]$

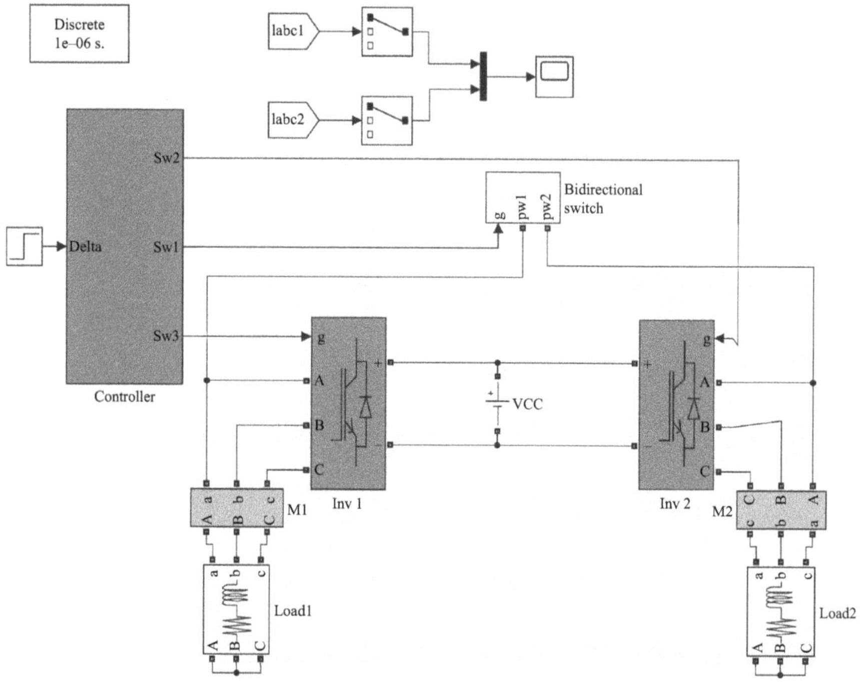

Figure 5.12 Simulink diagram of the back-to-back converter for fault-tolerant operation

Figure 5.12 shows the Simulink diagram for testing the fault-tolerant operation for the back-to-back configuration. Figure 5.13 shows the operation in the transition between healthy and fault-tolerant operation, with full DC-link utilization during healthy operation. In this case, to guarantee fault-tolerant operation in the linear range, the voltage drops to half its healthy operation value. Figure 5.14 shows the case for best DC-link use during fault-tolerant operation, maintaining voltage amplitude. In this case, DC-link use drops to half during healthy operation and has been adjusted to maintain constant voltage/current magnitude.

5.2.4 Three-level NPC converter

There are several strategies for fault-tolerant control in three-level neutral-point-clamped (NPC) converters [73,74]. For the unmodified standard NPC topology shown in Figure 5.15, there are alternatives for the fault-tolerant operation, as defined in [73]. For a fault in one clamping diode, only one set of states is affected, and the modulation strategy should be modified so that the faulty states are replaced or not used for the nonredundant states. In this case, the control algorithm should operate the faulty branch in two-level mode. Figure 5.16(a) shows the affected states for a fault in D_{aH}, when $i_a < 0$ or a fault in D_{aH} when $i_a > 0$. Therefore, limited three-level operation is possible.

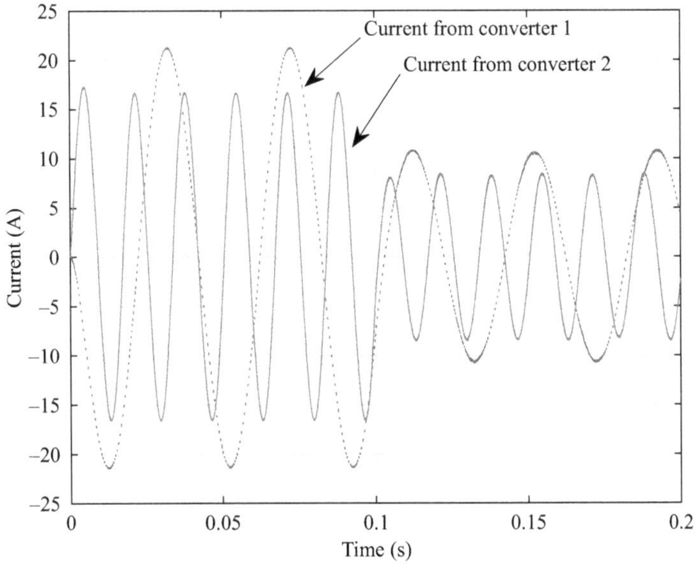

Figure 5.13 Transition between healthy and fault-tolerant operation with best DC-link use during healthy operation

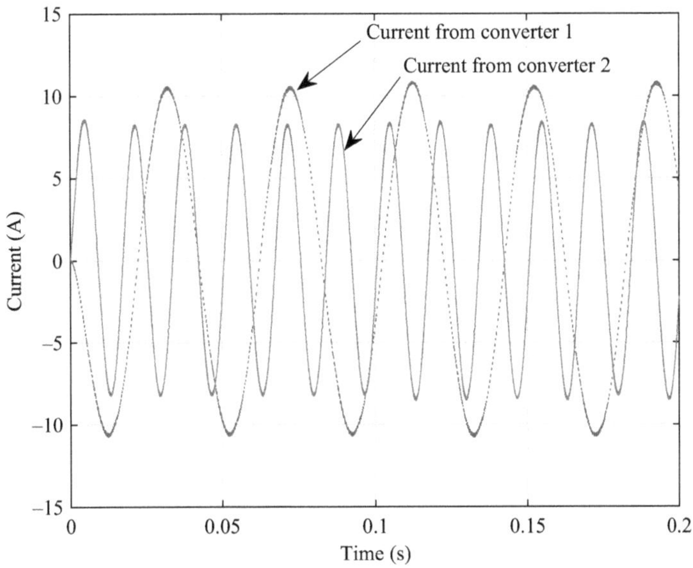

Figure 5.14 Transition between healthy and fault-tolerant operation with best DC-link use during fault-tolerant operation

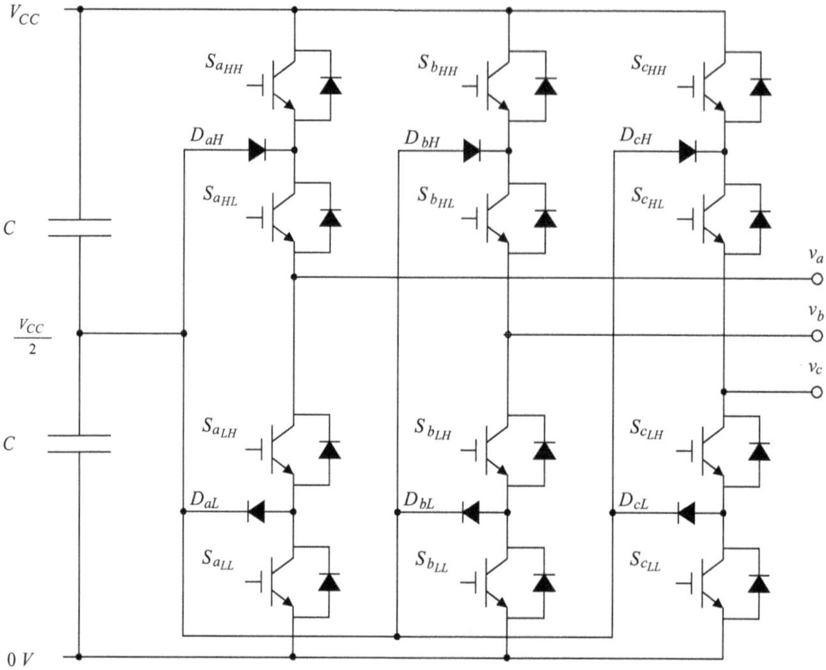

Figure 5.15 Standard NPC topology. Adapted from Reference [75]

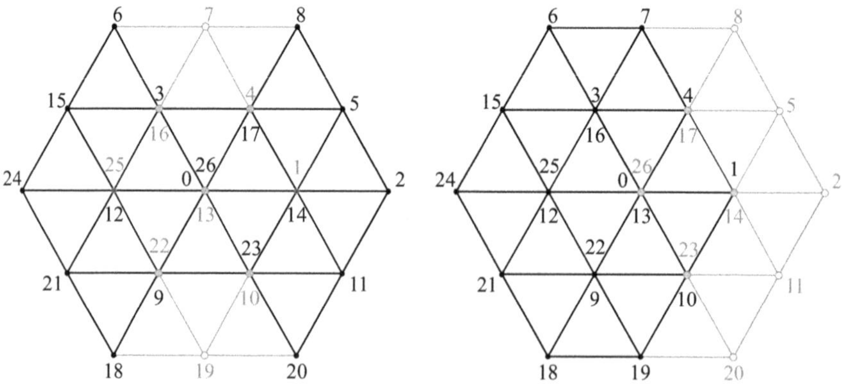

(a) (b)

Figure 5.16 Affected states for (a) fault in clamping diode D_{aH} (or in D_{aL}) and $i_a < 0$ or fault in D_{aL} and $i_a > 0$ and (b) fault in S_{aHH}

For a fault in the S_{xHH} or S_{xLL} switches, the voltage capabilities for the faulty phase reduce to half, and the controller behaves similarly to the two-level fault-tolerant operation described in Section 5.2.2. Figure 5.16(b) shows the states affected by a fault in S_{aHH}. Only the central sub-hexagon is available for control. Faults in the

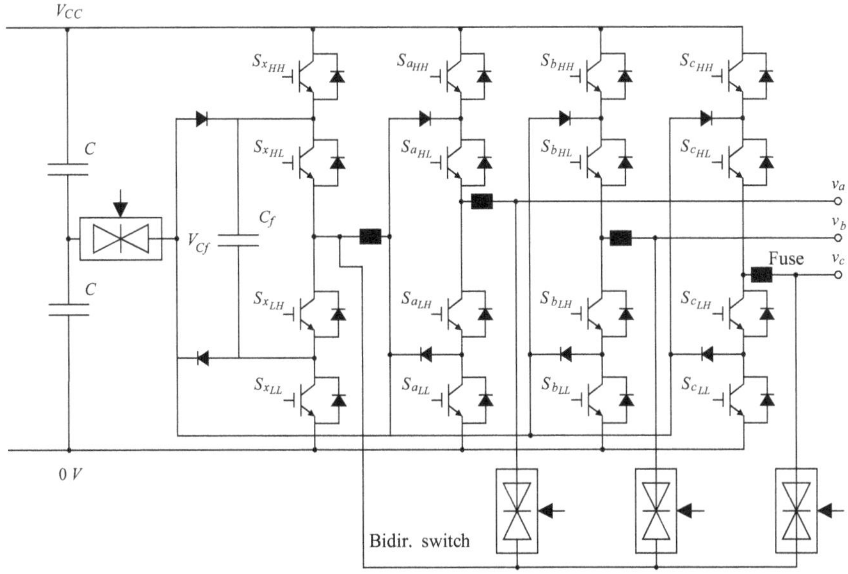

Figure 5.17 Multilevel NPC converter with flying capacitor branch for fault-tolerant operation. Adapted from References [74,76]

central power devices (S_{xHL} or S_{xLH}) leave the affected branch with current flow in one direction and with the capability to force only one voltage level out of three. In this case, the converter does not have a reasonable fault-tolerant capability, and the converter topology needs to be modified to allow continuous operation.

The topology for overcoming this problem, shown in Figure 5.17, uses three legs with standard NPC structure, together with a fourth leg with flying capacitor structure. During healthy operation, the fourth leg sets a solid middle voltage (neutral-point). During fault-tolerant operation, the fourth leg can replace the faulty leg by isolating the troublesome leg with the corresponding fuse and using the fourth leg through the bidirectional switch to take the task of the isolated leg.

5.3 Fault-tolerant operation of the open-end converter

Figure 5.18 shows a diagram of the dual converter, used here for switch suppression operation. This topology for the dual converter comprises two three-phase two-level inverters sharing the same DC link. In this topology, the dual converter can be considered as a six-phase converter, where the number of possible states are $4^6 = 4,096$, of which $4^6 - 3^6 = 3,367$ are forbidden states because they short circuit the DC-link; from the valid states, $3^6 - 2^6 = 665$ have at least one "fly-wheeling" floating branch, and only $2^6 = 64$, shown in Figure 5.19, are well established and in general considered to be usable states for normal operation, resulting in 19 different voltage space

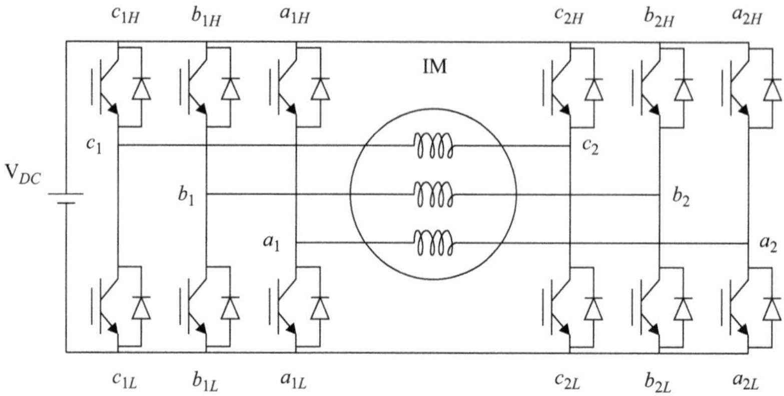

Figure 5.18 Diagram of the dual converter with one DC supply feeding an open-end induction machine. ©2018 IEEE. Reprinted, with permission, from Reference [79]

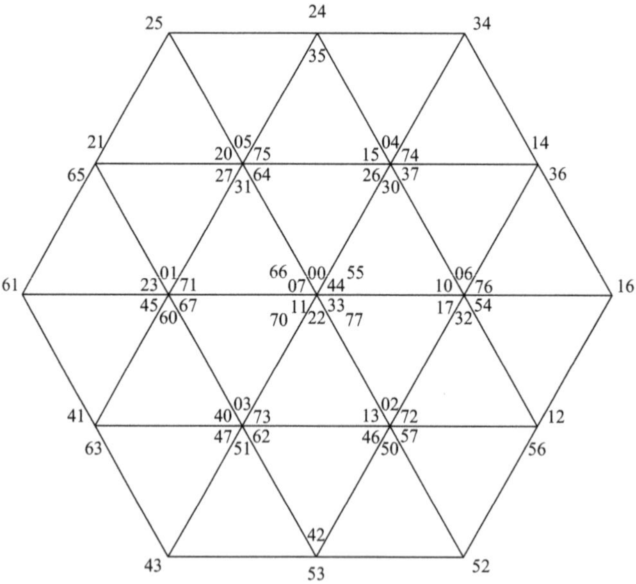

Figure 5.19 Voltage space vectors with their respective switching states in base 8. Adapted from [79]

vectors, shown in Figure 5.20. To simplify the state information, the connectivity in Figure 5.20 uses octal notation, where the most significant digit indicates the state for the left converter, and the least significant digit indicates the state for the right converter. For example, (O41 =B100001) indicates that c_{1H}, b_{1L}, a_{1L}, a_{2H}, b_{2L}, and c_{2L} are "ON" and all the other IGBTs are "OFF."

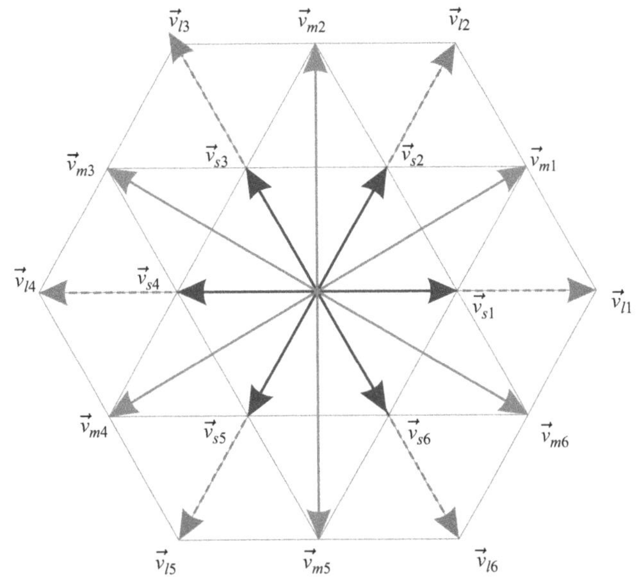

Figure 5.20 Sets of voltage space vectors in the dual converter. ©2018 IEEE. Reprinted, with permission, from Reference [79]

For simplicity, the power drives are sometimes analyzed by assuming switching with on and off states, and assuming the loads to be purely resistive or without regeneration capabilities. This generates what has been previously called usable states. However, when the load has regenerative capabilities or energy-storage properties such as inductances or electrical machines, the energy storage provides additional control alternatives expressed in additional transient usable states. In the case of fault-tolerant strategies, these additional states serve as an important support for optimization during trigger suppression. In this paper, these states are defined as "opportunistic states," since they can only be used when they become available, and the control algorithm can favor their occurrence since they are determined by the current circulating in the machine's winding under control. The presence of "opportunistic states" allows for the use of "regenerative vectors."

Since the two sub-converters making up the dual converter share a common DC-link, the modulation strategies used to synthesize the space vector in each sub-converter need to avoid the injection of zero-sequence components, as they will result in corresponding zero-sequence currents flowing through the load [77,78]. However, this zero sequence is allowed, with proper provisions, during fault-tolerant operation [79].

The dual converter topology depicted in Figure 5.18 has the following voltage levels in each phase of the induction machine:

- $+V_{DC}$, obtained when the associated branch in sub-converter 1 is in high level and its paired branch in sub-converter 2 is in low level; when there is enough energy stored in the machine inductances, the following voltage levels can be applied.

- 0, when both the branches feeding the phase winding are simultaneously in high or low level in both sub-converters.
- $-V_{DC}$, when the associated branch in sub-converter 1 is in low level, and its paired branch in sub-converter 2 is in high level.

In this way, the space vectors in the dual converter can be classified in the following sets:

- First, a set of six "positive" small-sized voltage space vectors obtained with the combination of voltage levels $\{0, +V_{DC}\}$.
- A second set of six "negative" small-sized voltage space vectors obtained with the combination of voltage levels $\{-V_{DC}, 0\}$.
- A third set of six medium-sized voltage space vectors obtained with one phase set to $-V_{DC}$, the other phase to 0, and the last phase to $+V_{DC}$ between each winding terminal.
- A fourth set of six large-sized voltage space vectors obtained using voltage levels V_{DC} and $+V_{DC}$ between the winding terminals.
- Finally, a set of three zero-magnitude voltage space vectors is obtained when the voltage in the three winding terminals are the same $\{(-V_{DC}, -V_{DC}, -V_{DC}), (0, 0, 0), (+V_{DC}, +V_{DC}, +V_{DC})\}$.

Figure 5.20 shows the resulting space vector locations. The zero-sequence voltage, responsible for the zero-sequence current in the induction machine phases, is defined as [77]

$$v_{ZS} = \frac{1}{3}\left(v_{a_1a_2} + v_{b_1b_2} + v_{c_1c_2}\right) \tag{5.4}$$

For testing, the open-end induction machine fed from the dual converter, as shown in Figure 5.18, uses the classical direct torque control (DTC) as proposed by Takahashi and Noguchi [80]. During normal operation, DTC can take advantage of the additional space vectors produced by the dual converter and produce less ripple in the system's state variables [81–83]. However, here the focus is on the machine's operation during switch suppression. Thus, a simplified DTC algorithm is used for healthy switching operation.

5.3.1 DTC algorithm

The classical DTC, originally proposed by Takahashi and Noguchi [80], uses the machine's instantaneous stator flux and electric torque calculation. The calculated values produce error signals when compared with the flux and torque references. That, with the angular stator flux location, serves to select the switching state for the inverter. The classical DTC algorithm divides the hexagonal space vector's area into six different zones. Table 5.2 summarizes the switching states in a two-level converter that is needed for the given flux and torque errors. In this work, the use of Clarke's or space vector transformation gives their x, y components:

$$\vec{f} = f_x + jf_y = \sqrt{\frac{2}{3}}\left(f_a\alpha^0 + f_b\alpha^1 + f_c\alpha^2\right) \tag{5.5}$$

Table 5.2 Switch selection for classical DTC [80]

Ψ Change	**τ Change**	Z_1	Z_2	Z_3	Z_4	Z_5	Z_6
Decreases	Decreases	$\vec{v}_5$	$\vec{v}_6$	$\vec{v}_1$	$\vec{v}_2$	$\vec{v}_3$	$\vec{v}_4$
	=	$\vec{v}_0$	$\vec{v}_0$	$\vec{v}_0$	$\vec{v}_0$	$\vec{v}_0$	$\vec{v}_0$
	Increases	$\vec{v}_3$	$\vec{v}_4$	$\vec{v}_5$	$\vec{v}_6$	$\vec{v}_1$	$\vec{v}_2$
Increases	Decreases	$\vec{v}_6$	$\vec{v}_1$	$\vec{v}_2$	$\vec{v}_3$	$\vec{v}_4$	$\vec{v}_5$
	=	$\vec{v}_0$	$\vec{v}_0$	$\vec{v}_0$	$\vec{v}_0$	$\vec{v}_0$	$\vec{v}_0$
	Increases	$\vec{v}_2$	$\vec{v}_3$	$\vec{v}_4$	$\vec{v}_5$	$\vec{v}_6$	$\vec{v}_1$

where $\alpha = e^{j2\pi/3}$. The stator flux linkage Ψ_s can be obtained by integrating the stator equation of the induction machine model [80]

$$\vec{\Psi}_s = \int \left(\vec{v}_s - R_s\vec{i}_s\right) dt \tag{5.6}$$

and the electric torque by multiplying the stator flux linkage and the stator current space vectors:

$$\tau_e = \vec{\Psi}_s \times \vec{i}_s \tag{5.7}$$

The DTC algorithm with the medium-sized space vectors has one branch at $-V_{DC}$, another branch at 0 V, and the remaining branch at V_{DC}. Therefore, according to (5.4), the voltage supplying the machine has a null zero-sequence component. The states that belong to the medium-sized space vectors are, according to Figs. 5.19 and 5.20, $\vec{v}_{m1} = \{14, 36\}, \vec{v}_{m2} = \{24, 35\}, \vec{v}_{m3} = \{21, 65\}, \vec{v}_{m4} = \{41, 63\}, \vec{v}_{m5} = \{42, 53\}, \vec{v}_{m6} = \{12, 56\}$, and the null vector $\vec{v}_0 = \{00, 11, 22, 33, 44, 55, 66, 77\}$. Figure 5.21 shows the medium-sized space vectors and the resulting hexagonal region covered by this set of space vectors. An advantage of applying the medium-sized space vectors to the dual converter is that they do not inject zero-sequence voltage; hence, there are no zero-sequence currents injected into the load. Any other selection of space vectors will inject a zero-sequence component to the load. So, special consideration of effects on the load is necessary before using any other space vectors than the medium-sized space vectors.

The classical DTC algorithm remains unchanged for healthy or trigger-suppressed operation, but for the dual converter, the healthy case uses the medium-sized space vectors versus using the small-sized space vectors in the trigger suppression mode. The medium-sized space vectors require a 30° rotation in the stator flux selection zones $(Z_1, Z_2, \ldots, Z_6)$, as shown in Figure 5.21.

5.3.2 DTC operation for switch trigger suppression

The number of switching states in the dual converter allows the implementation of different machine-control algorithms. For DTC, depending on the number of devices under trigger suppression, there are some voltage space vectors available to perform

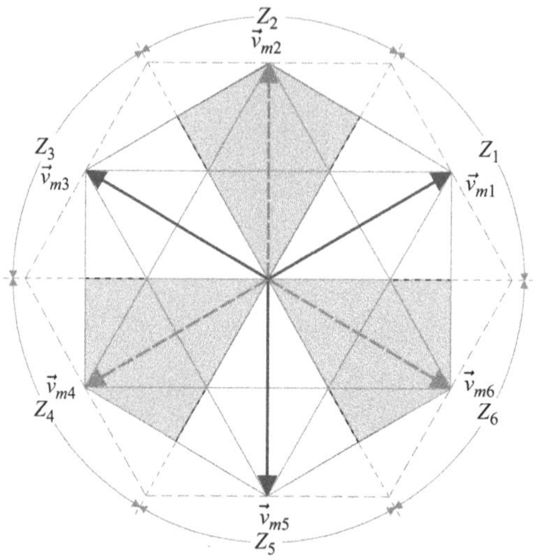

Figure 5.21 Medium-sized space vectors. ©2018 IEEE. Reprinted, with permission, from Reference [79]

the control algorithm on the machine. Also, depending on the number of switches under trigger suppression, it is possible to enforce the conditions for using all the space vectors thanks to the opportunistic states. Since DTC uses a computation of the stator flux linkage, (5.6) requires the voltage at the machine terminals. The use of the switching state of the dual inverter for estimating the stator voltage assumes the connection of each machine's terminal to $+V_{DC}$ or 0 V. This condition limits using the method to one switch's failure per branch. A fully asymmetric operation of the inverter results from six switches with trigger suppression, trigger suppression in three upper switches in one sub-converter, and trigger suppression in three lower switches in the other sub-converter.

The single-link dual converter is simulated for healthy operation with the set of medium-voltage space vectors.

Figure 5.22 shows the simulated stator currents, stator flux, and electric torque for a healthy operation of the dual converter. The advantage of this selection for healthy operation is that there is no zero-sequence injection at any time. It can be seen from Figure 5.22(a) that no zero-sequence component is present in the stator current. Also, since the classical DTC algorithm is employed, the torque shows the large ripple characteristic of this kind of controller. Figure 5.23 shows the experimental results obtained for healthy operation with DTC with the medium-sized set of space vectors; the experimental stator current waveforms confirm that for healthy operation with the selected set of space vectors, there is no zero-sequence injection (Table 5.3).

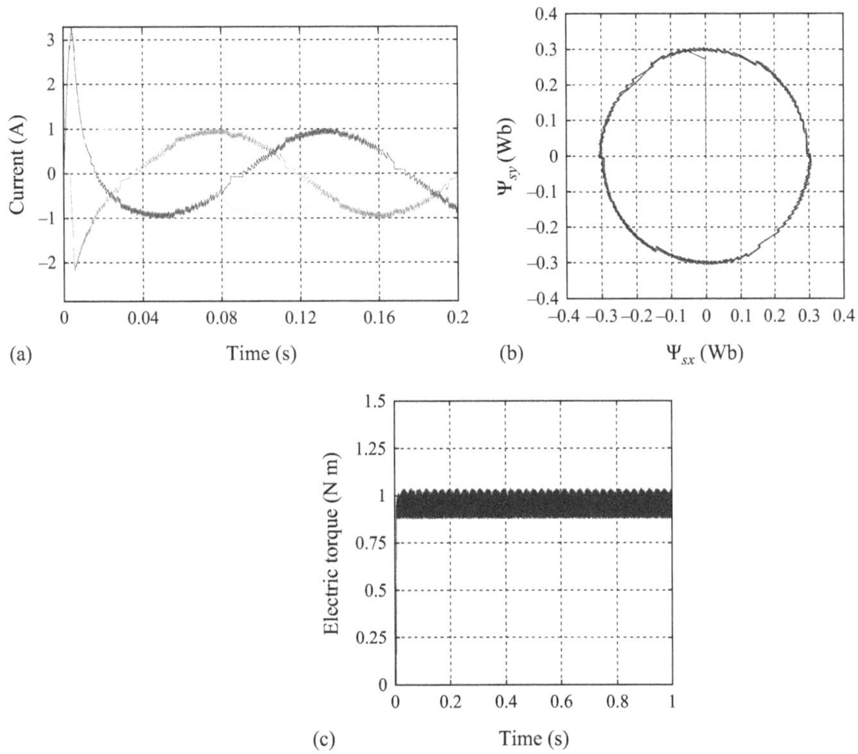

Figure 5.22 Simulated (a) stator currents, (b) stator flux, and (c) torque for normal operation with DTC using medium-sized voltage space vectors.

5.3.2.1 Trigger suppression in one and two devices

A healthy dual converter uses an H-bridge to feed each winding of an open-end machine. An H-bridge uses a couple of switches driving current in the positive direction and another couple of devices driving current in the opposite direction. Trigger suppression in one or two devices in the same phase, controlling current flow in one direction, affects several converter states. For trigger suppression in one switch, current does not flow in its controlled direction. However, if regeneration allows for the settling of valid states, all the original voltage space vectors are available for controlling the load. When the conditions do not allow for settling at a known state, trigger suppression in one device affects 32 out of the 64 usable states shown in Figures 5.24 and 5.25 and Table 5.4. This results in only five voltage space vectors out of 19 affected because of the dual converter's redundancy. A further suppression

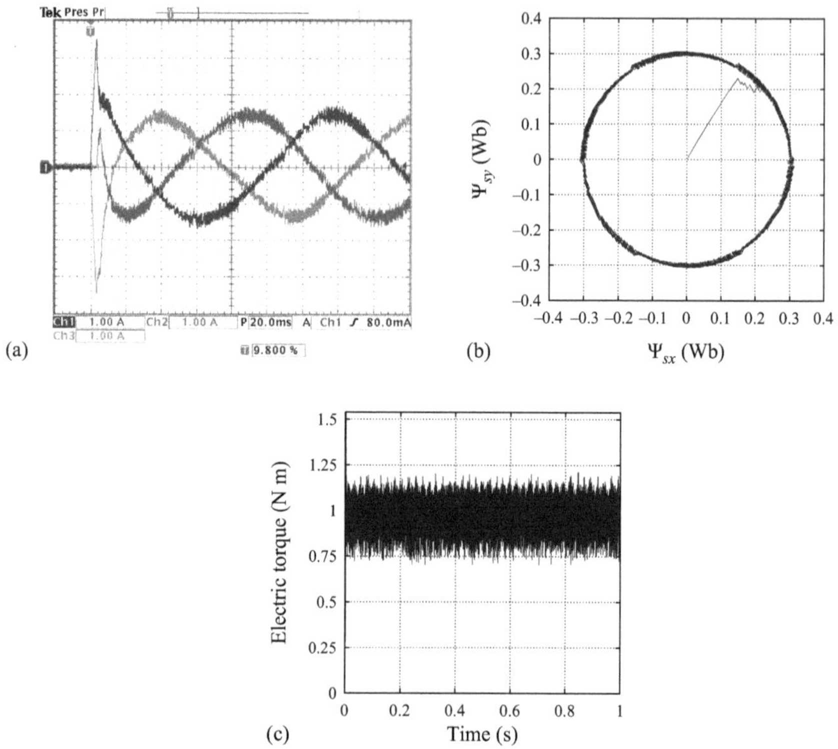

Figure 5.23 Experimental (a) stator currents, (b) stator flux, and (c) torque for normal operation with DTC using medium-sized voltage space vectors. ©2018 IEEE. Reprinted, with permission, from Reference [79]

in the second device, along the controlled direction, affects 16 extra states and five voltage space vectors. Figures 5.26 and 5.27 show the effects of trigger suppression for a_{1H} and a_{2L} in phase "*a*."

The devices under trigger suppression are sub-converter 1, low-side phase "*a*" (a_{1L}) and sub-converter 2, high-side phase "*a*" (a_{2H}). Trigger suppression and no regeneration in one of these devices prevent setting the outer space vectors in Figure 5.27 ($\vec{v}_{l2}$, $\vec{v}_{m1}$, $\vec{v}_{l1}$, $\vec{v}_{m6}$, $\vec{v}_{l6}$), and the additional trigger suppression in the second device prevent setting the space vectors ($\vec{v}_{m2}$, $\vec{v}_{s2}$, $\vec{v}_{s1}$, $\vec{v}_{s6}$, $\vec{v}_{m5}$). This also affects some states associated with voltage space vectors ($\vec{v}_0$, $\vec{v}_{s3}$, $\vec{v}_{s4}$, $\vec{v}_{s5}$), but the redundancy in dual converters' states allows for setting the associated voltage space vectors. If the second device with trigger suppression is (a_{2H}), the voltage space vectors affected are $\vec{v}_{l3}$, $\vec{v}_{m3}$, $\vec{v}_{l4}$, $\vec{v}_{m4}$, $\vec{v}_{l5}$. Injecting a zero-sequence component into the load current allows DTC on the load. All the switching states are available for control owing to regeneration. The proper selection of the switching state and its associated common mode voltage gives control over the zero-sequence injection.

Table 5.3 Common mode voltage for the positive and negative small voltage space vectors. ©2018 IEEE. Reprinted, with permission, from Reference [79]

v_a	v_b	v_c	Voltage space vector	v_{ZS}
V_{DC}	V_{DC}	V_{DC}	$\vec{v}_{s0}$	V_{DC}
0	0	0		0
$-V_{DC}$	$-V_{DC}$	$-V_{DC}$		$-V_{DC}$
V_{DC}	0	0	$\vec{v}_{s1}$	$(1/3)V_{DC}$
0	$-V_{DC}$	$-V_{DC}$		$-(2/3)V_{DC}$
V_{DC}	V_{DC}	0	$\vec{v}_{s2}$	$(2/3)V_{DC}$
0	0	$-V_{DC}$		$-(1/3)V_{DC}$
0	V_{DC}	0	$\vec{v}_{s3}$	$(1/3)V_{DC}$
$-V_{DC}$	0	$-V_{DC}$		$-(2/3)V_{DC}$
0	V_{DC}	V_{DC}	$\vec{v}_{s4}$	$(2/3)V_{DC}$
$-V_{DC}$	0	0		$-(1/3)V_{DC}$
0	0	V_{DC}	$\vec{v}_{s5}$	$(1/3)V_{DC}$
$-V_{DC}$	$-V_{DC}$	0		$-(2/3)V_{DC}$
V_{DC}	0	V_{DC}	$\vec{v}_{s6}$	$(2/3)V_{DC}$
0	$-V_{DC}$	0		$-(1/3)V_{DC}$

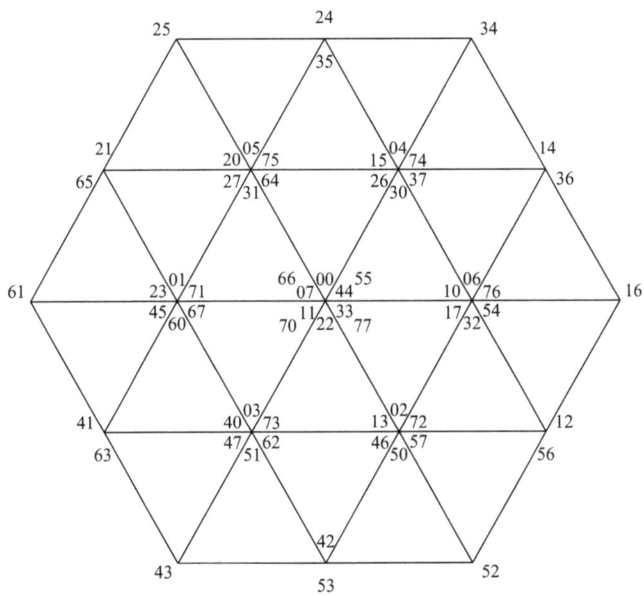

Figure 5.24 Switching states affected during trigger suppression in a_{1H}

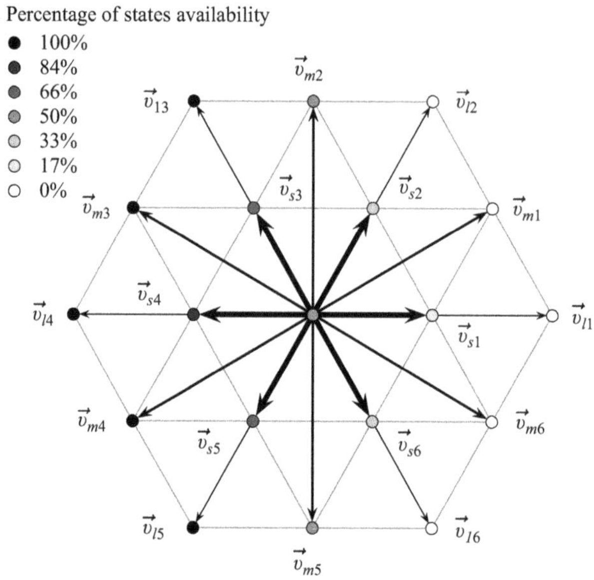

Figure 5.25 Space vectors availability during trigger suppression in a_{1H} without regenerative states

Table 5.4 States available for trigger suppression in two switches of phase "a" as presented in Figure 5.24

Voltage space vector	v_{ZS}
$\vec{v}_{s0}$	$-V_{DC}$
$\vec{v}_{s5}$	$-(2/3)V_{DC}$
$\vec{v}_{l5}$	$-(1/3)V_{DC}$
$\vec{v}_{s3}$	$-(2/3)V_{DC}$
$\vec{v}_{s4}$	$-(1/3)V_{DC}$
$\vec{v}_{m4}$	0
$\vec{v}_{l3}$	$-(1/3)V_{DC}$
$\vec{v}_{m3}$	0
$\vec{v}_{l4}$	$(1/3)V_{DC}$

Thanks to symmetry in the dual converter, the behavior of the other phases equals a rotated version of phase "*a*." Figure 5.28 shows the patterns for two devices in different phases, with trigger suppression.

For trigger suppression in "a_{1H}," shown in Figure 5.25, the voltage space vectors available for DTC are those with phase "*a* at $-V_{DC}$, shown in Table 5.4.

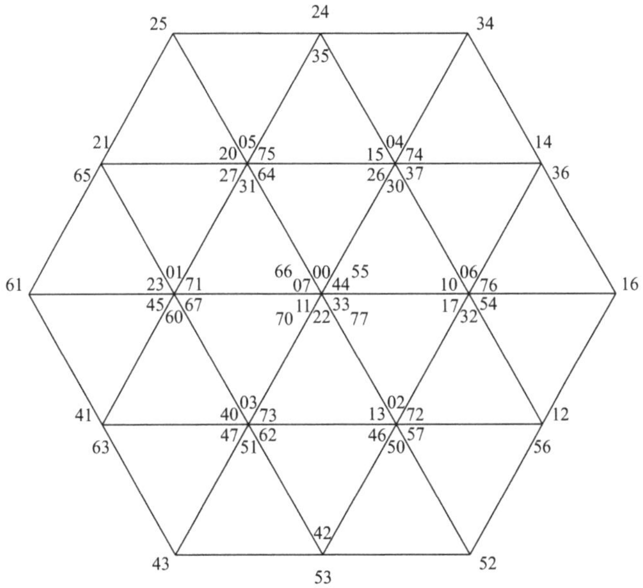

Figure 5.26 Switching states affected during trigger suppression in a_{1H} and a_{2L}

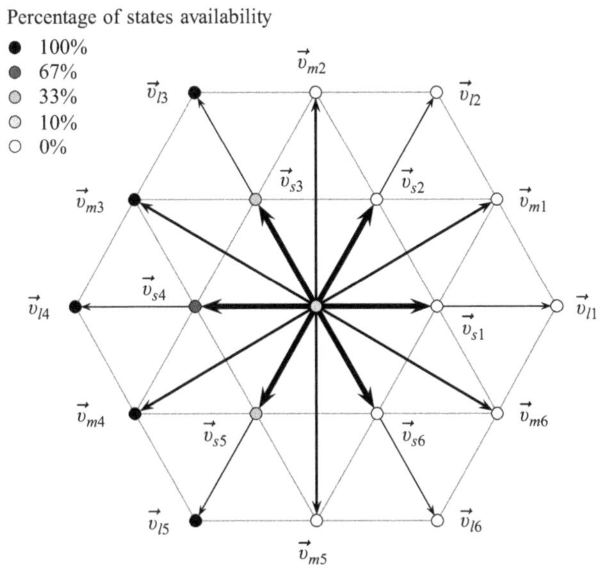

Figure 5.27 Space vectors availability during trigger suppression in a_{1H} and a_{2L} without regenerative states. Adapted from Reference [79]

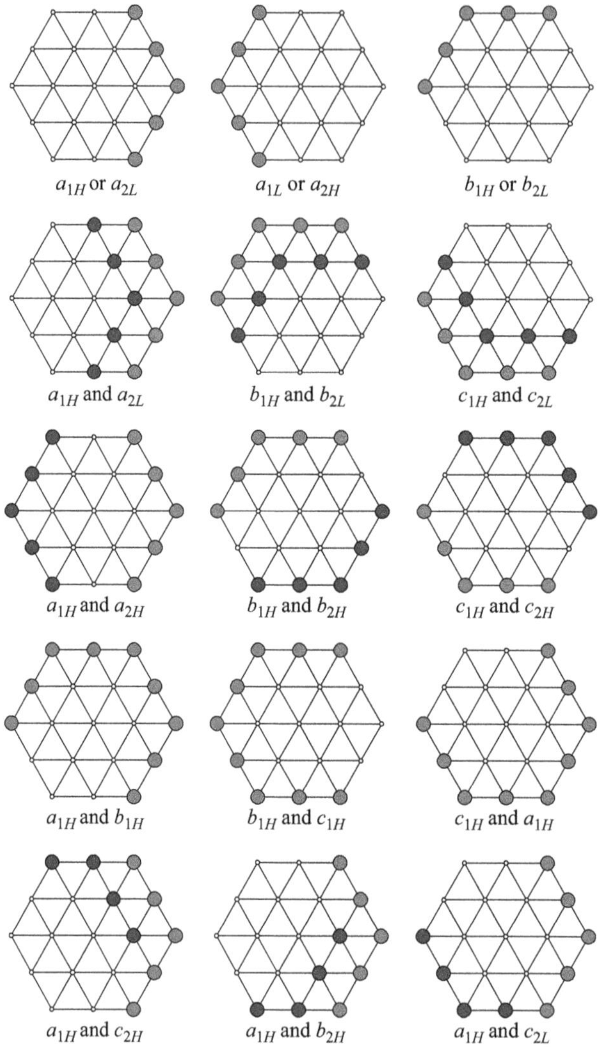

Figure 5.28 Example of different space vectors affected in the case of trigger suppression in one and two power devices. Adapted from Reference [79]

For two devices associated with the same phase and sub-converter, there is no possible combination of remaining switching devices that can impress the desired current value in that phase. Therefore, no opportunistic states are available for control, and a different control strategy should be considered, taking into account the phase where current does not flow.

Figure 5.29 shows the corresponding simulated state variables for trigger suppression in one or two switches associated to one-phase winding (a_{1L} and a_{2H}); in

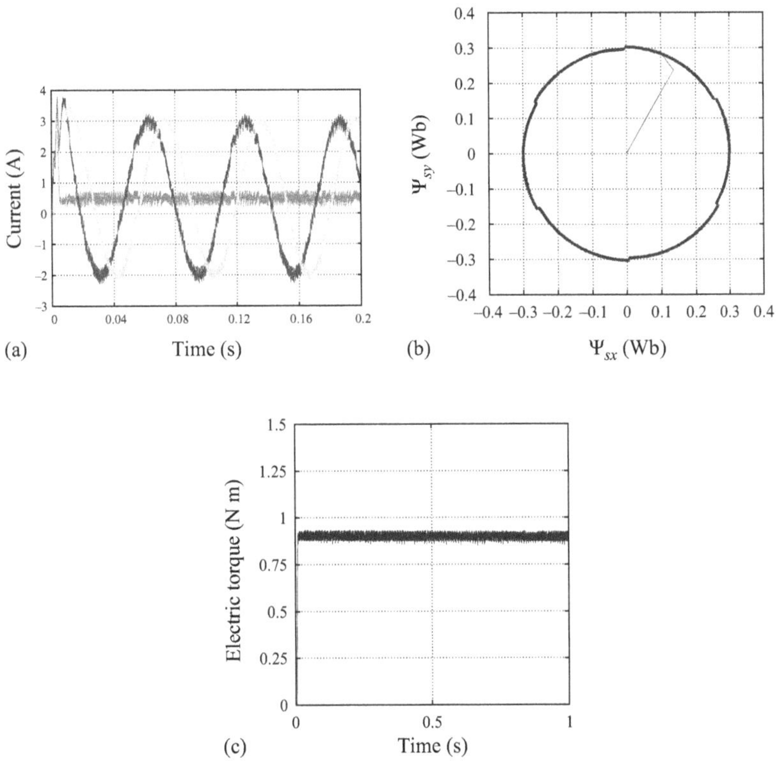

Figure 5.29 Simulated (a) stator currents, (b) stator flux, and (c) torque for normal operation with DTC using medium-sized voltage space vectors. ©2018 IEEE. Reprinted, with permission, from Reference [79]

this case, the current flowing through phase "*a*" can take only positive values, and to ensure known switching states in the dual converter, the phase current has to be positive at all times ($i_{sa} > 0$). If the trigger suppression is in the complementary switches (a_{1H} and a_{2L}), the current has to be negative at all times to ensure operation of the dual converter under fault with known states. The control strategy in this case is to use the set of regenerative space vectors whenever the current flowing through the affected phase is above a certain threshold that ensures the use of opportunistic states; otherwise, use the positive small set of space vectors to increase the stored energy in the affected winding to enable the use of opportunistic states in subsequent control cycles. Figure 5.30 shows the experimental state variables for trigger suppression in one or two switches associated to one-phase winding (a_{1L} and a_{2H}). The phase in phase "*a*" is maintained at just above zero to provide the opportunistic states.

5.3.2.2 Trigger suppression in six switches (asymmetric operation)

For trigger suppression in six devices, only 50% of power switches are available for controlling the current in the load. In this case, the resulting power stage topology

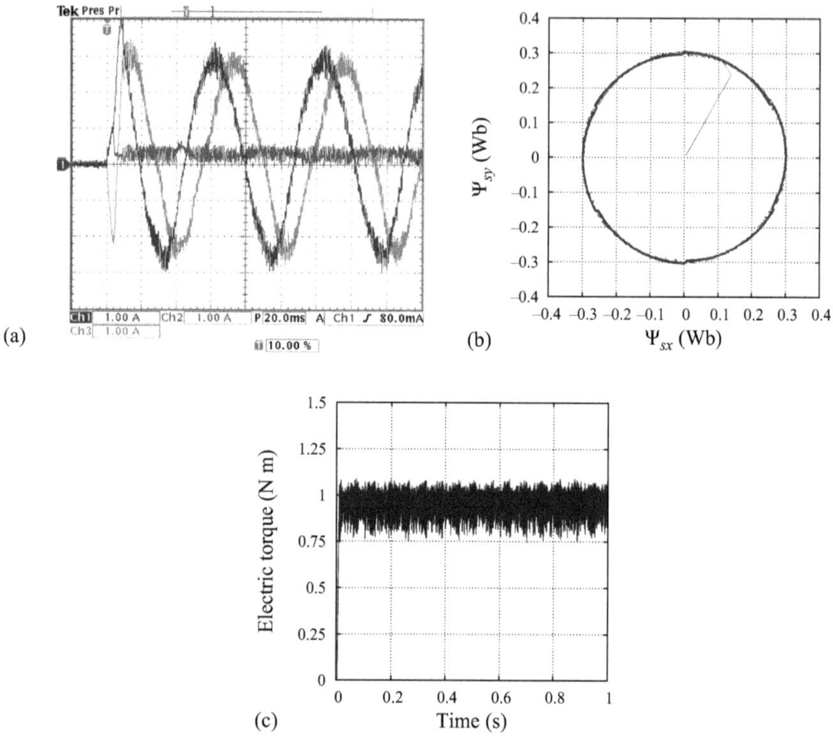

Figure 5.30 Experimental (a) stator currents, (b) stator flux, and (c) torque for trigger suppression in a_{1L} and a_{2H}. ©2018 IEEE. Reprinted, with permission, from Reference [79]

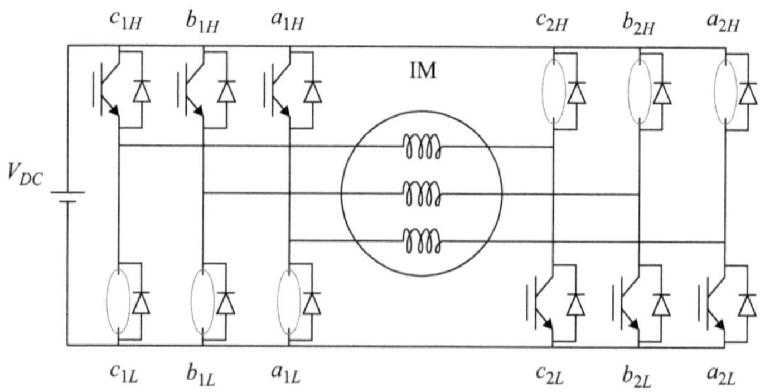

Figure 5.31 Diagram of the dual converter under six switches trigger suppression. Adapted from Reference [79]

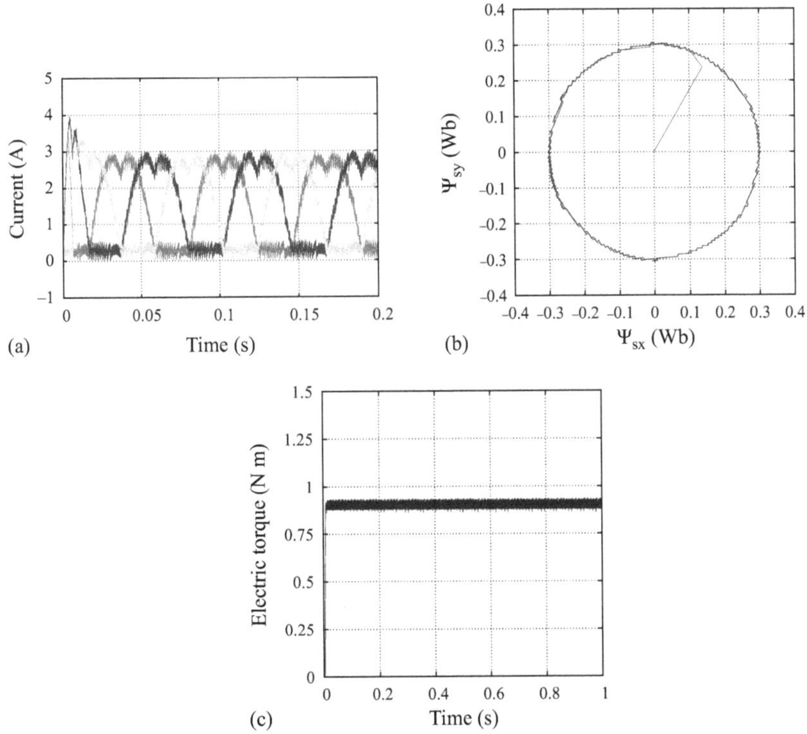

Figure 5.32 Simulated (a) stator currents, (b) stator flux, and (c) torque for trigger suppression in six switches, $\{a_{1L}, b_{1L}, c_{1L}\}$ *and* $\{a_{2H}, b_{2H}, c_{2H}\}$*.*

is similar to the asymmetric converter employed in the control of switched reluctance machines [84]. The term "asymmetric" applies for trigger suppression on the complementary devices belonging to the same phase, as shown in Figure 5.31. The authors in [85] show different switching strategies for the induction machine fed by an asymmetric converter. In this work, the DTC algorithm uses the positive and negative or regenerative set of space vectors. Under these conditions, all the switching states become available thanks to the opportunistic states.

For trigger suppression in the devices shown in Figure 5.31, the only stable state available to feed the machine is $\{1, 1, 1\}$ or null vector. The remaining states need current flow in each branch to operate with regenerative states.

Once the 64 states and 19 voltage space vectors become available, through regeneration, the DTC algorithm should use those states that preserve the zero sequence injection to the load.

The strategy used to control the zero-sequence injection is similar to the one employed in the case of trigger suppression with one and two devices. The opportunistic states (space vectors synthesized with 0 and -1 states) are used until a predefined

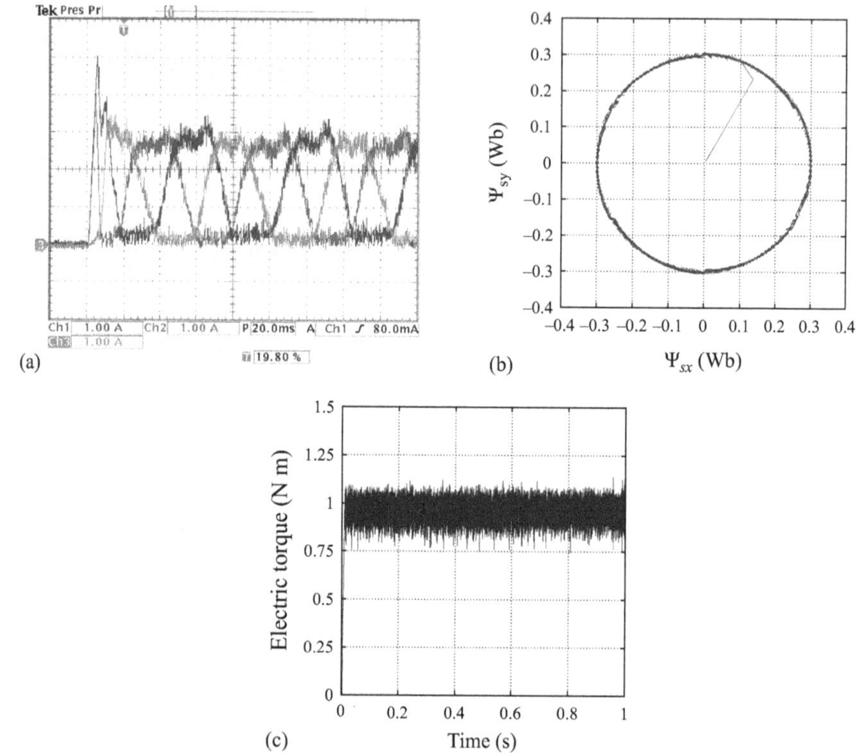

Figure 5.33 Experimental (a) stator currents, (b) stator flux, and (c) torque for trigger suppression in six switches, $\{a_{1L}, b_{1L}, c_{1L}\}$ *and* $\{a_{2H}, b_{2H}, c_{2H}\}$*. ©2018 IEEE. Reprinted, with permission, from Reference [79]*

threshold is reached. Then a space vector is synthesized using states 0 and 1, which increases the energy in the machine leakage reactances and goes back to the use of opportunistic states.

Although trigger suppression in six switches will affect all the space vectors synthesized by the dual converter, there are still known states available for storing energy in the machine inductances. This brings the possibility of achieving proper control of the dual converter, if enough energy can be stored in the machine inductances so that through regeneration, the opportunistic states become available. In this way, all the space vectors can be used, and the switching state of the dual converter is known. Figure 5.32 shows the simulated state variables for the dual converter operating under six switches with trigger suppression. The switches with trigger suppression are $(a, b, c)_{1L}$ and $(a, b, c)_{2H}$. The control strategy used in this case is similar to the one used previously and consists of applying the opportunistic states as long as there is enough energy stored in the machine-leakage inductances, and then switching to the complementary states to increase the stored energy when the energy is not sufficient to use the opportunistic states. For the case under test, the opportunistic states can be

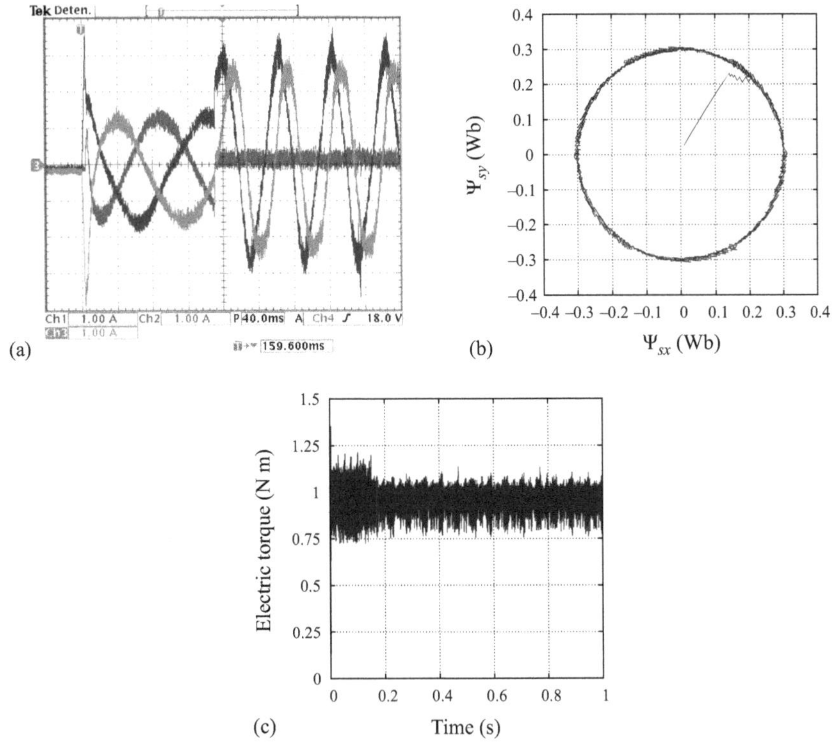

Figure 5.34 Experimental (a) stator currents, (b) stator flux, and (c) torque for the transition between healthy operation and trigger suppression in a_{1H} and a_{2L}. ©2018 IEEE. Reprinted, with permission, from Reference [79]

used if the phase currents are above a predefined minimum value, i.e., each phase circuit is operating in a continuous switching mode. Figure 5.33 shows the experimental results that confirm the ones obtained previously in the simulation. Although there is a high zero-sequence voltage and current component injected into the machine terminals, this zero sequence does not produce any braking torque because there is no zero-sequence component in the stator flux. The disadvantage of this operating mode is that for the same electrical torque, the machine is operating with higher rms currents, hence with increased power losses and reduced system efficiency. This results in higher thermal stress and reduced life span for the operating power components.

Figure 5.34 shows the instantaneous transition between healthy operation and switch trigger suppression in two power devices. Since the stator flux and torque are maintained as constant before and after the change in operating mode, the transition of the remaining state variables toward steady state is fast.

Table 5.5 shows the rms value of the stator currents and the total harmonic distortion (THD) for the dual converter when operating under a healthy condition, as

Table 5.5 RMS current and THD for healthy operation, two switches trigger suppression and six switches trigger suppression. ©2018 IEEE. Reprinted, with permission, from Reference [79]

		RMS **(*A*)**	***THD***
Healthy	i_a	0.9731	0.0647
	i_b	0.9397	0.0574
	i_c	1.0160	0.0608
2sts	i_a	0.2913	–
	i_b	1.7797	0.2143
	i_c	1.8834	0.2325
6sts	i_a	1.8883	1.4729
	i_b	1.7342	1.2646
	i_c	1.7967	1.4013

well as with two switches under trigger suppression and also with six switches under trigger suppression. From these results, it is clear that there is an increase in both rms current and the THD in most phases, which in turn results in an increase in the machine's temperature for the same output torque.

5.4 Summary

The widespread use of power electronics-based converters has resulted in continuous improvement in the power switches' reliability. Although the authors in [86] show a failure rate of medium-power IGBT modules of 0.1 failures per 10^6 h, many critical applications need further assurance of operation. This chapter has shown a sample of the techniques used to increase the system operating range under adverse conditions. Some techniques are applied before the onset of a device's fault, by allowing the device to move out of line for "recovery." Other techniques modify the topology, either by disabling the faulty devices or by also including extra elements to replace the faulty ones. In most cases, this operation needs a corresponding change in the control strategy, as well as in the amount of power that the system can handle during this fault-tolerant operation.

References

[1] D. Kastha and B. K. Bose, "Investigation of fault modes of voltage-fed inverter system for induction motor drive," *IEEE Transactions on Industry Applications*, vol. 30, no. 4, pp. 1028–1038, 1994.

[2] F. W. Fuchs, "Some diagnosis methods for voltage source inverters in variable speed drives with induction machines – a survey," in *Industrial Electronics Society, 2003. IECON'03. The 29th Annual Conference of the IEEE*, vol. 2, 2003, pp. 1378–1385.

[3] M. Baybutt, C. Minnella, A. E. Ginart, P. W. Kalgren, and M. J. Roemer, "Improving digital system diagnostics through prognostic and health management (PHM) technology," *IEEE Transactions on Instrumentation and Measurement*, vol. 58, no. 2, pp. 255–262, 2009.

[4] Q. T. An, L. Z. Sun, K. Zhao, and L. Sun, "Switching function model-based fast-diagnostic method of open-switch faults in inverters without sensors," *IEEE Transactions on Power Electronics*, vol. 26, no. 1, pp. 119–126, 2011.

[5] B. Mirafzal, "Survey of fault-tolerance techniques for three-phase voltage source inverters," *IEEE Transactions on Industrial Electronics*, vol. 61, no. 10, pp. 5192–5202, 2014.

[6] B. Lu and S. K. Sharma, "A literature review of IGBT fault diagnostic and protection methods for power inverters," *IEEE Transactions on Industry Applications*, vol. 45, no. 5, pp. 1770–1777, 2009.

[7] R. Wu, F. Blaabjerg, H. Wang, M. Liserre, and F. Iannuzzo, "Catastrophic failure and fault-tolerant design of IGBT power electronic converters – an overview," in *IECON 2013 – 39th Annual Conference of the IEEE Industrial Electronics Society*, 2013, pp. 507–513.

[8] K. Rothenhagen and F. W. Fuchs, "Performance of diagnosis methods for IGBT open circuit faults in three phase voltage source inverters for AC variable speed drives," in *2005 European Conference on Power Electronics and Applications*, 2005.

[9] P. O'Connor, R. W. Cox, and J. M. Anderson, "Near real-time incipient fault detection in IGBT switches," in *IECON 2014 – 40th Annual Conference of the IEEE Industrial Electronics Society*, 2014, pp. 4484–4491.

[10] A. M. S. Mendes and A. J. M. Cardoso, "Voltage source inverter fault diagnosis in variable speed AC drives, by the average current Park's vector approach," in *Electric Machines and Drives, 1999. International Conference IEMD'99*, 1999, pp. 704–706.

[11] K. Rothenhagen and F. W. Fuchs, "Performance of diagnosis methods for IGBT open circuit faults in voltage source active rectifiers," in *2004 IEEE 35th Annual Power Electronics Specialists Conference (IEEE Cat. No.04CH37551)*, vol. 6, 2004, pp. 4348–4354.

[12] R. Peuget, S. Courtine, and J. P. Rognon, "Fault detection and isolation on a PWM inverter by knowledge-based model," *IEEE Transactions on Industry Applications*, vol. 34, no. 6, pp. 1318–1326, 1998.

[13] R. L. de Araujo Ribeiro, C. B. Jacobina, E. R. C. da Silva, and A. M. N. Lima, "Fault detection of open-switch damage in voltage-fed PWM motor drive systems," *IEEE Transactions on Power Electronics*, vol. 18, no. 2, pp. 587–593, 2003.

[14] O. S. Yu, N. J. Park, and D. S. Hyun, "A novel fault detection scheme for voltage fed PWM inverter," in *IECON 2006 – 32nd Annual Conference on IEEE Industrial Electronics*, 2006, pp. 2654–2659.

[15] P. Gilreath and B. N. Singh, "A new centroid based fault detection method for 3-phase inverter-fed induction motors," in *2005 IEEE 36th Power Electronics Specialists Conference*, 2005, pp. 2664–2669.

[16] L. Zhang, I. B. Aris, and L. N. Hulley, "A knowledge-based system for online fault diagnosis of power inverter circuits for AC machines," in *Proc. Eur. Power Electron. Conf.*, 1995, pp. 334–339.

[17] R. Szczesny, H. Piquet, and P. Kurzynski, "Fault detection and diagnosis in the electric drives," in *Proc. Eur. Power Electron. Appl. Conf.*, 1997, pp. 995–1000.

[18] B. Raison, G. Rostaing, and J. P. Rognon, "Towards a global monitoring scheme for induction motor drives," in *Proc. Int. Power Electron. Conf.*, 2000, pp. 1183–1188.

[19] C. Kral and K. Kafka, "Power electronics monitoring for a controlled voltage source inverter drive with induction machines," in *2000 IEEE 31st Annual Power Electronics Specialists Conference. Conference Proceedings (Cat. No.00CH37018)*, vol. 1, 2000, pp. 213–217.

[20] Y. Guan, D. Sun, and Y. He, "Mean current vector based online real-time fault diagnosis for voltage source inverter fed induction motor drives," in *2007 IEEE International Electric Machines Drives Conference*, vol. 2, 2007, pp. 1114–1118.

[21] M. R. Mamat, M. Rizon, and M. Khanniche, "Fault detection of 3-phase VSI using wavelet-fuzzy algorithm," *American Journal of Applied Sciences*, vol. 3, no. 1, pp. 1642–1648, 2006.

[22] F. Charfi, F. Sellami, and K. Al-Haddad, "Fault diagnostic in power system using wavelet transforms and neural networks," in *2006 IEEE International Symposium on Industrial Electronics*, vol. 2, 2006, pp. 1143–1148.

[23] M. A. Awadallah and M. M. Morcos, "Diagnosis of switch open-circuit fault in PM brushless DC motor drives," in *Large Engineering Systems Conference on Power Engineering, 2003*, 2003, pp. 69–73.

[24] J.-H. Park, D.-H. Kim, S.-S. Kim, D.-J. Lee, and M.-G. Chun, "C-ANFIS based fault diagnosis for voltage-fed PWM motor drive systems," in *Fuzzy Information, 2004. Processing NAFIPS'04. IEEE Annual Meeting of the*, vol. 1, 2004, pp. 379–383.

[25] M. A. Masrur, Z. Chen, B. Zhang, and Y. L. Murphey, "Model-based fault diagnosis in electric drive inverters using artificial neural network," in *2007 IEEE Power Engineering Society General Meeting*, 2007, pp. 1–7.

[26] K. Debebe, V. Rajagopalan, and T. S. Sankar, "Expert systems for fault diagnosis of VSI fed AC drives," in *Conference Record of the 1991 IEEE Industry Applications Society Annual Meeting*, vol. 1, 1991, pp. 368–373.

[27] A. Kopta, M. Rahimo, U. Schlapbach, N. Kaminski, and D. Silber, "Limitation of the short-circuit ruggedness of high-voltage IGBTs," in *2009 21st International Symposium on Power Semiconductor Devices IC's*, 2009, pp. 33–36.

[28] H. J. Schulze, F. J. Niedernostheide, F. Pfirsch, and R. Baburske, "Limiting factors of the safe operating area for power devices," *IEEE Transactions on Electron Devices*, vol. 60, no. 2, pp. 551–562, 2013.

[29] F. Yang, J. Tan, S. Lu, and Y. Zhu, "Investigation on the robustness during short-circuit turn-off and its tradeoff characteristics with performance in IGBTs," *IEEE Transactions on Electron Devices*, vol. 64, no. 8, pp. 3293–3297, 2017.

[30] T. Basler, R. Bhojani, J. Lutz, and R. Jakob, "Dynamic self-clamping at short-circuit turn-off of high-voltage IGBTs," in *2013 25th International Symposium on Power Semiconductor Devices IC's (ISPSD)*, 2013, pp. 277–280.

[31] I. Takata, "Non thermal destruction mechanisms of IGBTs in short circuit operation," in *Proceedings of the 14th International Symposium on Power Semiconductor Devices and ICs*, 2002, pp. 173–176.

[32] S. Milady, D. Silber, F. Pfirsch, and F. J. Niedernostheide, "Simulation studies and modeling of short circuit current oscillations in IGBTs," in *2009 21st International Symposium on Power Semiconductor Devices IC's*, 2009, pp. 37–40.

[33] F. Hille, F. Umbach, T. Raker, and R. Roth, "Failure mechanism and improvement potential of IGBT's short circuit operation," in *2010 22nd International Symposium on Power Semiconductor Devices IC's (ISPSD)*, 2010, pp. 33–36.

[34] T. Laska, G. Miller, M. Pfaffenlehner, *et al.*, "Short circuit properties of trench-/field-stop-IGBTs-design aspects for a superior robustness," in *Power Semiconductor Devices and ICs, 2003. Proceedings. ISPSD'03. 2003 IEEE 15th International Symposium on*, 2003, pp. 152–155.

[35] M. Otsuki, Y. Onozawa, H. Kanemaru, Y. Seki, and T. Matsumoto, "A study on the short-circuit capability of field-stop IGBTs," *IEEE Transactions on Electron Devices*, vol. 50, no. 6, pp. 1525–1531, 2003.

[36] X. Perpiña, J. F. Serviere, X. Jorda, *et al.*, "Over-current turn-off failure in high voltage IGBT modules under clamped inductive load," in *2009 13th European Conference on Power Electronics and Applications*, 2009, pp. 1–10.

[37] B. C. Kok, M. S. Looi, and H. H. Goh, "Insulated gate bipolar transistor failure analysis in overvoltage condition," in *International Conference on Renewable Energies and Power Quality*, 2012.

[38] J. G. Fossum and Y.-S. Kim, "Static and dynamic latchup in the LIGBT," *IEEE Transactions on Electron Devices*, vol. 35, no. 11, pp. 1977–1985, 1988.

[39] K. Heumann and M. Quenum, "Second breakdown and latch-up behavior of IGBTs," in *1993 Fifth European Conference on Power Electronics and Applications*, vol. 2, 1993, pp. 301–305.

[40] K.-H. Oh, Y. C. Kim, K. H. Lee, and C. M. Yun, "Investigation of short-circuit failure limited by dynamic-avalanche capability in 600-V punchthrough IGBTs," *IEEE Transactions on Device and Materials Reliability*, vol. 6, no. 1, pp. 2–8, 2006.

[41] L. Benbahouche, A. Merabet, and A. Zegadi, "A comprehensive analysis of failure mechanisms: latch up and second breakdown in IGBT(IXYS) and

improvement," in *2012 19th International Conference on Microwaves, Radar Wireless Communications*, vol. 1, 2012, pp. 190–192.

[42] P. L. Hower and V. G. K. Reddi, "Avalanche injection and second breakdown in transistors," *IEEE Transactions on Electron Devices*, vol. 17, no. 4, pp. 320–335, 1970.

[43] C.-C. Shen, A. R. Kefner, D. W. Berning, and J. B. Bernstein, "Failure dynamics of the IGBT during turn-off for unclamped inductive loading conditions," in *Conference Record of 1998 IEEE Industry Applications Conference. Thirty-Third IAS Annual Meeting (Cat. No.98CH36242)*, vol. 2, 1998, pp. 831–839.

[44] G. Breglio, A. Irace, E. Napoli, M. Riccio, and P. Spirito, "Experimental detection and numerical validation of different failure mechanisms in IGBTs during unclamped inductive switching," *IEEE Transactions on Electron Devices*, vol. 60, no. 2, pp. 563–570, 2013.

[45] M. Riccio, E. Napoli, A. Irace, G. Breglio, and P. Spirito, "Energy and current crowding limits in avalanche operation of IGBTs," in *2013 25th International Symposium on Power Semiconductor Devices IC's (ISPSD)*, 2013, pp. 273–276.

[46] J. Lutz, H. Schlangenotto, U. Scheuermann, and R. De Doncker, *Semiconductor Power Devices: Physics, Characteristics, Reliability*. Springer, Berlin Heidelberg, 2014.

[47] F. Saint-Eve, S. Lefebvre, and Z. Khatir, "Influence of repetitions of short-circuit conditions on IGBT lifetime," *EPE Journal*, vol. 15, no. 4, pp. 7–12, 2005. [Online]. Available: https://doi.org/10.1080/09398368.2005.11463598.

[48] J. Lutz, R. Dobler, J. Mari, and M. Menzel, "Short circuit III in high power IGBTs," in *2009 13th European Conference on Power Electronics and Applications*, 2009, pp. 1–8.

[49] T. Shoji, M. Ishiko, S. Kawaji, T. Sugiyama, K. Hotta, T. Fukami, and K. Hamada, "Investigation of short-circuit capability of IGBT under high applied voltage conditions," *R&D Review of Toyota CRDL*, vol. 39, no. 4, pp. 22–26, 2004.

[50] H. Y. Long, N. Luther-King, M. R. Sweet, and E. M. S. Narayanan, "Numerical evaluation of the short-circuit performance of 3.3-kV CIGBT in field-stop technology," *IEEE Transactions on Power Electronics*, vol. 27, no. 5, pp. 2673–2679, 2012.

[51] T. Basler, J. Lutz, T. Brückner, and R. Jakob, "IGBT self-turn-off under short-circuit condition," in *10th International Seminar on Power Semiconductors*, 2010.

[52] J. Lutz and T. Basler, "Short-circuit ruggedness of high-voltage IGBTs," in *2012 28th International Conference on Microelectronics Proceedings*, May 2012, pp. 243–250.

[53] Y. Chen, W. Li, F. Iannuzzo, H. Luo, X. He, and F. Blaabjerg, "Investigation and classification of short-circuit failure modes based on three-dimensional safe operating area for high-power IGBT modules," *IEEE Transactions on Power Electronics*, vol. 33, no. 2, pp. 1075–1086, 2018.

[54] M. Naidu, S. Gopalakrishnan, and T. W. Nehl, "Fault-tolerant permanent magnet motor drive topologies for automotive X-By-Wire systems," *IEEE Transactions on Industry Applications*, vol. 46, no. 2, pp. 841–848, 2010.

[55] M. Pecht, *Prognostics and Health Management of Electronics*. Hoboken, NJ: Wiley, 2008.

[56] S. Yang, D. Xiang, A. Bryant, P. Mawby, L. Ran, and P. Tavner, "Condition monitoring for device reliability in power electronic converters: a review," *IEEE Transactions on Power Electronics*, vol. 25, no. 11, pp. 2734–2752, 2010.

[57] Y. Avenas, L. Dupont, N. Baker, H. Zara, and F. Barruel, "Condition monitoring: a decade of proposed techniques," *IEEE Industrial Electronics Magazine*, vol. 9, no. 4, pp. 22–36, 2015.

[58] B. Ji, V. Pickert, B. Zahawi, and M. Zhang, "In-situ bond wire health monitoring circuit for IGBT power modules," in *6th IET International Conference on Power Electronics, Machines and Drives (PEMD 2012)*, 2012, pp. 1–6.

[59] V. Smet, F. Forest, J. J. Huselstein, A. Rashed, and F. Richardeau, "Evaluation of v_{rmce} monitoring as a real-time method to estimate aging of bond wire-IGBT modules stressed by power cycling," *IEEE Transactions on Industrial Electronics*, vol. 60, no. 7, pp. 2760–2770, 2013.

[60] J. Sun, H. Xu, X. Wu, and K. Sheng, "Comparison and analysis of short circuit capability of 1200 V single-chip SiC MOSFET and Si IGBT," in *2016 13th China International Forum on Solid State Lighting: International Forum on Wide Bandgap Semiconductors China (SSLChina: IFWS)*, 2016, pp. 42–45.

[61] S. Hong, Z. Zhou, C. Lv, and H. Guo, "Prognosis for insulated gate bipolar transistor based on Gaussian Process Regression," in *2013 IEEE Conference on Prognostics and Health Management (PHM)*, 2013, pp. 1–5.

[62] B. Liu, D. Liu, Y. Tang, and M. Chen, "The investigation on the lifetime prediction model of IGBT module," *Energy Procedia*, vol. 12, pp. 394–402, 2011. the Proceedings of International Conference on Smart Grid and Clean Energy Technologies (ICSGCE) 2011. [Online]. Available: http://www.sciencedirect.com/science/article/pii/S1876610211018790.

[63] S. Lefebvre, Z. Khatir, and F. Saint-Eve, "Experimental behavior of single-chip IGBT and COOLMOS devices under repetitive short-circuit conditions," *IEEE Transactions on Electron Devices*, vol. 52, no. 2, pp. 276–283, 2005.

[64] A. E. Ginart, P. W. Kalgren, M. J. Roemer, D. W. Brown, and M. Abbas, "Transistor diagnostic strategies and extended operation under one-transistor trigger suppression in inverter power drives," *IEEE Transactions on Power Electronics*, vol. 25, no. 2, pp. 499–506, 2010.

[65] S. Yang, A. Bryant, P. Mawby, D. Xiang, L. Ran, and P. Tavner, "An industry-based survey of reliability in power electronic converters," *IEEE Transactions on Industry Applications*, vol. 47, no. 3, pp. 1441–1451, 2011.

[66] F. Wu, J. Zhao, Y. Liu, D. Zhou, and H. Luo, "Primary source inductive energy analysis based real-time multiple open-circuit fault diagnosis in two-level

three-phase PWM boost rectifier," *IEEE Transactions on Power Electronics*, vol. 33, no. 4, pp. 3411–3423, 2018.

[67] J. Restrepo, V. Guzmán, M. Giménez, A. Bueno, and J. Aller, "Parallelogram based method for space vector pulse width modulation," *Revista Facultad de Ingeniería Universidad de Antioquia*, no. 52, pp. 161–171, 2010.

[68] J. Restrepo, J. M. Aller, A. Bueno, J. C. Viola, A. Berzoy, and T. Habetler, "Direct power control of a dual converter operating as synchronous rectifier," in *2010 Twenty-Fifth Annual IEEE Applied Power Electronics Conference and Exposition (APEC)*, 2010, pp. 343–348.

[69] J. Restrepo, J. M. Aller, A. Bueno, V. M. Guzmán, and M. I. Giménez, "Generalized algorithm for pulse width modulation using a two-vectors based technique," *EPE Journal*, vol. 21, no. 2, pp. 30–39, 2011.

[70] C. B. Jacobina, R. L. de Araujo Ribeiro, A. M. N. Lima, and E. R. C. da Silva, "Fault-tolerant reversible ac motor drive system," *IEEE Transactions on Industry Applications*, vol. 39, no. 4, pp. 1077–1084, 2003.

[71] M. Jones, S. N. Vukosavic, D. Dujic, E. Levi, and P. Wright, "Five-leg inverter PWM technique for reduced switch count two-motor constant power applications," *IET Electric Power Applications*, vol. 2, no. 5, pp. 275–287, 2008.

[72] J. Restrepo, J. Viola, and F. Quizhpi, "Simplified FPGA implementation of the generalized space vector pulse width modulation (GSVPWM) for three wire three-phase inverters," in *2014 Australasian Universities Power Engineering Conference (AUPEC)*, 2014, pp. 1–6.

[73] U. M. Choi, J. S. Lee, F. Blaabjerg, and K. B. Lee, "Open-circuit fault diagnosis and fault-tolerant control for a grid-connected NPC inverter," *IEEE Transactions on Power Electronics*, vol. 31, no. 10, pp. 7234–7247, 2016.

[74] S. Ceballos, J. Pou, E. Robles, *et al.*, "Three-level converter topologies with switch breakdown fault-tolerance capability," *IEEE Transactions on Industrial Electronics*, vol. 55, no. 3, pp. 982–995, 2008.

[75] A. Nabae, I. Takahashi, and H. Akagi, "A new neutral-point-clamped PWM inverter," *IEEE Transactions on Industry Applications*, vol. IA-17, no. 5, pp. 518–523, 1981.

[76] S. Ceballos, J. Pou, J. Zaragoza, *et al.*, "Efficient modulation technique for a four-leg fault-tolerant neutral-point-clamped inverter," *IEEE Transactions on Industrial Electronics*, vol. 55, no. 3, pp. 1067–1074, 2008.

[77] V. T. Somasekhar, S. Srinivas, B. Prakash Reddy, C. Nagarjuna Reddy, and K. Sivakumar, "Pulse width-modulated switching strategy for the dynamic balancing of zero-sequence current for a dual-inverter fed open-end winding induction motor drive," *IET Electric Power Applications*, vol. 1, pp. 591–600, 2007.

[78] V. T. Somasekhar, S. Srinivas, and K. K. Kumar, "Effect of zero-vector placement in a dual-inverter fed open-end winding induction-motor drive with a decoupled space-vector PWM strategy," *IEEE Transactions on Industrial Electronics*, vol. 55, no. 6, pp. 2497–2505, 2008.

[79] J. A. Restrepo, A. Berzoy, A. E. Ginart, J. M. Aller, R. G. Harley, and T. G. Habetler, "Switching strategies for fault tolerant operation of single DC-link

dual converters," *IEEE Transactions on Power Electronics*, vol. 27, no. 2, pp. 509–518, 2012.

[80] I. Takahashi and T. Noguchi, "A new quick-response and high-efficiency control strategy of an induction motor," *IEEE Transactions on Industry Applications*, vol. IA-22, no. 5, pp. 820–827, 1986.

[81] A. Shukla, A. Ghosh, and A. Joshi, "Improved multilevel hysteresis current regulation and capacitor voltage balancing schemes for flying capacitor multilevel inverter," *IEEE Transactions on Power Electronics*, vol. 23, pp. 518–529, 2008.

[82] M. F. Escalante, J. C. Vannier, and A. Arzande, "Flying capacitor multilevel inverters and DTC motor drive applications," *IEEE Transactions on Industrial Electronics*, vol. 49, no. 4, pp. 809–815, 2002.

[83] F. Khoucha, S. M. Lagoun, K. Marouani, A. Kheloui, and M. El Hachemi Benbouzid, "Hybrid cascaded H-bridge multilevel-inverter induction-motor-drive direct torque control for automotive applications," *IEEE Transactions on Industrial Electronics*, vol. 57, no. 3, pp. 892–899, 2010.

[84] S. Vukosavic and V. Stefanovic, "SRM inverter topologies: a comparative evaluation," *IEEE Transactions on Industry Applications*, vol. 27, no. 6, pp. 1034–1047, 1991.

[85] J. Restrepo, J. M. Aller, V. M. Guzman, M. I. Gimenez, and J. C. Viola, "Switching strategies for DTC on asymmetric converters," in *2009 13th European Conference on Power Electronics and Applications*, 2009, pp. 1–9.

[86] R. Burgos, G. Chen, F. Wang, D. Boroyevich, W. G. Odendaal, and J. D. V. Wyk, "Reliability-oriented design of three-phase power converters for aircraft applications," *IEEE Transactions on Aerospace and Electronic Systems*, vol. 48, no. 2, pp. 1249–1263, 2012.

Chapter 6

Motor diagnostics and protection using inverter capabilities

Stefan Grubic[1], José M. Aller[2,3], and Thomas G. Habetler[4]

This chapter outlines the state of the art on motor protection, monitoring, and failure prediction using inverter capabilities (sensing, data processing, and signal application) while feeding the electric motor. The methods included in this comprehensive review are thermal monitoring and protection schemes, monitoring schemes for insulation-related issues, and methods that analyze the mechanical health of the motor or an asset connected to the motor. Other topics of interest include thermal modeling of electric machines, spectral analysis, expert systems, neural networks, and parameter estimation. The techniques described in this chapter are applied to induction machines, brushless direct current (DC) machines, and reluctance and synchronous motors.

6.1 Introduction

The field of monitoring and diagnostics (M&D) of electrical machines has been a popular research topic over the last three decades. There are several reasons for the growing interest in this field. The health of electrical assets is critical for the flawless operation of industrial processes and plants, and the failure of those assets might have undesired consequences. The most obvious ones are financial losses associated with the downtime of the process and the costs required to replace or repair the faulty asset. An estimation for the average downtime cost for different industries is given in Figure 6.1 [1]. Another critical aspect of equipment failure is a potential threat to the health of the operator or others in the vicinity of the faulty asset.

M&D of electrical assets is an essential tool for improving the reliability, availability, and safety of processes that are driven by electrical equipment. The benefit of the M&D is usually diminished by the additional cost that is incurred for the

[1]Evelozcity, USA
[2]Electrical Engineering Department, Universidad Politécnica Salesiana – Sede Cuenca, Ecuador
[3]Electrical Engineering Department, Simon Bolivar University, Venezuela
[4]School of Electrical and Computer Engineering, Georgia Institute of Technology, USA

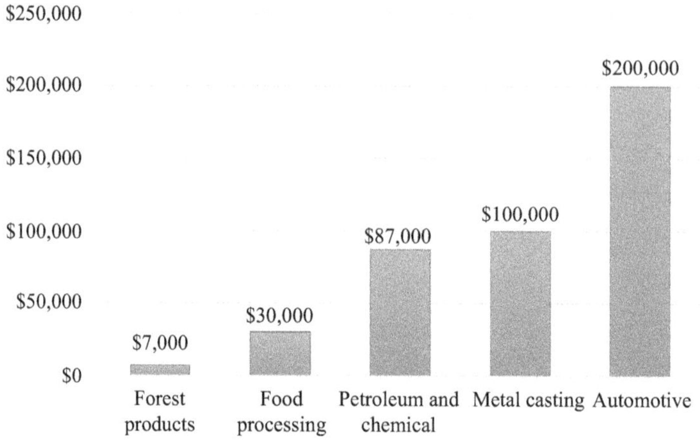

Figure 6.1 Average downtime cost of different industries due to failures of electric motors

monitoring equipment. To keep the cost of the monitoring process low, it is therefore an advantage to utilize existing assets to perform the monitoring duties. The fundamental components required for the M&D process are (1) an excitation, (2) a data-acquisition system, and (3) some data-processing capability.

In some cases, the excitation is provided by the regular operation of the asset. For example, in an online capacitance and dissipation factor (C&DF) measurement, the excitation is provided by the fundamental of the operating voltage [2] (see Section 6.3.2). In other cases, a specific signal can be incorporated in the operating sequence. An example for this is the thermal monitoring approach using a closed-loop drive presented in [3] (see Section 6.2.2). Some methods, such as the online surge test, might even require an additional source of excitation [4]. The same applies to signal acquisition and data processing. For some applications, the on-board sensors and data-acquisition system can be used, while other applications might require additional sensors and a specific data-acquisition system. Often the data processing is too complex to be performed by the inverter and instead has to be carried out by a more powerful asset such as an external server.

There are several different types of failures in electric machines and several types of methods that are used to diagnose the problem. The most common failures are [5,6]

1. gearbox or bearing failures caused by thermal, mechanical, or electrical stresses;
2. stator faults due to an insulation failure (phase-to-phase, phase-to-ground, and turn-to-turn);
3. rotor faults such as broken bars or end rings on induction machines, winding faults for wound synchronous machines, or demagnetization faults in permanent magnet machines; and
4. static or/and dynamic eccentricities or irregularities.

There are several surveys that give estimates on the relative likelihood of the different failure mechanisms. In [7], the failure rate in small and medium-sized motors due to a malfunction of the bearings or the stator winding faults is reported to be around 45% each. In large-sized AC motors, bearing issues seem to be less significant (16%), and the amount of stator winding failures increases (66%) [6]. Broken rotor bars and/or end rings and rotor eccentricity are around 10% in each of the referenced reports.

Various surveys on different aspects of M&D of electrical machines have been published. Here are some of the most relevant review papers in this area and a brief comment on the main topic of the respective review:

- "A Review of Induction Motors Signature Analysis as a Medium for Faults Detection" (2000) [8]. This review gives an overview of current signature analysis as well as of frequency and time-frequency processing methods that are used for current signature-based analysis for induction machines.
- "Recent Developments of Induction Motor Drives Fault Diagnosis Using AI Techniques" (2000) [9]. An overview on the use of artificial intelligence (AI) algorithms such as neural networks or fuzzy logic applied to the detection of induction motor faults is given in this review.
- "Induction Machine Drive Condition Monitoring and Diagnostic Research – A Survey" (2003) [10]. This work gives a comprehensive review of online monitoring techniques applicable to induction machines and also addresses signal processing and decision-making via AI methods.
- "Condition Monitoring and Fault Diagnosis of Electrical Motors – A Review" (2005) [5]. This is a review that touches on a variety of topics with a strong focus on current signature analysis.
- "A Review of Stator Fault Monitoring Techniques of Induction Motors" (2005) [11]. The focus of this work is on stator insulation-related monitoring techniques, especially on methods relevant for induction machines.
- "Condition Monitoring of Rotating Electrical Machines" (2008) [12]. A comprehensive and in-depth overview on the condition monitoring of electrical machines that covers a large variety of topics.
- "A Survey on Testing and Monitoring Methods for Stator Insulation Systems of Low-Voltage Induction Machines Focusing on Turn Insulation Problems" (2008) [13]. This publication deals with turn insulation monitoring in low-voltage induction machines. It also offers a comprehensive overview of stator insulation monitoring and related topics.
- "Recent Advances in Modeling and Online Detection of Stator Interturn Faults in Electrical Motors" (2011) [14]. This paper has a similar topic as the previous paper but is focused on the application of turn fault detection in permanent magnet synchronous machines (PMSM).
- "A Survey of Condition Monitoring and Protection Methods for Medium-Voltage Induction Motors" (2011) [6]. Much of the research conducted by universities has been focused on smaller low-voltage machines. This survey assesses the applicability of popular monitoring methods to larger medium-voltage machines.

- "A Review of Condition Monitoring and Fault Diagnosis for Permanent Magnet Motors" (2012) [15]. Even though induction machines are still the most commonly used machines in industrial application, permanent magnet machines are getting more attention due to their popularity in traction applications and the increased electrification of traction applications. This paper gives a brief overview of the state of condition monitoring for permanent magnet machines.

These reviews provide references to several hundred papers related to M&D of electrical machines. The research area has been very active since the 1970s and has received increased attention since the 1990s. There is a plethora of publications on approaches for M&D algorithms. Most of the methods can be grouped by the basic underlying concepts. For example, one popular group of methods is based on diagnosing an imbalance in the electric machine that occurs due to a fault. Those methods can be generally labeled as sequence component-based methods. A very common application for this type of methods is the diagnosis of stator turn faults [16]. Another popular group of methods is based on the spectral analysis of the electrical signals. The fault produces specific frequency components that can be found via frequency or time-frequency analysis of the machine currents or voltages. Typically, the spectral methods are applied to mechanical faults such as broken rotor bars, eccentricities, bearing faults [7], or even faults in the equipment connected to the machine [17]. Model-based approaches can be used for a large variety of problems, such as thermal problems. Vibration-based analysis can be applied to mechanical faults but can also serve to identify electrical problems. The application of signal-processing techniques is of great importance. Some of the more popular methods are fast Fourier transform (FFT), power spectral density (PSD), short time Fourier transform (STFT), and wavelet transform (WT). In recent years, more sophisticated methods that address the drawbacks of the traditional approaches have been introduced into M&D applications. The use of AI, especially for the automation of the monitoring process, has also been explored by using neural networks, fuzzy logic, or genetic algorithms [9,16,18–22]. A general differentiation can be made between methods that can be performed online with the motor operating and those that can be performed when the motor is off-line, i.e., when it is not in service. Both types will be introduced in this chapter since both can be facilitated through the inverter.

A large number of the monitoring techniques are based on measurements of instantaneous voltages and currents at the machine terminals, which are conducted online (during normal operation). These methods are nonintrusive, and usually their implementation cost is low. These diagnostic methods can easily be incorporated in an electronic power converter (soft starter, inverter, cycloconverter, etc.). The converter provides the sensors for the data acquisition of current and/or voltages and the A/D conversion. It usually contains a powerful DSP that controls the operation of the converter (Figure 6.2). This platform can be used for the data processing. Adding some diagnostic and monitoring capability only marginally increases the production cost, yet the maintenance expenses related to electrical machine failures can be reduced significantly.

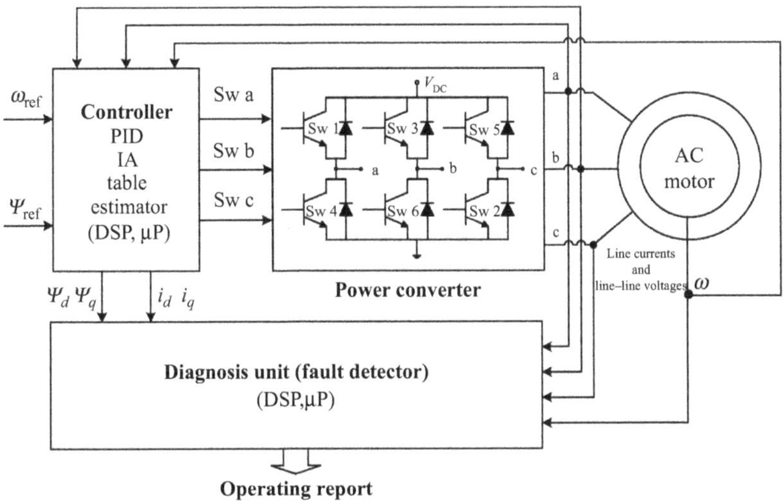

Figure 6.2 Power converter with online electrical diagnostic and monitoring features [23]

This chapter focuses on the analysis of the health of electrical machines facilitated through the power converter (inverter). For example, the inverter can be used to execute the monitoring algorithm on the inverter platform. In some cases, the fundamental excitation also serves as the excitation of the monitoring process. In other cases, the inverter generates a specific excitation that has to be incorporated into the fundamental excitation. The inverter can be used for the data acquisition, and the data processing can be executed on the inverter platform as well. It is feasible for the inverter to execute either all of the tasks required for the monitoring process or only some of them. Several applications of monitoring technologies that involve the inverter are discussed, and a comprehensive overview of the technologies that have been developed in this area is given. We begin with a discussion of thermal monitoring and protection schemes that mainly deals with methods to estimate the stator temperature of the electric machine. Then, an overview over stator-related monitoring is given, which largely is a discussion of insulation monitoring. This is followed by a review of rotor-related issues such as broken bars and rotor eccentricities and by a discussion on methods to detect mechanical problems such as bearing or gearbox faults. An extensive list of references is given along the way.

6.2 Thermal monitoring and protection

Thermal monitoring is an essential tool to protect the motor from thermal overload. Thermal overload can have several root causes and can occur in different operating modes. The overload can be of a short duration when the motor operates in a transient condition (startup, stall), or when an insulation fault occurs. The thermal overloading can also happen during steady state. It can occur when the motor is overloaded or operated with an unbalanced voltage supply. Impaired cooling mechanisms or a high ambient temperature also contribute to thermal overloading. Some of the consequences of thermal overloading include the deterioration of important motor components such as the winding insulation, the bearings, the conductors, and the core. The thermal protection of the motor requires measurement or estimation of the motor temperature. In general, the methods for obtaining the temperature can be divided into three groups:

1. Direct measurement of the temperature, which is usually achieved by using some form of embedded temperature sensor [e.g., thermocouple (TC), resistance temperature detector] or infrared temperature sensor;
2. Temperature estimation based on a thermal model, which usually includes the estimation of the motor losses and some sort of heat-transfer model; and
3. Parameter-based temperature estimation, which is often based on the measurement of a temperature-sensitive quantity such as the stator winding resistance.

6.2.1 Thermal models

The most commonly used technique in protection relays is the model-based temperature estimation. A thermal model requires the estimation of the power losses in the electrical machine and of the thermal parameters of the model. Several thermal models of the induction machine have been proposed and analyzed [24]. A first-order model does not always have the accuracy required by the application. Increasing the model order can increase the accuracy of the prediction but also requires more computational power and comes at the cost of an increased number of parameters that have to be estimated. A second-order model is often used as a compromise between complexity and accuracy [25].

The mathematical expression for a second-order thermal model can be written as

$$\begin{bmatrix} \dot{\theta}_s \\ \dot{\theta}_r \end{bmatrix} = \begin{bmatrix} -\frac{R_{\theta s}+R_{\theta sr}}{R_{\theta s}R_{\theta sr}C_{\theta s}} & \frac{1}{R_{\theta sr}C_{\theta s}} \\ \frac{1}{R_{\theta sr}C_{\theta r}} & -\frac{R_{\theta r}+R_{\theta sr}}{R_{\theta r}R_{\theta sr}C_{\theta r}} \end{bmatrix} \begin{bmatrix} \theta_s \\ \theta_r \end{bmatrix} + \begin{bmatrix} \frac{P_{s,loss}}{C_{\theta s}} \\ \frac{P_{r,loss}}{C_{\theta r}} \end{bmatrix} \tag{6.1}$$

where θ_s and θ_r are the stator and rotor winding temperatures; $R_{\theta s}$ and $R_{\theta r}$ are the stator and rotor thermal resistance representing the heat dissipation capability to the ambient; $R_{\theta sr}$ is the thermal resistance between stator and rotor; $C_{\theta s}$ and $C_{\theta r}$ are the stator and rotor thermal capacitance representing the capability of the system to store thermal energy; and $P_{s,loss}$ and $P_{r,loss}$ represent the losses in stator and rotor, respectively. The equation for a first-order model is

$$\dot{\theta}_i = -\frac{1}{R_\theta C_\theta}\theta_i + \frac{P_{loss}}{C_\theta} \tag{6.2}$$

where θ_i is the estimated stator winding temperature; R_θ is the thermal resistance between the stator winding and the ambient; C_θ is the thermal capacitance between the stator winding and the ambient; and P_{loss} is the total motor losses.

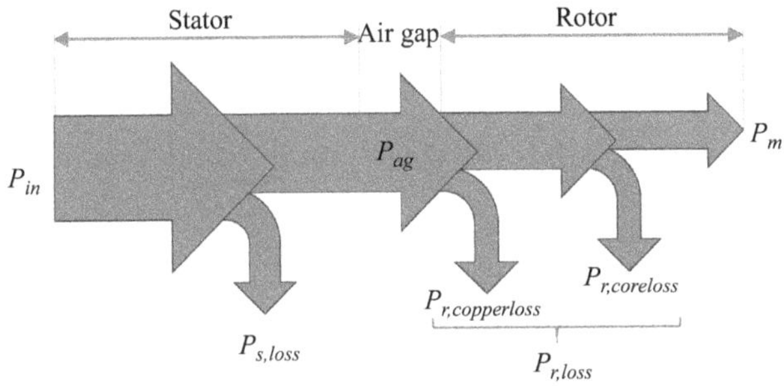

Figure 6.3 Power flow and losses for induction motors

6.2.1.1 Estimation of the motor losses

In [26,27], the nonintrusive air-gap torque method for an online efficiency estimation of an induction machine is proposed. Figure 6.3 shows a power flowchart including the machine losses for induction motors. Each of the losses can be obtained by a nonintrusive method.

The active input power P_{in} is calculated as [28]

$$P_{in} = \Re e\left(\vec{v}_s \cdot \vec{i}_s^*\right) = v_{ab}i_a + v_{cb}i_c \tag{6.3}$$

and the air-gap power

$$P_{ag} = T_e\,\omega_s \tag{6.4}$$

The electric torque in the air gap (T_e) is given as [28]

$$T_e = p\vec{\lambda}_s \times \vec{i}_s = p\left(\lambda_{s\alpha}i_{s\beta} - \lambda_{s\beta}i_{s\alpha}\right) \tag{6.5}$$

where

$$\vec{\lambda}_s = \int_0^t \left(\vec{v}_s - R_s\vec{i}_s\right) d\tau$$

$$\vec{v}_s = \sqrt{\frac{2}{3}}\frac{1}{1-\alpha^2}\left(v_{ab} + \alpha v_{bc} + \alpha^2 v_{ca}\right)\ ;\ \vec{i}_s = \sqrt{\frac{2}{3}}\left(i_a + \alpha i_b + \alpha^2 i_c\right)\ ;\ \alpha = e^{j(2\pi/3)}$$

The angular speed (mechanical angular speed $=\omega_m$, electrical angular speed $=\omega_r$) can be estimated based on harmonics in the stator current that arise from rotor slotting and eccentricity effects. From [26,29],

$$\omega_m = \frac{2}{p}\omega_r = \frac{2}{p}\omega_s(1-s) = \frac{2\pi\left(f_{sh} - f_s n_w\right)}{kN_R + n_d} \tag{6.6}$$

where s is the rotor slip, p is the number of poles; f_{sh} is the frequency of the rotor slot harmonic; f_s is the fundamental frequency; N_R is the number of rotor slots; $n_w = \pm 1, \pm 3, \ldots$; $n_d = 0, \pm 1, \ldots$ is the eccentricity order; and $k \in \mathbb{N}$. The most suitable choice for the parameters n_w, n_d, and k depends on the motor and can be determined during a commissioning process. During the monitoring phase, f_{sh} is

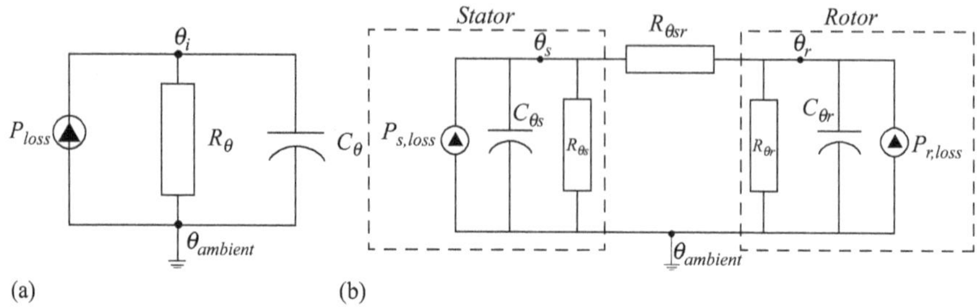

Figure 6.4 Induction machine thermal model: (a) first-order thermal model and (b) second-order thermal model

found via frequency analysis from the stator current and is then used to calculate the mechanical angular speed ω_m by using (6.6). Once the angular speed is determined, the developed mechanical power P_m can be calculated as

$$P_m = T_e\,\omega_m \tag{6.7}$$

The difference between the input power P_{in} (6.3) and the air-gap power P_{ag} obtained by (6.4) yields the stator losses:

$$P_{s,loss} = P_{in} - P_{ag} \tag{6.8}$$

The rotor losses are obtained by calculating the difference between the air-gap power P_{ag} (6.4) and the developed mechanical power P_m (6.7):

$$P_{r,loss} = P_{ag} - P_m \tag{6.9}$$

The expressions (6.8) and (6.9) are the power sources of a second-order thermal model illustrated in Figure 6.4 [30].

6.2.1.2 Thermal model parameter estimation

The parameters of the second-order thermal model as defined in (6.1) (see Figure 6.4), $P_{s,loss}$, $P_{r,loss}$, $R_{\theta s}$, $R_{\theta r}$, $R_{\theta sr}$, $C_{\theta s}$, and $C_{\theta r}$, have to be identified. First, the dependence of the thermal parameters on the operating variables has to be established. The dependence of the stator and rotor losses on the phase currents (i_s and i_r), the operating frequency (ω_s), and the slip (s) can be approximated by [30]

$$P_{s,loss} = 3\,R_s i_s^2 + k_{Hs}\omega_s + k_{Es}\omega_s^2 \tag{6.10}$$

$$P_{r,loss} = 3\,R_r i_r^2 + k_{Hr} s\omega_s + k_{Er} s^2\omega_s^2 \tag{6.11}$$

where R_s and R_r are the stator and rotor electrical resistance, and k_{Hs}, k_{Es}, k_{Hr}, and k_{Er} are the loss coefficients for hysteresis and eddy current losses (core losses). Further, the resistance values depend on the operating temperatures.

$$R_s = R_{s0}[1 + \alpha_s(\theta_s - \theta_0)] \tag{6.12}$$

$$R_r = R_{r0}[1 + \alpha_r(\theta_r - \theta_0)] \tag{6.13}$$

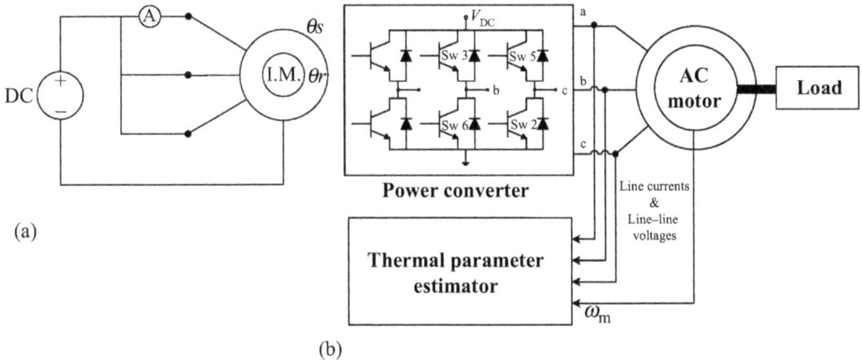

Figure 6.5 Test setup for thermal resistances estimation [30]: (a) DC test and (b) AC test

where R_{s0} and R_{r0} are the electrical resistance values at the reference temperature θ_0, and α_s and α_r are the temperature coefficients of the stator and rotor winding, respectively. R_{s0} and R_{r0} can easily be determined by using an ohmmeter and the temperature coefficients depend on the material used for the windings and can be found in literature. The thermal conductance values depend linearly on the rotor speed:

$$G_{\theta s} = \frac{1}{R_{\theta s}} = G_{\theta s0}(1 + b_s\omega) \tag{6.14}$$

$$G_{\theta r} = \frac{1}{R_{\theta r}} = G_{\theta r0}(1 + b_r\omega) \tag{6.15}$$

$$G_{\theta sr} = \frac{1}{R_{\theta sr}} = G_{\theta sr0}(1 + b_{sr}\omega) \tag{6.16}$$

where $G_{\theta s0}$, $G_{\theta r0}$, and $G_{\theta sr0}$ are the thermal conductance values at zero speed, and b_s, b_r, and b_{sr} are the coefficients with respect to the variation of the motor speed.

Plugging (6.10)–(6.16) into (6.1) and using thermal conductance instead of thermal resistance, one obtains

$$\begin{aligned} &3\,R_{s0}[1 + \alpha_s(\theta_s - \theta_0)]\,i_s^2 + k_{Hs}\omega_s + k_{Es}\omega_s^2 = \\ &\quad = G_{\theta s0}(1 + b_s\omega)\,\theta_s + C_{\theta s}\dot{\theta}_s - G_{\theta sr0}(1 + b_{sr}\omega)(\theta_r - \theta_s) \end{aligned} \tag{6.17}$$

$$\begin{aligned} &3\,R_{r0}[1 + \alpha_r(\theta_r - \theta_0)]\,i_r^2 + k_{Hr}s\omega_s + k_{Er}s^2\omega_s^2 = \\ &\quad = G_{\theta r0}(1 + b_s\omega)\,\theta_r + C_{\theta r}\dot{\theta}_r - G_{\theta sr0}(1 + b_{sr}\omega)(\theta_s - \theta_r) \end{aligned} \tag{6.18}$$

According to [30], the estimation of the parameters $P_{s,loss}$, $P_{r,loss}$, $R_{s\theta}$, $R_{\theta r}$, $R_{\theta sr}$ requires six tests:

1. Two tests in which the machine is supplied with different DC voltages (see Figure 6.5(a)),

2. Two measurements obtained for an alternating current (AC) excitation at different operating frequencies while the motor speed is zero (locked rotor, see Figure 6.5(b)), and
3. A set of two measurements under AC excitation with the inverter supplying two different operating frequencies and load levels (see Figure 6.5(b)).

The DC test requires access to the neutral of the stator winding. A temperature sensor (e.g., TCs, infrared) is required to measure the rotor and stator winding temperature. Furthermore, the instantaneous stator currents and voltages, the fundamental stator frequency and the angular speed of the rotor have to be measured. The test signals have to be applied until a thermal steady state is reached, where the temperatures θ_s and θ_r are constant and the derivative terms $\dot{\theta}_s$ and $\dot{\theta}_r$ are zero. The steady-state operating points are then plugged into (6.17) and (6.18) to calculate the thermal parameters. In the first step, results from the DC test are used to compute the thermal conductance values $G_{\theta s0}$, $G_{\theta r0}$, and $G_{\theta sr0}$. Then, the coefficients related to the core loss k_{Hs}, k_{Es}, k_{Hr}, and k_{Er} are calculated from the results of the AC test at zero speed and the parameters obtained from the previous step. In a third step, the parameters b_s, b_r, and b_{sr} are derived from the test results of the AC test at different motor speeds.

Finally, the thermal capacitances $C_{\theta s}$ and $C_{\theta r}$ are estimated. Rearranging (6.1) with respect to $C_{\theta s}$ and $C_{\theta r}$ yields

$$C_{\theta s} = \left(\frac{d\theta_s}{dt}\right)^{-1} \left[P_{s,loss} - (G_{\theta s} + G_{\theta sr})\,\theta_s + G_{\theta sr}\theta_r\right] \tag{6.19}$$

$$C_{\theta r} = \left(\frac{d\theta_r}{dt}\right)^{-1} \left[P_{r,loss} - (G_{\theta r} + G_{\theta sr})\,\theta_r + G_{\theta sr}\theta_s\right] \tag{6.20}$$

The thermal capacitances can be estimated by curve fitting the temperature data from any of the measurements performed above (DC, AC locked rotor, and AC rotating) and by using the parameters determined from the previous steps and plugging the results into (6.19) and (6.20).

6.2.2 Parameter-based temperature estimation

The temperature can also be estimated by using an indirect measurement. A popular approach is to measure the stator resistance. The stator temperature is then estimated from the resistance measurement since the relationship between the resistance and the temperature is well known. One such method that is applicable to line-fed induction machines is introduced in [31]. The estimation is based on the following relation:

$$\hat{T}_s = T_{s0} + \frac{(\hat{R}_s - R_{s0})}{\alpha R_{s0}} \tag{6.21}$$

where T_{s0} is the reference temperature, R_{s0} is the electrical resistance at the reference temperature; $\hat{T}_s$ and $\hat{R}_s$ are the estimated temperature and resistance at the desired operating point and α the temperature coefficient of the resistivity. The basic idea is to inject a small DC voltage that is superimposed to the regular operating voltage and to measure the resulting DC current. The DC component is small, and the injection is

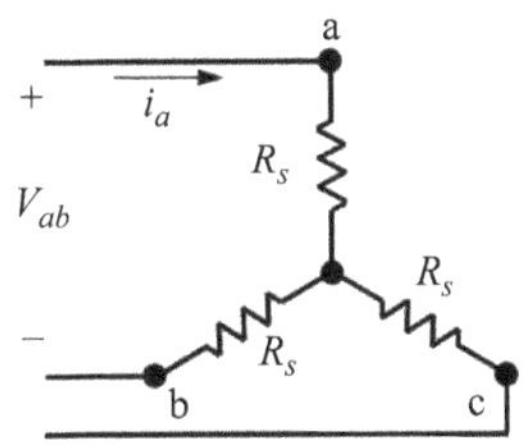

Figure 6.6 DC equivalent circuit of three-phase AC machine

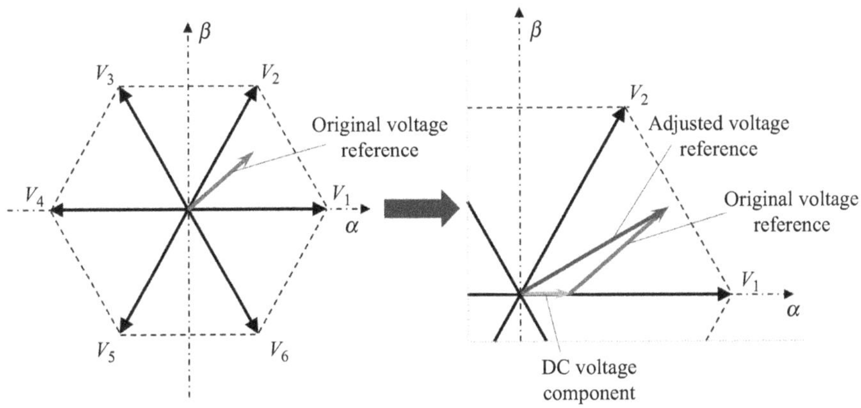

Figure 6.7 Modification of SVPWM for DC-signal injection in an open-loop application using a two-level inverter

performed for a short period of time every few minutes so that the additional heating due to the DC component is negligible. The DC equivalent circuit of a three-phase AC machine is shown in Figure 6.6, and the resistance can be calculated as

$$\hat{R}_s = \frac{2}{3}\frac{v_{ab}^{DC}}{i_a^{DC}} \tag{6.22}$$

where v_{ab}^{DC} is the DC voltage applied and i_a^{DC} the DC current measured.

The approach suggested in [31] uses additional circuitry to inject the DC voltage and is applied to line-fed machines only. Instead of installing additional hardware, Zhang *et al.* suggest performing the DC signal injection via a soft starter [32] and further expanding the application to open-loop inverter-fed induction machines in [33]. In the inverter-based application, the DC voltage is superimposed on the regular operating voltage during the test as illustrated in Figure 6.7.

Some factors that need to be considered in the estimation are the other resistive elements in the circuit that cause a voltage drop such as the cable resistance or the voltage drop across the power electronic devices (see Figure 6.8).

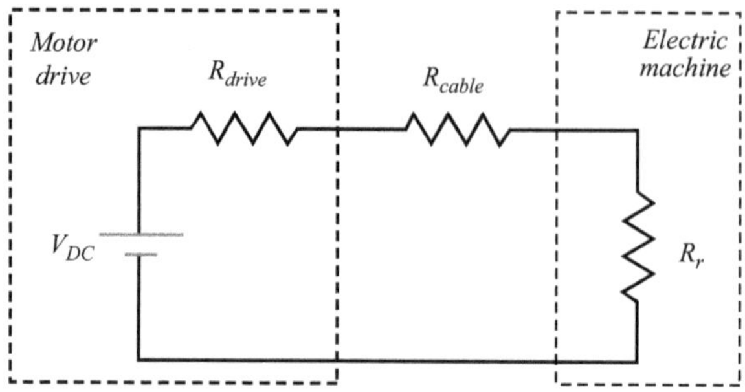

Figure 6.8 Single-phase DC equivalent circuit of three-phase AC machine supplied by inverter drive

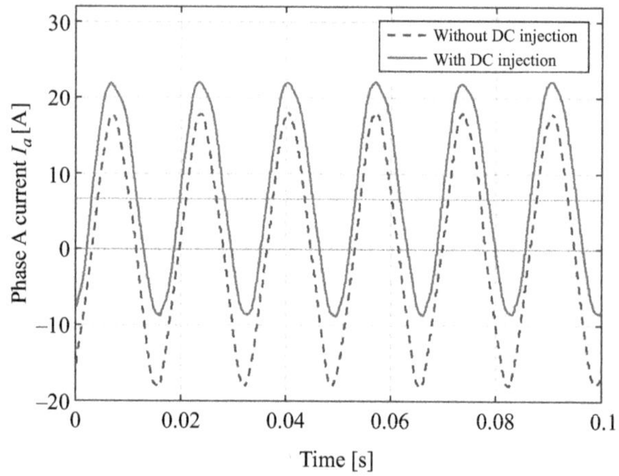

Figure 6.9 Stator current with and without DC injection

The superimposed DC voltage induces a torque pulsation that should be kept as small as possible to avoid any negative impact on the operation. The magnitude of the torque pulsation with respect to the torque applied is

$$T_{ripple} \approx \frac{I_{a,DC}}{I_{a,peak}\cos(\varphi)}$$

where $\cos(\varphi)$ is the power factor, and $I_{a,DC}$ and $I_{a,peak}$ are the magnitude of the DC current component and the peak value of the fundamental AC current component of phase A, respectively. An example for the stator current without DC injection and with a DC injection of 5 V in phase A is shown in Figure 6.9. A comparison of a stator temperature estimation using the DC injection and a measurement using TCs is shown in Figure 6.10. The experiments were conducted with a four-pole induction motor rated for 7.5 hp that were operated at no load, and at 100%, 50%, and 75% of the rated load, respectively.

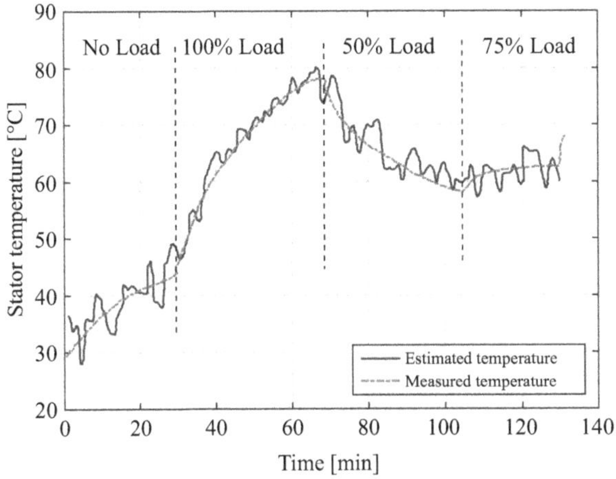

Figure 6.10 Stator temperature estimation for varied operating conditions

The application to an induction machine supplied by inverters that are operated in closed-loop via field-oriented control is investigated in [3]. Instead of superimposing a voltage component on the A-phase as proposed in the open-loop application (see [33]), the DC component is introduced via the current command.

$$\Delta i_{d,cmd} = I_{DC}\cos\theta,$$
$$\Delta i_{q,cmd} = I_{DC}\sin\theta.$$

where θ is the transformation angle of the synchronous dq-reference frame and I_{DC} is the magnitude of the injected DC current component. This results in the following additional current command components in the stationary abc-reference frame.

$$\Delta i_a = I_{DC}, \quad \Delta i_b = -0.5\,I_{DC}, \quad \Delta i_c = -0.5\,I_{DC}.$$

The stator resistance is estimated by using (6.22). The DC current can be measured by using the current sensors of the inverter. The voltage can be obtained from the voltage commands or from the duty cycles commanded and the DC bus voltage measurement. Several nonidealities such as dead time, the voltage drop across the devices, and device turn-on and turn-off delays have to be taken into account to provide an accurate calculation of the DC voltage. Some additional signal-processing steps are recommended to extract the DC component of the stator current and voltage. The most important step is the removal of the fundamental operating component, which is large compared to the DC component injected. A block diagram of the processing steps is shown in Figure 6.11.

An experimental result obtained with a 5 hp, four-pole induction machine operated by an inverter switching at 10 kHz, is given in Figures 6.12 and 6.13. The torque and the speed before and during the DC injection are shown in Figure 6.12. While the torque ripple can be spotted easily, there is no noticeable impact on the

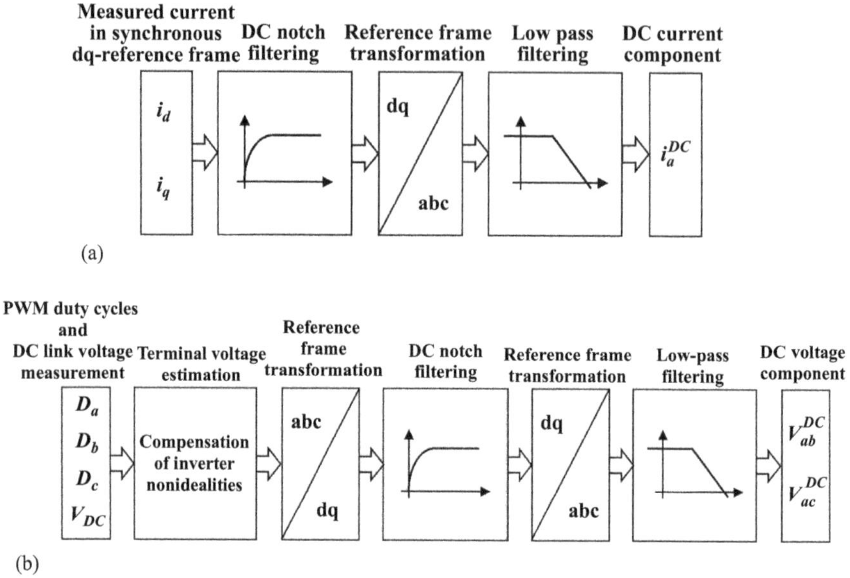

Figure 6.11 Block diagram of the signal-processing steps to extract DC components: (a) DC current extraction and (b) DC voltage extraction

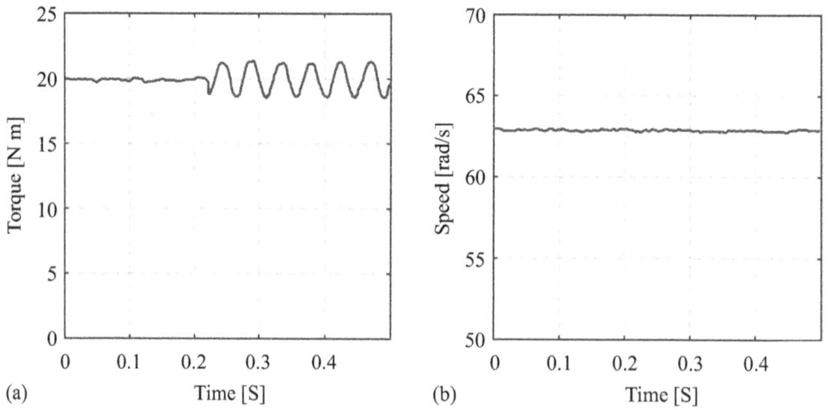

Figure 6.12 Motor torque and speed before and during a DC signal injection: (a) torque and (b) speed

motor speed. A result for the temperature estimation with and without considering the inverter nonidealities is given for varying operating conditions. The result shows that the consideration of the nonidealities is crucial for an accurate result.

Matic *et al.* focus on the compensation of the inverter nonlinearities when applying the DC injection to induction machines operated under field-oriented control [34].

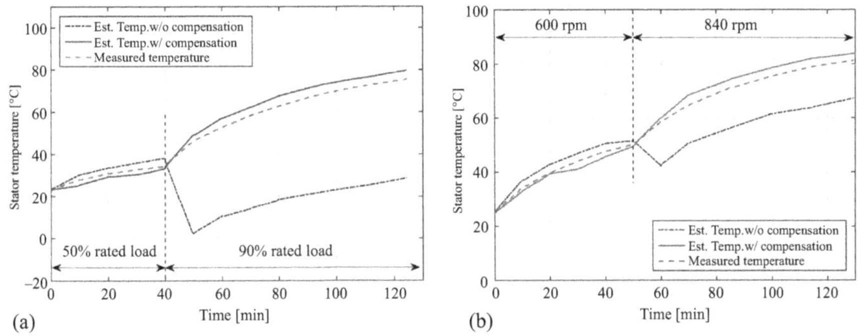

Figure 6.13 Experimental results for a temperature estimation using DC-signal injection under varying operating conditions: (a) varying load level and (b) varying operating speed and load level

The application of the DC injection method to induction machines operated under direct torque control (DTC) is introduced in [35].

A slightly different approach for stator temperature monitoring from the DC injection is suggested by Popov *et al.* [36]. To avoid the torque ripple caused by the DC injection and the necessary compensation of inverter nonidealities, the use of high-frequency signatures due to the PWM excitation is suggested. The temperature estimation is based on the spectral analysis of the input impedance of the machine and requires a reference measurement. An experimental evaluation shows a good accuracy of the proposed method. Despite the obvious advantages over the DC injection, the method requires a high sampling rate that is usually not available in standard inverter drives. Therefore, the practical application of this approach is currently limited to high performance drives with high sampling rates and sufficient computational resources. The authors also anticipate a reduced applicability of the approach to certain machine designs. A variation of this approach for estimating the rotor temperature in induction machines is suggested in [37].

6.3 Monitoring and protection of stator-related issues

Stator-related faults account for a large percentage of electric machine failures, and the percentage of stator faults increases with the machine size [6]. Stator failures can be classified into winding/insulation- and lamination/stator core-related faults [11]. The most common stator problems arise from the insulation system. Failures due to the stator core are rather rare. According to [11,13,38], the root cause for stator winding faults can be divided into four types of stresses the windings are exposed to:

1. thermal stress due to high temperatures and thermal cycling;
2. mechanical stress due to vibration, excessive forces, and abrasion;

3. electrical stress due to the operating voltages and voltage disturbances; and
4. environmental factors such as contamination and humidity.

Other factors that have a strong influence on the integrity of the insulation are the quality and the rating of the insulation usually labeled as insulation class, and the specific application the motor is used in. The sequence of stator winding faults often begins with a short in-between one or more turns of the stator winding. The fault current in the shorted windings usually is significantly larger than the rated stator current. This short-circuit current results in severe localized heating and further deterioration of the adjacent insulation. The turn fault can then propagate and evolve into a more severe failure of the primary insulation system, which is either the failure of the phase-to-phase insulation or the failure of the groundwall insulation. This type of catastrophic failure usually requires the shutdown and repair of the machine. In the worst case, the large ground currents that occur during a groundwall fault can damage the stator core. A detailed overview of stator-related faults and monitoring techniques is given in [11,39] and a survey with a focus on turn insulation is presented in [13]. This section will first go over several approaches for monitoring the turn insulation system and then go over some techniques for assessing the primary insulation system. The section is concluded by a short discussion on open-winding faults and open inverter switches, as well as a brief look at the diagnosis of stator core faults.

6.3.1 *Turn insulation*

Shorted turns tend to occur more frequently in random-wound machines than in form-wound machines. However, the consequence of a shorted turn in form-wound machines can be more severe because the of the size and power involved in this kind of equipment. A very popular method for detecting shorted turns in the coil of a motor winding is based on the calculation of the negative sequence current. Negative sequence components in general are an indicator of the imbalance in a three-phase system. Shorted turns are introducing an imbalance and therefore increase the negative sequence current. The fundamentals of this technique were introduced by [40]. The negative sequence current I_- can be calculated by using the phase currents as follows:

$$\bar{I}_- = \frac{1}{3}\left(\bar{I}_a + e^{j(4\pi/3)}\bar{I}_b + e^{j(2\pi/3)}\bar{I}_c\right). \tag{6.23}$$

Equation (6.23) requires the measurement of the instantaneous currents and the calculation of the phase current phasors $\bar{I}_a$, $\bar{I}_b$, and $\bar{I}_c$. The problem is that there are numerous factors that can contribute to the negative sequence currents, such as an unbalance in the voltage supply or inherent asymmetries in the machine. Those components need to be accounted and compensated for to avoid a fault detection by error (false positive). Hence, the main challenge of any sequence component-based method usually is that other sources of asymmetry (e.g., inherent winding asymmetry, supply imbalance, sensor imbalance, etc.) need to be taken into account and compensated for. Many of the publications related to sequence component-based monitoring of turn faults are focused on finding indicators that have a high sensitivity to the fault-producing component and are insensitive to other factors that increase the asymmetry of the system. For example, the monitoring of the motor impedance yields more robust results [41–44].

In [44], an off-diagonal term of the sequence impedance matrix is used for turn fault detection since it is highly susceptible to an increase in machine asymmetry due to turn fault. The sequence component voltage equation is defined as follows:

$$\begin{bmatrix} \tilde{V}_p \\ \tilde{V}_n \end{bmatrix} = \begin{bmatrix} Z_{pp} & Z_{pn} \\ Z_{np} & Z_{nn} \end{bmatrix} \begin{bmatrix} \tilde{I}_p \\ \tilde{I}_n \end{bmatrix} \tag{6.24}$$

where the subscripts p and n stand for positive and negative sequence components and the fault indicator ΔZ_{np} is defined as follows:

$$\Delta Z_{np} = Z_{np} - Z_{np0} \tag{6.25}$$

Z_{np} is obtained during the monitoring phase and Z_{np0} is obtained during a commissioning or learning phase. The authors point out that a learning phase is required by any algorithm that has the ability to account for nonidealities during the monitoring stage. To calculate Z_{np}, two sets of linearly independent data sets are required. This is achieved by using one data set that is obtained under balanced supply voltage conditions and another one that is obtained under (intentionally) unbalanced supply voltages. Z_{np} is calculated as follows from the two data sets

$$\begin{aligned} \begin{bmatrix} Z_{np} \\ Z_{nn} \end{bmatrix} &= \begin{bmatrix} \tilde{I}_{p1} & \tilde{I}_{n1} \\ \tilde{I}_{p2} & \tilde{I}_{n2} \end{bmatrix}^{-1} \begin{bmatrix} \tilde{V}_{n1} \\ \tilde{V}_{n2} \end{bmatrix} \\ Z_{np} &= \frac{\tilde{I}_{n2}\tilde{V}_{n1} - \tilde{I}_{n1}\tilde{V}_{n2}}{\tilde{I}_{p1}\tilde{I}_{n2} - \tilde{I}_{p2}\tilde{I}_{n1}} \end{aligned} \tag{6.26}$$

where the subscripts 1 and 2 stand for the first and for the second measurement, respectively. For example, during the learning phase, data set one is taken under balanced voltage conditions, and set two is acquired under unbalanced voltage conditions. Equation (6.26) is used to calculate the baseline value Z_{np0} from the baseline measurements. During the monitoring phase, the newly acquired data set and the data set under the unbalanced condition are inserted into (6.26) to obtain Z_{np} and then into (6.25) to compute the fault indicator ΔZ_{np}. The authors conduct a series of simulations and experiments to show that the method is immune to supply unbalance, inherent motor asymmetries, and measurement errors. An example for the detection of one, two, and three shorted turns near no-load condition is shown in Figure 6.14. The experiments are conducted with a 5 hp, 460 V, four-pole induction machine that has 216 turns per phase. Some modifications allow the application of between one and three shorted turns. The fault indicator $|\Delta Z_{np}|$ increases with the fault severity, and the faulty condition can be distinguished from the healthy motor even if just one out of 216 turns is shorted. A similar result is achieved under load conditions close to the rated load.

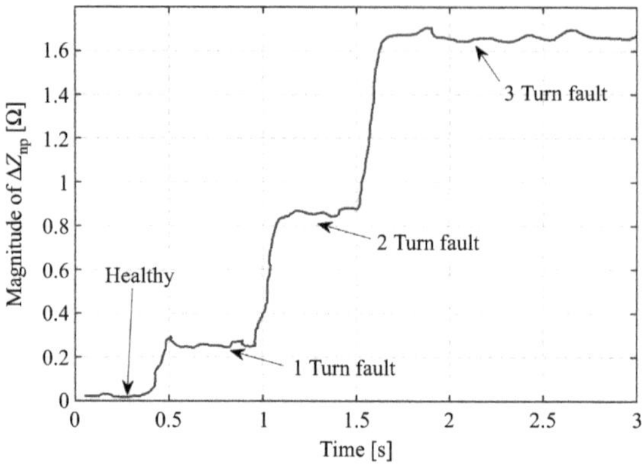

Figure 6.14 $|\Delta Z_{np}|$ *for a developing turn fault (no turn fault up to three shorted turns) at low slip condition*

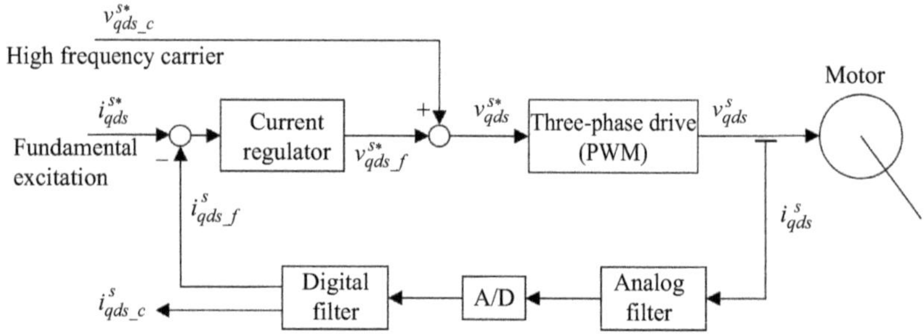

Figure 6.15 Diagram of HF injection scheme for the monitoring of the motor condition

Another focus of recent research in this area is the automation of the monitoring process and the decision-making process. For example, Tallam *et al.* deploy neural networks [16,45] to achieve the automation of the process. The approaches described above can be applied to mains-fed as well as inverter-fed motors.

A slightly modified approach based on sequence component analysis that uses high-frequency (HF) signal injection is presented in [46,47]. This method is specifically designed for inverter-fed machines since a test signal is generated by the supply. A rotating balanced high-frequency carrier signal (6.27) is superimposed on the controller output. The schematic for this approach is shown in Figure 6.15.

With the excitation frequency ω_c set significantly higher than the stator dynamics, the resulting high-frequency machine model can be written as (6.28)

$$v^s_{qds,c} = V_c e^{j\omega_c t} \tag{6.27}$$

$$v^s_{qds,c} \cong j\omega_c L_{\sigma s} i^s_{qds,c} \tag{6.28}$$

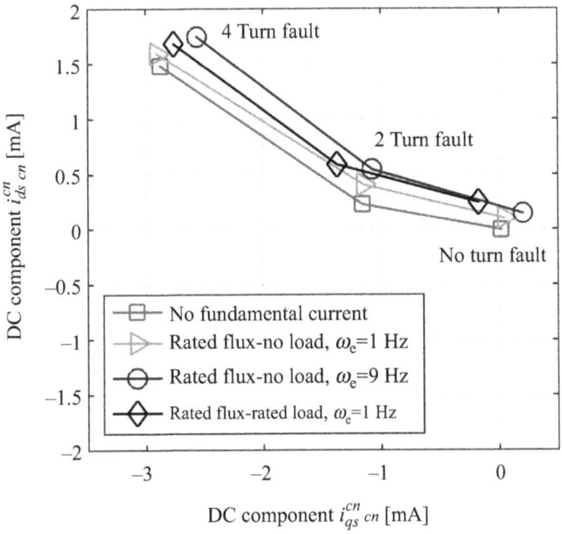

Figure 6.16 Experimental result for HF injection-based turn fault detection applied to a motor with zero-, two-, and four-turn faults under different operating conditions

where V_c is the magnitude of the excitation and $L_{\sigma s}^{s}$ is the stator transient inductance in the synchronous reference frame. The resulting high-frequency current (6.29) can be split into a positive and negative sequence component.

$$i_{qds,c}^{s} = i_{qds,c+}^{s} + i_{qds,c-}^{s} = -jI_{c+}e^{j\omega_c t} - jI_{c-}e^{j(h\theta_e - \omega_c t)} \tag{6.29}$$

The negative sequence component contains the desired saliency spatial information that depends on the level of asymmetry in the machine, which will increase if a turn fault is present. The turn fault indicator proposed is the DC component of the negative sequence current in the carrier-signal reference frame, i.e., the fundamental component of the negative sequence current that is induced by the injected high-frequency voltage. The robustness of the indicator is tested with respect to varying operating conditions (load level and fundamental frequency), variations in the resistive phase imbalance and temperature changes. An experimental result for various operating conditions and a motor that has no turn fault, two shorted and four shorted turns are shown in Figure 6.16. The fault indicator proposed can separate the faulty from the healthy windings, and the separation between the faulty and the healthy case increases with an increasing fault severity. For the implementation on an inverter drive platform, a simple method for extracting the fault indicator is presented as well. First, the measured phase current is demodulated with the carrier frequency. This moves the positive sequence component to $2\omega_c$ and the negative sequence component to DC. Then, a simple second-order low-pass filter is used to remove any operating-related

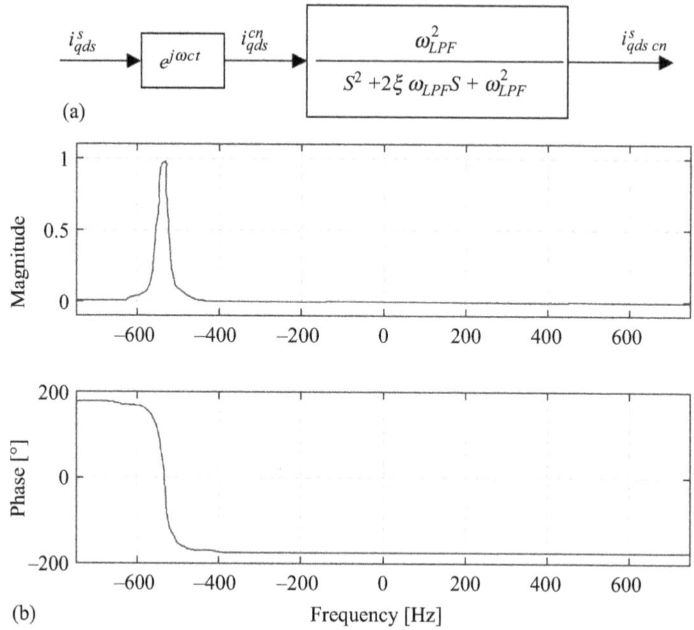

Figure 6.17 Digital processing of current to extract turn insulation fault indicator. (a) Block diagram of the coordinate transformation and second-order low-pass filter to extract the carrier-signal current from the overall stator current. (b) Frequency plot of the filter used for the fault indicator extraction according to part (a) in the stationary reference frame. $\omega_c = 535.7$ *Hz,* ω_{LPF} *corresponding to a cutoff frequency of 10 Hz, and the damping factor* $\xi = 0.707$

frequencies, as well as the positive sequence component due to the signal injection. The block diagram of the filter and the frequency response in the stationary reference frame are displayed in Figure 6.17(a) and 6.17(b), respectively.

In [48], the method is adopted for use with soft starters. Briz *et al.* show that the fundamental frequency and lower order harmonic components of the negative sequence current are highly sensitive to a turn fault condition during the start-up transient that is facilitated by the soft starter. Experimental results show that the sensitivity of these harmonics to the turn fault condition decreases significantly once the operation transits from the start-up phase to the steady state (see Figure 6.18). Therefore, in contrast to the application with inverter-fed machines, the method can be used during the start-up transient only. The signal-processing steps required to extract the fault-indicating components are the same as in the inverter-based application described above. Instead of using the carrier frequency ω_c for the reference frame transformation, the respective fault-indicating frequency has to be used (e.g., $-\omega_e$, $5\omega_e$, $-7\omega_e$).

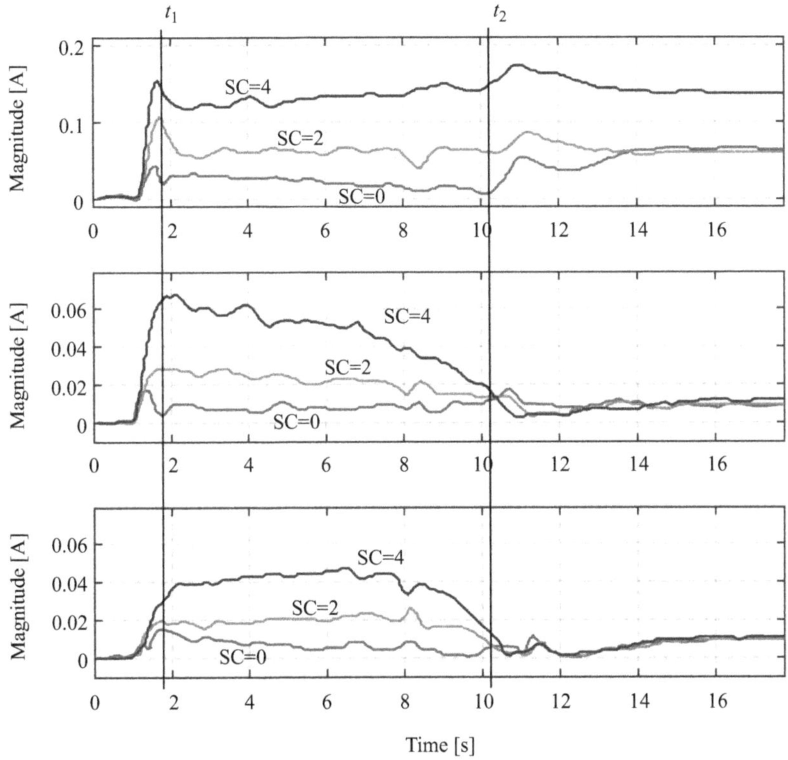

Figure 6.18 Magnitude of the negative sequence stator current components at $-\omega_e$ (upper), $5\omega_e$ (middle), and $-7\omega_e$ (lower) during startup transient, for a healthy machine (SC = 0) and machines with two- (SC = 2) and four- (SC = 4) short-circuited turns

Research on the detection of shorted stator turns in motors that are operated with DTC is presented in [23,49]. Some researchers (see [50,51]) suggest approaches that are based on the monitoring of specific stator current harmonics. This is a very popular method for the detection of mechanical issues such as rotor eccentricity, broken rotor bars, or bearing faults (see Sections 6.4 and 6.5). An approach that utilizes spectral analysis and is applied to the terminal voltages after switch-off is suggested by Nandi *et al.* [52]. Du *et al.* [53] propose a signal injection-based approach for turn fault detection in permanent magnet motors. The method is supposed to be efficient enough to be implemented on platforms typically used for electric vehicles. The detection of shorted turns based on Concordia patterns in combination with neural networks for decision-making is discussed by El-Arabawy *et al.* [54].

An entirely different group of methods to monitor the condition of the turn insulation is based on the excitation of the turn-to-turn impedance. A well-established test method that stresses the turn-to-turn impedance is the surge test [38]. The surge test applies a short current pulse to the machine windings so that a high-voltage stress between the turns is induced. The monitoring and evaluation of the voltage response can be used to detect the early stages of turn insulation failure. If there is any form of arcing between turns, the voltage response is modified and the insulation problem can be identified at an early stage even before a turn short develops. Some attempts have been made to apply the surge test online [4,55], but some additional hardware is required here, and the application to inverter-fed machines has not been explored so far. Furthermore, the surge test has never been used to predict remaining insulation life, and there are concerns about the level of stress that is applied to the turn insulation. Another method that monitors the turn insulation condition before solid turn short develops will inject a low-voltage, high-frequency signal into the machine [56,57]. Instead of monitoring a negative sequence component of the stator current as in [46,47] or stressing the insulation and observing the onset of an insulation breakdown as shown in the surge test, this method observes the turn-to-turn insulation condition via measuring the line-to-neutral impedance. A shift in specific frequency components of this impedance is used as an indicator for turn-insulation deterioration. In other words, this method performs an online spectroscopy of the motor impedance. Unfortunately, this method also requires additional hardware for the monitoring process and therefore does not fit the mold of low-cost methods that can easily be integrated within the inverter drive. However, a promising new idea, that provides similar capabilities as the online impedance spectroscopy, bridges this gap and is designed specifically for the use with inverter-fed machines [58,59]. To assess the insulation health, the authors suggest utilizing the characteristic high-frequency oscillation of the motor current when the motor is excited by a voltage step. The excitation of the insulation system is simply provided by the inverter switching. The current response due to this voltage "step excitation" has to be measured with a sufficient resolution to capture the high-frequency "ringing." Changes in the insulation condition induced by aging impact the high-frequency current waveform (see Figure 6.19).

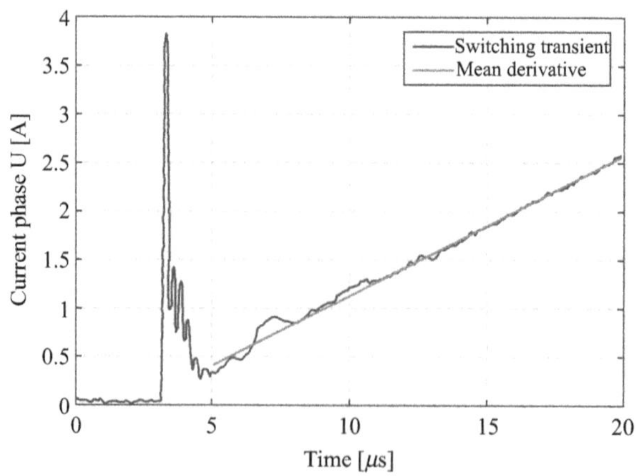

Figure 6.19 Stator current during switching instant and mean derivative of stator current right after switching instant

The authors point out that environmental factors such as temperature, moisture, or even a change in the cabling between inverter and motor have little to no impact on the method, and they refer to some of their previous publications for more details [60,61]. The first essential preprocessing step is the removal of the "mean derivative," which simply refers to the current ramp due to the step excitation of the voltage. This can be achieved by calculating

$$y(t) = i(t) - \frac{i(t_2) - i(t_1)}{t_2 - t_1}\, t$$

where $i(t)$ is the measured current and $(i(t_2) - i(t_1))/(t_2 - t_1)$ is the approximation of the derivative of the current. To avoid a constant offset, t_1 is set to the instant of the application of the voltage step and t_2 to a suitable time ($t_2 > t_1$) that allows an accurate calculation of the current slope. After the removal of the "mean derivative," an FFT is applied to the "ringing" portion of the preprocessed current waveform (see Figure 6.20(a) and 6.20(b)).

The frequency domain waveform is then used to derive the proposed insulation state indicator (ISI) for each phase as the root-mean-square deviation (RMSD) of a portion of the current spectrum:

$$ISI_{p,k} = RMSD_{p,k}(x_1, x_2) = \frac{\sqrt{\sum_{g=n_{low}}^{n_{high}} \left(|Y_{ref,p}(f_d)| - |Y_{con,p,k}(f_d)|\right)^2}}{n_{high} - n_{low}} \tag{6.30}$$

where Y_{ref} and Y_{con} are the frequency domain representation of the "ringing" portion of the waveform for the reference measurement and the condition monitoring measurement obtained as described above; n_{low} and n_{high} define the frequency range for the calculation of the fault indicator. The parameter p indicates which phase the voltage was applied to (A, B, or C); the parameter k stands for the test or measurement number; and the variable f_d is the discrete frequency. The discrete frequency is defined as an integer multiple of the frequency resolution that is calculated as the inverse of the measurement length $f_d = (g/t_{win})$ with t_{win} being the measurement length and

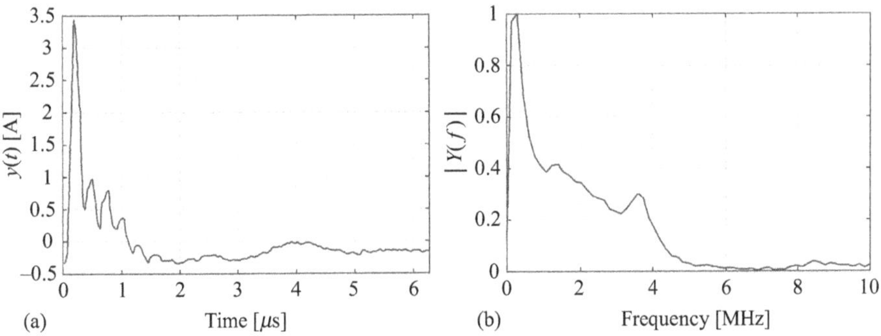

Figure 6.20 Time and frequency representation of stator current switching transient after subtraction of the mean derivative: (a) time-domain waveform and (b) frequency-domain waveform

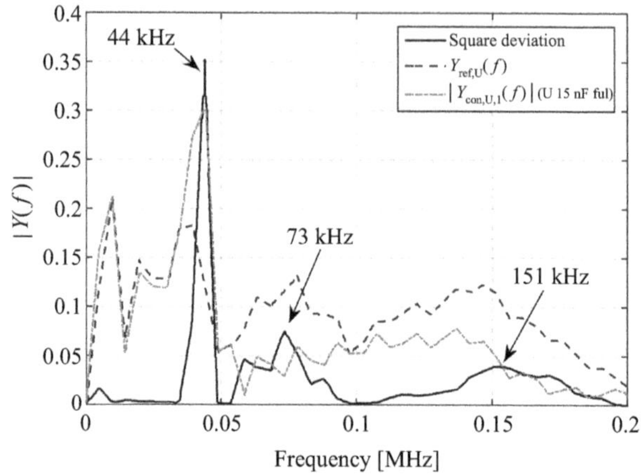

Figure 6.21 Reference waveform for baseline case $|Y_{ref,U}(f)|$ (dashed), condition assessment waveform $|Y_{con,U,1}(f)|$ (dash-dot) for 15-nF capacitor inserted in parallel to the entire winding of phase U, and square deviation (solid) between the two

$g = 0, 1, 2, 3, \ldots$. Instead of having an insulation health indicator for each phase, the three indicators can be combined to yield a three phase or spatial ISI (SISI):

$$SISI = ISI_A + ISI_B e^{j(2\pi/3)} + ISI_C e^{j(4\pi/3)} \tag{6.31}$$

Insulation aging rarely happens in exactly the same way in each of the phases, i.e., the insulation condition changes are rather asymmetrical with respect to the three phases. Usually there are local mechanical or environmental stresses that will affect certain portions of the insulation more than others. On the other hand, side effects such as the variation in temperature or cabling are symmetrical changes that affect all three phases in the same way. The SISI indicator eliminates the influence of symmetrical changes in the insulation, because the zero sequence component is canceled out in (6.31). Experimental studies with a 5.5 kW and a 1.4 MW squirrel cage induction machine are presented to validate the effectiveness of the method. The sampling rate is set to 40 MHz. Both machines have access to portions of the windings via the terminal box, which allows the user to modify the capacitance or resistance of the winding. A measurement result for the 1.4 MW motor is shown in Figure 6.21 for a scenario where a 15 nF capacitor is connected in parallel to the entire phase A. The change in capacitance can clearly be picked up by the ISI indicator (square deviation). A larger number of test cases for various modifications of the winding capacitance are shown in Figure 6.22(a).

The larger capacitance modifications and the capacitance modification across a larger portion of the winding can clearly be distinguished from the baseline case (healthy). For some test cases, the change in the SISI fault indicator is subtle. To get a better idea of whether the more subtle deviations can still be separated from the reference case, a close-up of Figure 6.22(a) is given in Figure 6.22(b). The close-up

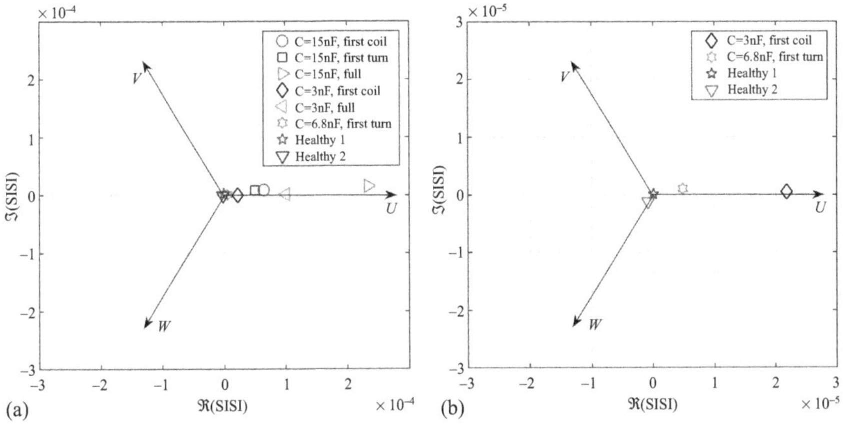

Figure 6.22 SISI for 1.4 MW induction machine for emulated changed of hf-behavior and a sampling rate of 40 MHz: (a) full data set and (b) close up of part (a)

shows that even small capacitance modifications across a small portion of the winding can be detected with this method. Even though the method seems to be very effective, the large sampling frequency of 40 MHz presents a challenge for the adaptation in regular inverter drives. To show that the method can still be applied at lower sampling frequencies, the sampling rate is reduced from 40 to 2.6 MHz without a noticeable impact on the detection accuracy. A further reduction of the sampling frequency would be required for a broader application of the approach, which is not restricted to the application with high performance drives only. Another possible limitation in practical application is the bandwidth of the current sensors. Broadband current sensors that have a flat gain over a wide frequency range were used for this investigation. The application of sensors commonly used in inverter drives calls for further investigations.

6.3.2 Primary insulation system

The primary insulation system consists of the insulation between the windings and the stator core laminations (groundwall insulation) and the insulation between the windings of different phases (phase-to-phase insulation). The breakdown of the primary insulation system usually results in a catastrophic failure. Immediate termination of the operation is required to protect the stator core from severe damage. It is preferable to avoid such a fault condition, to detect the onset of insulation deterioration and to service the machine before the insulation fault fully develops. There are several tests capable of providing such a monitoring function. Some of the most popular methods are based on the detection of partial discharges [38,62] or the measurement of the insulation C&DF [38,63]. Traditionally, partial discharge has only been applied to medium- or high-voltage machines. Over the last decade, there have been investigations on the applicability to inverter-fed low-voltage machines as well [64,65].

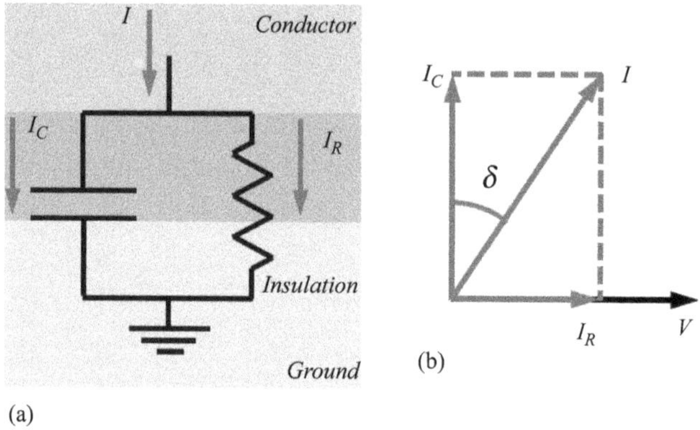

Figure 6.23 First-order model of primary insulation system: (a) first-order equivalent circuit of electrical insulation system and (b) phasor diagram for circuit in part (a)

Usually, partial discharge requires expensive hardware, which can only be justified if the application is of utmost importance and outages have to be avoided at all costs. The C&DF test is a traditional off-line test. It is based on the measurement of the leakage current due to a high voltage that is applied between the machine terminal and the stator core. The insulation can be modeled as a simple parallel circuit that consists of a capacitor and a resistor as shown in Figure 6.23.

In [66,67], Lee *et al.* show the applicability of the C&DF test to operate machines. The basic idea is illustrated in Figure 6.24(a) and 6.24(b), respectively.

A change in the insulation resistance (IR) and capacitance measured at the fundamental operating frequency is recorded over time and used as an indicator of the insulation condition. The applicability of this method to inverter-fed machines has been investigated by Zhang *et al.* [68]. In addition to the analysis of the impedance at the fundamental frequency, the authors suggest using additional frequency harmonics to enhance the sensitivity of the method. However, this approach requires the installation of an expensive specialized current transformer to accurately measure the insulation leakage current and does not utilize the inverter.

Assessment of the insulation condition using the inverter is suggested by Yang *et al.* [69,70]. The authors propose applying several standard insulation tests during periods when the machine is not operating. Besides the C&DF test, the IR test, which is essentially a DC version of the C&DF test [38,71], is applied as well. Some small low-cost modifications are required to enable the inverter drive to perform these standard off-line tests. Both the required circuit modification to conduct a C&DF AC insulation test and a DC IR measurement are illustrated in Figure 6.25.

Some contactors and a sensing element need to be added to the inverter circuit to facilitate the insulation tests. For the C&DF test (see Figure 6.25(a)), the AC source line to neutral voltage is applied to the groundwall insulation by closing contact 1 and opening contacts 2 and 3. The leakage current through the groundwall insulation, which is equivalent to the common mode three-phase current, can be measured by

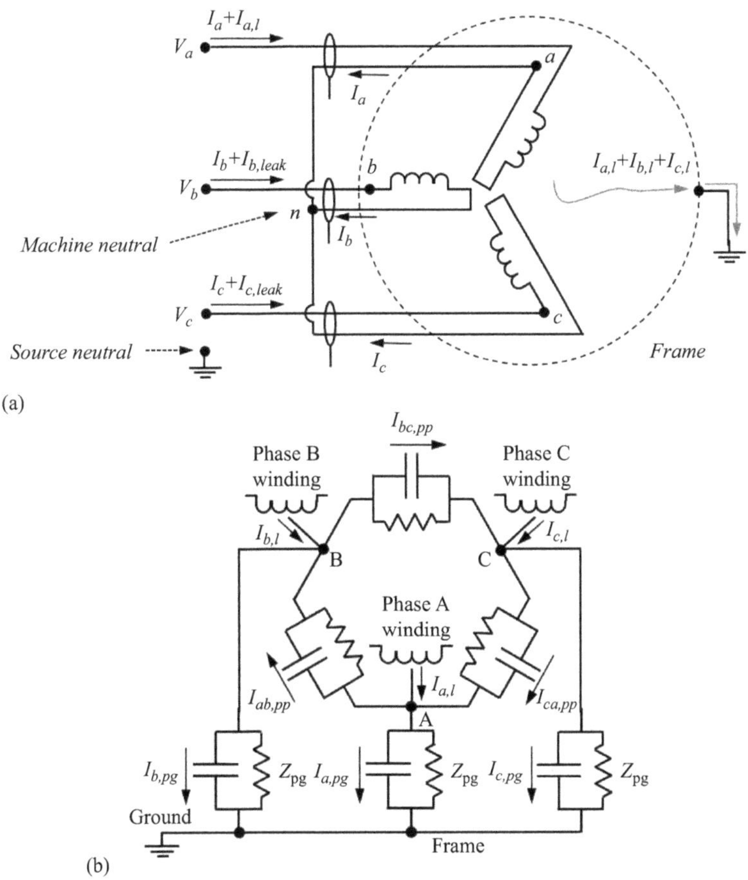

Figure 6.24 Online insulation monitoring scheme for three-phase AC machines. (a) Differential leakage current measurement scheme for online assessment of stator winding insulation condition. (b) Simplified equivalent circuit for an online insulation monitoring scheme for a three-phase AC machine

using a ground fault CT (if available) or by using a sensing resistor. A slightly different configuration enables the application of the IR test by applying the rectified DC voltage to the groundwall insulation (see Figure 6.25(b)). To conduct the test, contact 1 is closed and contacts 2, 3, and 4 are opened. The leakage current is measured via a sensing resistor. A simulation result for the measurement of the leakage current of a machine that is excited with a rectified DC voltage is shown in Figure 6.26.

One advantage of the proposed approach is that the method can be implemented with small, low-cost modifications to the inverter topology. Furthermore, it enables the user to monitor the insulation condition on a regular basis, and it eliminates the impact of the operating condition since the test is applied during off-intervals.

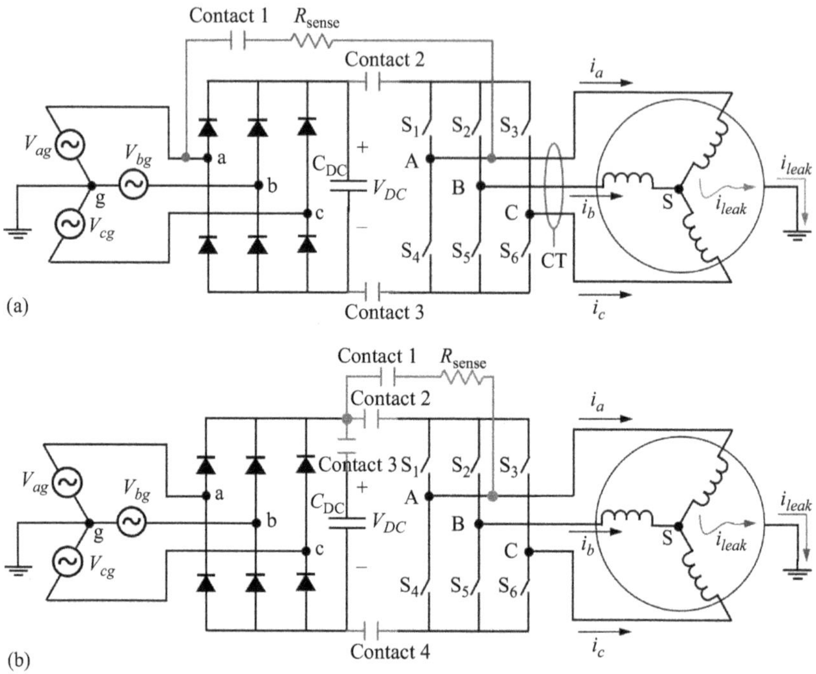

Figure 6.25 Circuit modifications to conduct standard insulation tests during periods when the drive unit is not operating. (a) Groundwall insulation test using the inverter and the AC source line-neutral voltage (C&DF) and (b) groundwall insulation test using the inverter and the rectified DC voltage (insulation resistance)

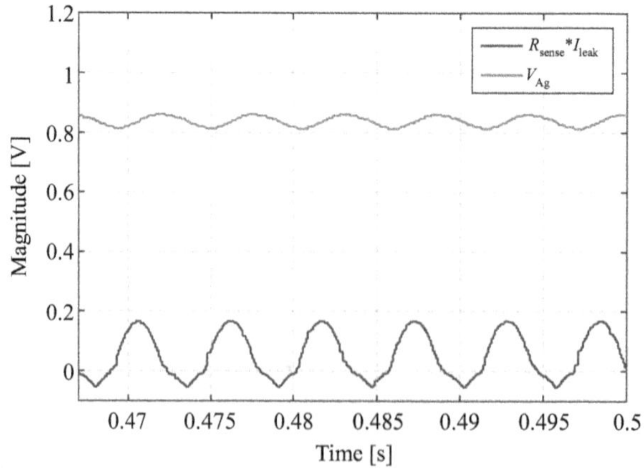

Figure 6.26 Simulated result for the method shown in Figure 6.25(b) with $R_{eq} = 20\ M\Omega$

6.3.3 Open stator winding faults and open-switch faults

Another type of stator fault is an open circuit (OC) fault. This type of failure can occur due to problems with the joints or due to the mechanical failure of the conductor (open winding), which can be a partial or a full break of the winding. An inverter-based monitoring of open-winding faults is presented in [72]. The authors suggest running a model and comparing the model output with the measurements. The difference between the model and the measurement (I_{mM}) is compared to a threshold (I_{th}) and used as a fault indicator (residue r_1):

$$\begin{aligned} I_{mM} &= i_{measurement} - i_{model} \\ r_1 &= |I_{mM}| - I_{th} \end{aligned}$$

For the open-winding fault, the authors use the current residue of each phase to detect the OC. Experimental results for a single-phase induction machine are presented for validation. The authors also suggest using this method for the detection of shorted coils.

A more common scenario is that the OC occurs in one or multiple switches of the inverter instead of the machine. Even though this is not a machine fault, this issue is briefly discussed here, and several references are given. A major difference between the open switch and the open winding is that in the case of the open IGBT switch, there is a diode that can conduct current when the phase current is negative. In [73], a method applicable to closed-loop controlled permanent magnet motors is presented that is based on the current residual vector (CRV). Load effects and the impact of the main controller are canceled, and the monitoring process can be executed by the existing infrastructure of the inverter drive. The method is based on modeling the system including the electric motor and the inverter with a mixed logical dynamical model and comparing the predicted currents to the model output. The basic concept is depicted in Figure 6.27.

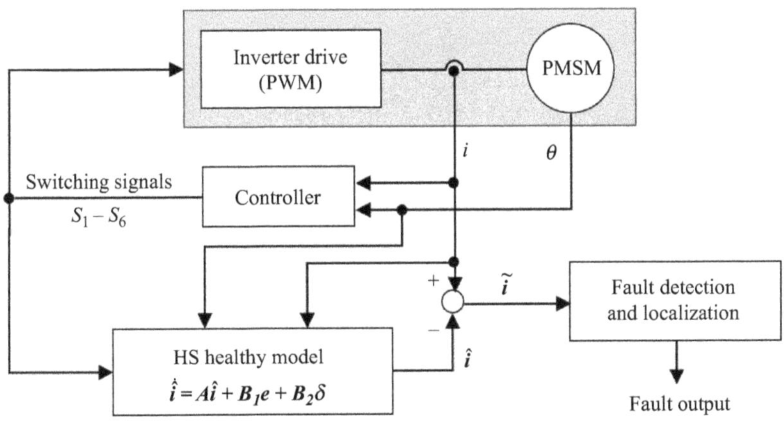

Figure 6.27 Block diagram of CRV-based open-switch diagnostic method

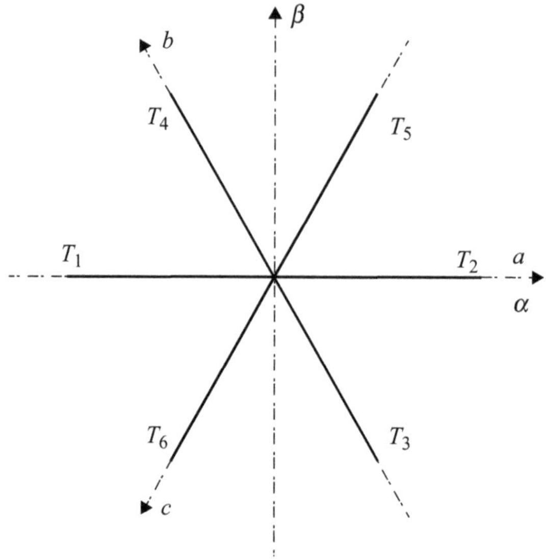

Figure 6.28 Trajectories of the CRV for various open-switch conditions

The inputs to the model are the switching states of the inverter and the sign of the phase currents as well as the rotor position. The difference between the currents obtained from the model and from the measurements is the residual phase currents $\tilde{i}_a, \tilde{i}_b, \tilde{i}_c$. These can be transformed into the $\alpha\beta$-frame via Clarke's transformation. The CRV is then defined as

$$\tilde{I} = \tilde{i}_\alpha + j\tilde{i}_\beta \tag{6.32}$$

and can be used as an indicator for open-switch faults. The CRV trajectories indicating an open switch are along the *a*,*b*,*c*-vectors of the hexagon in the complex plane as shown in Figure 6.28.

Simulation and experimental results for different fault scenarios are given to verify the feasibility of the technique proposed. The CRV trajectories obtained from an experiment with a healthy inverter, with an OC fault of either switch T_1 or T_2, and with an inverter that has two faulty switches, T_1 and T_2, are displayed in Figure 6.29.

The resulting CRV trajectories match the theoretical trajectories shown in Figure 6.28. A more detailed look at the fault currents for an OC fault of T_1 is given in Figure 6.30. The A-phase currents from the model and from the plant for an OC fault of T_1 are shown in Figure 6.30(a). The phase A current is always negative after the upper switch in the A-phase fails to open, while the model current maintains the sinusoidal pattern that the current would follow if all switches were healthy. The resulting current residues in the stationary abc-frame are displayed in Figure 6.30(b), and the corresponding current residues in the $\alpha\beta$-frame are shown in Figure 6.30(c). As predicted by the theoretical analysis of an OC fault of T_1, the current residue $\tilde{i}_\beta$ is approximately zero, while the current residue for $\tilde{i}_\alpha$ varies between

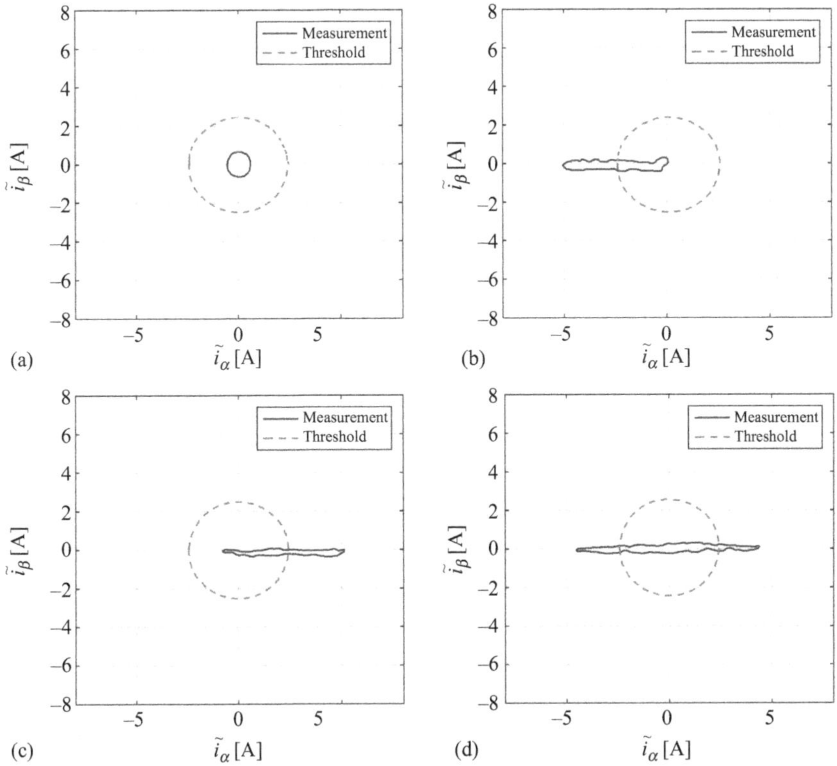

Figure 6.29 Experimental CRV trajectories for various test cases: (a) healthy inverter, (b) OC fault of T_1, (c) OC fault of T_2, and (d) OC fault of T_1 and T_2

zero and some minimum value. In either fault case (T_1, T_2, T_1&T_2), the OC fault can be identified for the suggested threshold setting ($I_{thres} = 2.5$ A).

A concept based on the calculation of the averaged residual voltage is presented in [74]. The authors propose measuring the pole voltage and subtracting it from the reference voltage to obtain the instantaneous residual voltage. The instantaneous residual is then averaged, and a lookup table is used to determine whether the averaged residual voltage is exceeding a threshold that indicates an open switch. For validation, the authors present experimental results for several fault scenarios in which up to four switches fail to open. An obvious disadvantage of this method is the required measurement of the pole voltages.

In [75], the authors suggest using the trajectory of the current vector obtained by Concordia's/Clarke's transformation or the instantaneous frequency (fundamental electrical frequency) for an open-switch detection. Both methods are able to detect an open switch. However, the current vector trajectory can identify which of the switches is faulty, while the instantaneous frequency method merely detects that one

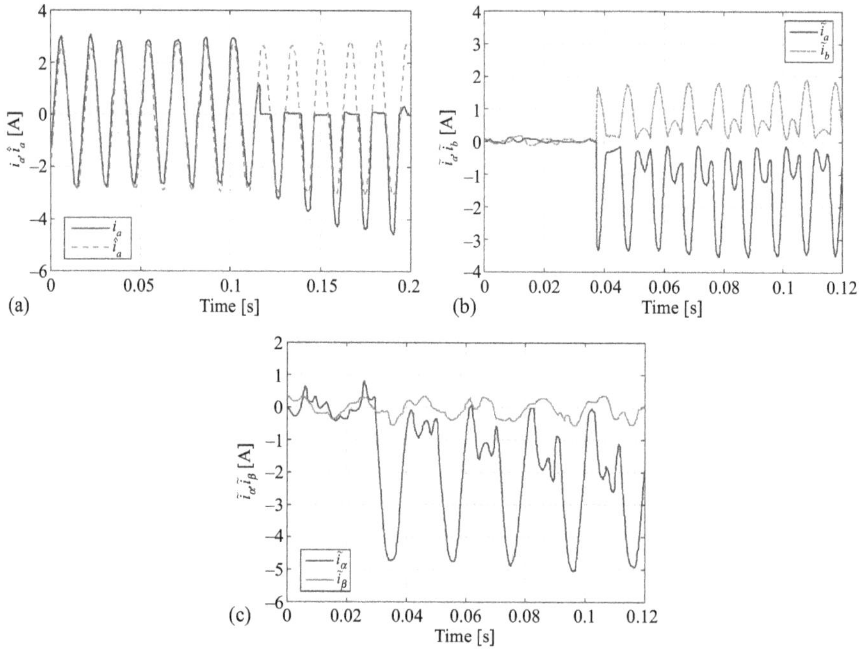

Figure 6.30 Phase currents in stationary abc-frame and αβ-frame for OC fault of T_1 (see Figure 6.29(b)): (a) measured and estimated phase A current in stationary reference frame, (b) current residuals of phases A and B in stationary reference frame, and (c) current residuals in stationary αβ-frame

of the switches is defective without being able to determine which switch is causing the issue. A benefit of the instantaneous frequency method is that it is comparatively easy to implement. Experimental results are provided to verify either of the approaches suggested.

A method based on the normalized average current Park vector that is applicable to three-level inverters is suggested in [76]. A few more interesting references related to the diagnosis of open-switch inverter faults are given at this point without going into anymore detail, since the monitoring of the inverter itself is not within the main scope of this chapter [77–85].

6.3.4 Stator core monitoring

A defective stator core leads to increased core losses and excessive heating of the machine. Localized heating can also deteriorate the stator insulation and eventually lead to a ground failure (see Section 6.3.2). In the worst case, a small defect such as the loss of insulation between a few laminations can eventually evolve into a molten stator core. A method to finding problems with the stator core in inverter-fed induction

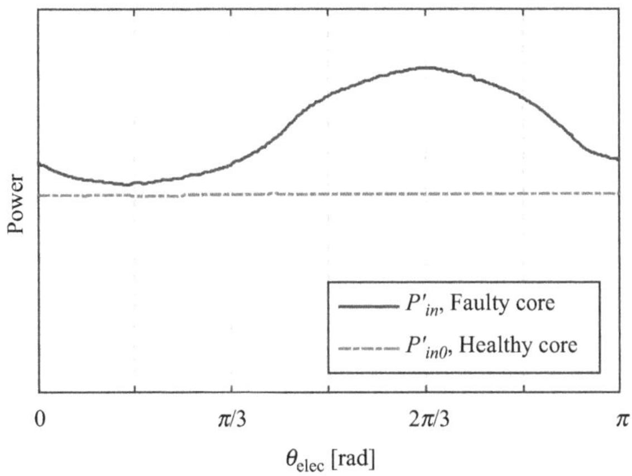

Figure 6.31 P'_{in} *as a function of the flux-vector electrical angle for a motor with a healthy and a faulty stator core. The faulty stator laminations are located at the 210° position.*

machines is presented in [86,87]. Lee *et al.* suggest applying a set of test signals to the motor via the inverter when the motor is at standstill. The basic idea is to apply a pulsating field at different angles. This is equivalent to applying the test signal at different stator circumferential positions. The power loss is used as an indicator of the integrity of the stator core. The loss components for an AC excitation when the machine is at standstill are the winding losses in stator $P_{w,s}$ and rotor $P_{w,r}$ and the core losses in stator $P_{c,s}$ and rotor $P_{c,r}$.

$$P_{in} = P_{w,s} + P_{w,r} + P_{c,s} + P_{c,r} \tag{6.33}$$

where the fault-indicating component is the stator core loss. While the rotor losses and the stator core loss are difficult to separate, the loss in the stator windings can be calculated and subtracted from P_{in}, which yields

$$P'_{in} = P_{in} - \hat{R}_s\left(i_{as}^2 + i_{bs}^2 + i_{cs}^2\right) \tag{6.34}$$

where $\hat{R}_s$ is the estimated stator resistance. If there are no faults present in the machine, the loss does not depend on the excitation angle, i.e., P'_{in} is a constant with respect to the excitation angle θ. If a fault is present, there will be an increase in the stator core losses, as depicted in Figure 6.31.

The fault location is at 210° and the loss power peaks at 120°, i.e., there is a −90° offset of the fault location with respect to the fault indicator. For the excitation, the authors recommend using as high a frequency as possible to maximize the impact of the stator core losses with respect to the other loss components. The limiting factors for the excitation frequency and magnitude are the maximum sampling rate and the processing capability, as well as the DC-link voltage of the inverter that is

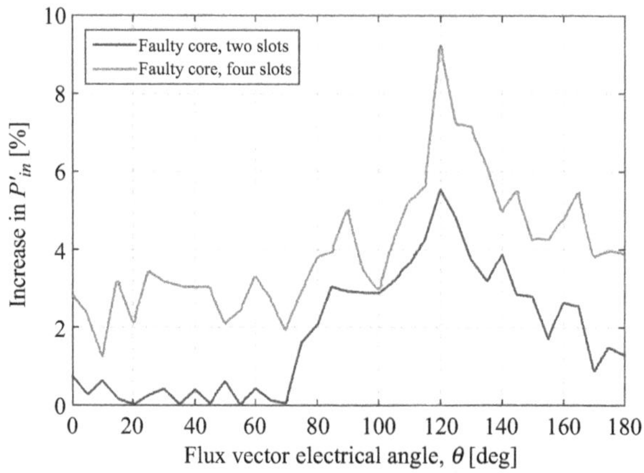

Figure 6.32 Experimental result for $\%\Delta P'_{in}$ using a motor with an emulated stator core fault and a 500-Hz excitation

driving the motor. For convenience, the computation of the percentage increase of P'_{in} with respect to the healthy machine is suggested. This requires that some initial measurements for a healthy stator core (P'_{in0}) are available.

$$\%\Delta P'_{in}(\theta) = \frac{P'_{in} - P'_{in0}}{P'_{in0}} \times 100\%$$

An experimental result with a machine that has two and four welded stator laminations (at the 210° position), respectively, is presented in Figure 6.32. The excitation frequency is 500 Hz. Both fault cases can be identified at the expected excitation angles (210° − 90° = 120°). An advantage of the method is the possibility of frequently assessing the stator core condition without having to operate or disassemble the motor. No additional test equipment is required since the excitation and data acquisition can be performed by the inverter. The fault can be located by observing at which excitation angle the input power (i.e., the loss) increases.

6.4 Rotor-related issues

Stator-related issues are similar for different types of machines (IM, SM, PMSM). Rotor issues, however, can be specific to the machine type. While induction machines can suffer from broken rotor bars, permanent magnet synchronous machines are at risk of demagnetization, and wound rotor machines (induction or synchronous machines) can develop problems with the insulation of the rotor windings. Rotor eccentricity is a problem that is common to all machine types, but not every fault indicator works for the different machine types.

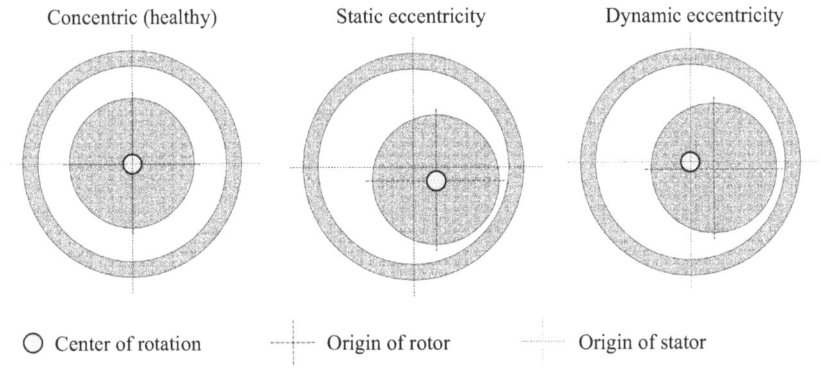

Figure 6.33 Different types of rotor eccentricity

6.4.1 Rotor eccentricity

Rotor eccentricity is a nonuniformity of the air gap. If the eccentricity is not changing with respect to the rotor position, it is labeled as static eccentricity (see Figure 6.33, middle).

For this type of eccentricity, the rotor rotates around its origin, and the origin of the rotor is not aligned with the origin of the stator, which creates a nonuniform air gap. In the case of a dynamic eccentricity, the nonuniformity of the air gap varies with rotor position (see Figure 6.33, right). In this case, the rotor's center of rotation coincides with the origin of the stator, but the origin of the rotor is not aligned with the center of rotation. A mixed eccentricity refers to a machine that has both types of eccentricities. The most commonly used method to detect a rotor eccentricity is the monitoring of specific harmonics induced by the eccentricities in the stator current. Some of the earlier research on the detection of rotor eccentricity in induction machines can be found in [18,88]. If sufficient information on the machine parameters is available, the following equation can be used to calculate the frequency components caused by rotor eccentricities [88,89]:

$$f_{ecc,slot} = f_s \left[(nR \pm n_e) \left(\frac{1-s}{p} \right) \pm n_{ws} \right] \tag{6.35}$$

where f_s is the fundamental operating frequency; $n = 0, 1, 2, \ldots$; R is the number of rotor slots; p is the number of pole pairs; n_e is the eccentricity order number; n_{ws} is the order number of the stator magnetomotive force (MMF) time harmonics. For a static eccentricity, $n_e = 0$, and for dynamic eccentricity, $n_e = 1, 2, \ldots$.

Another method of monitoring the sidebands of the fundamental operating frequency that does not require any detailed information about the construction of the machine uses the following stator current frequencies for monitoring [88,89]:

$$f_{ecc,IM} = f_s \left(1 \pm k \frac{1-s}{p} \right) \tag{6.36}$$

where f_s is the fundamental operating frequency, $k = 1, 2, 3, \ldots$, and p is the number of pole pairs.

While [88] and some of the earlier publications focus on the fundamentals of the detection method, some of the more recent publications focus on the signal processing to improve the reliability of the approach [8,89] or to improve the computational efficiency [90]. In particular, the work described in [90] is relevant for the application with inverter drives since the limited computational resources of the inverter platform require an efficient implementation of the monitoring algorithm. More details on the method implemented in [90] are given in Section 6.5.1.

An investigation on the applicability of (6.36) to an induction machine that is operated under DTC is conducted in [91]. The authors suggest a detection algorithm based on a method called multiple reference frames. This strategy uses a positive sequence component that is induced by stator and rotor faults for the monitoring process. With small variations of the processing steps required for line-fed machines, the method is applicable for finding stator turn faults, broken rotor bars, and rotor eccentricities in inverter-fed machines. The main difference in the DTC case is some additional filtering. An interesting observation from the experimental results is that additional frequency components not predicted by (6.36) occur in the stator current.

In [92], an eccentricity detection in PMSM based on the detection of the fundamental sideband frequencies is suggested. The equivalent frequencies caused by rotor eccentricity in PMSMs are

$$f_{ecc,PMSM} = f_s\left(\frac{1 \pm k}{p}\right). \tag{6.37}$$

with p equal to the number of pole pairs and $k = 1, 2, 3, \ldots$. The concept is evaluated on a machine with emulated dynamic and static eccentricities. By using the frequency components given by (6.37), the dynamic eccentricity is clearly identified, while the static eccentricity case cannot be distinguished from the healthy machine. The authors suggest using an analysis of the negative sequence current to identify the static eccentricity instead. Unlike the dynamic eccentricity, the static eccentricity introduces an asymmetry in the impedance of the stator phases, which can be picked up by using sequence component analysis. The data analysis confirms that the static eccentricity can be identified successfully by using the negative sequence current.

A series of research papers [93–95] proposes a novel eccentricity-detection method using the inverter drive during off-cycles, i.e., when the machine is not operating. In [93,95], the detection is applied to induction machines, and in [94], a method for permanent magnet motors is introduced. In [93], it is shown that the leakage inductance, and therefore the imaginary part of the impedance of the motor, is a function of the level of rotor eccentricity. Since the analysis is performed at standstill, the slip is equal to one and the leakage inductance has a large impact on the overall motor impedance. The equivalent circuit of an induction machine at standstill is shown in Figure 6.34.

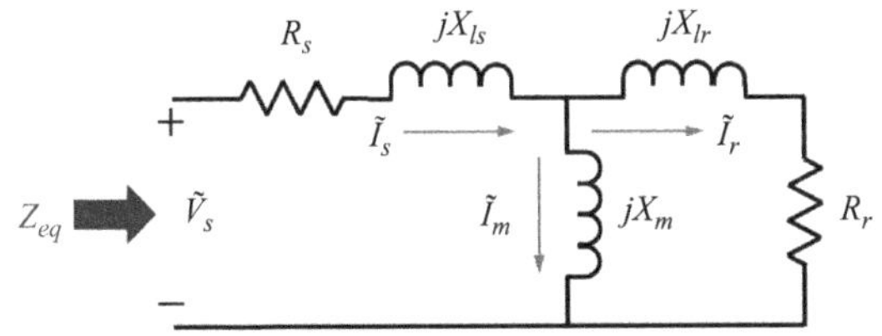

Figure 6.34 Electrical equivalent circuit model of induction machine at standstill (s = 1)

To measure the impedance, a pulsating voltage is injected and the current response is measured. The suggested excitation has the following form:

$$\begin{aligned} v_{as}^*(\theta, \omega t) &= V \cos(\theta)\ \mathrm{squ}(\omega t) \\ v_{bs}^*(\theta, \omega t) &= V \cos\left(\frac{\theta - 2\pi}{3}\right)\ \mathrm{squ}(\omega t) \\ v_{cs}^*(\theta, \omega t) &= V \cos\left(\frac{\theta + 2\pi}{3}\right)\ \mathrm{squ}(\omega t) \end{aligned} \tag{6.38}$$

where θ is the angle; V is the magnitude; and ω is the frequency of the injected square-wave voltage. By using the injected voltage and the measured current, the equivalent impedance can be calculated as follows:

$$Z_{eq} = \frac{\bar{V}_\theta}{\bar{I}_\theta} = R_{eq} + jX_{eq} \approx (R_s + R_r) + j(X_{ls} + X_{lr}) \tag{6.39}$$

where R_s and R_r are the stator and rotor resistance, and X_{ls} and X_{lr} are the stator and rotor leakage inductance. A finite element analysis (FEA) shows that the imaginary part of the impedance(X_{eq}) depends on the angle of the injected voltage θ as displayed in Figure 6.35.

Despite the dependence on the injection angle, the average impedance $X_{eq,avg}$ strictly increases when the degree of eccentricity is raised. Therefore, the average impedance is a reliable indicator of the degree of eccentricity. The analysis also shows that the relationship between the degree of the rotor eccentricity and the change in the motor impedance is strongly nonlinear (see Figure 6.35). The sensitivity of the equivalent impedance with respect to the degree of eccentricity increases sharply as the level of eccentricity rises (Figure 6.36). An experimental evaluation with emulated rotor eccentricities of 20%, 40%, 60%, and 80% for excitation frequencies of 50 and 250 Hz is shown in Figure 6.37.

Even though small degrees of eccentricity (<20%) have little impact on the fault indicator, the authors point out that consistent monitoring during off-intervals will still allow for a detection of the problem before a critical level of eccentricity is reached. The method is easy to implement and automate. Additional hardware is not required, and the machine parameters need not be known either. Since the method is

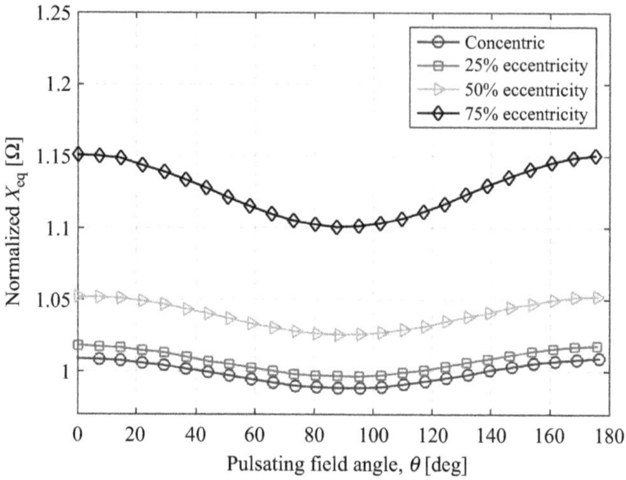

Figure 6.35 FEA simulation results of normalized X_{eq} for a four-pole motor with 0%, 25%, 50%, and 75% eccentricity under 50-Hz excitation

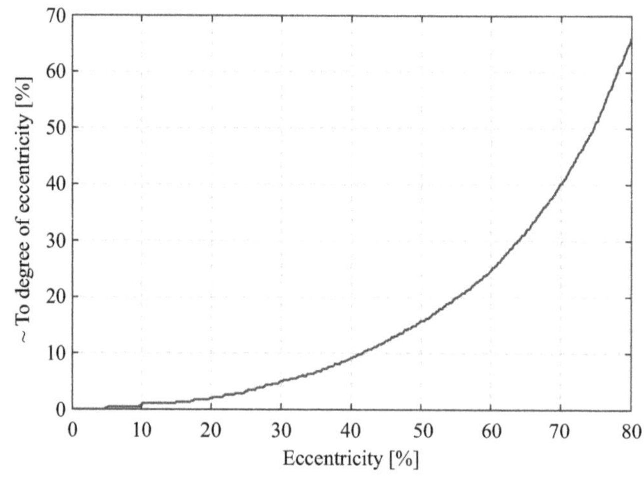

Figure 6.36 Percentage increase of the average of inverse air-gap length with respect to the degree of eccentricity

performed at standstill during off-intervals, the machine operation has no influence on the results.

To improve the sensitivity of the method in [93], Hong *et al.* suggest an approach based on the monitoring of the differential inductance [95]. The authors show that the differential inductance, which is defined as $L' = d\lambda/di$, has a high sensitivity to a change in rotor eccentricity when operating at the knee point of the BH-curve (see Figure 6.38).

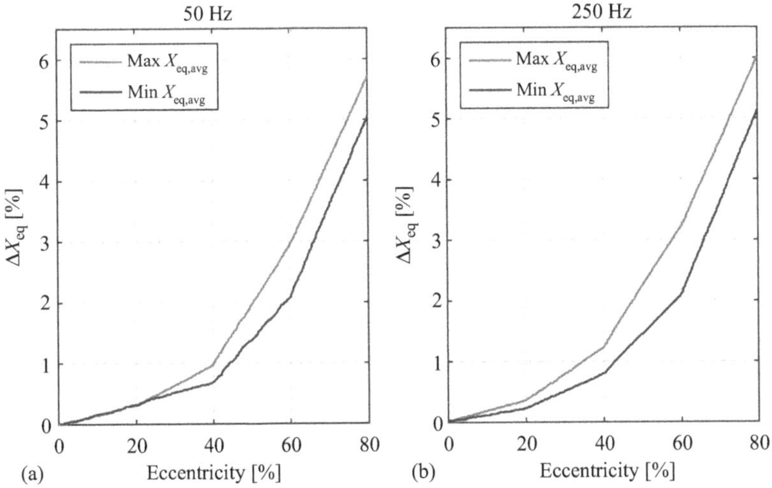

Figure 6.37 Experimental results for the minimum and maximum values of the increase of $X_{eq,avg}$ for 0%, 20%, 40%, 60%, and 80% eccentricity: for (a) 50-Hz excitation and (b) 250-Hz excitation

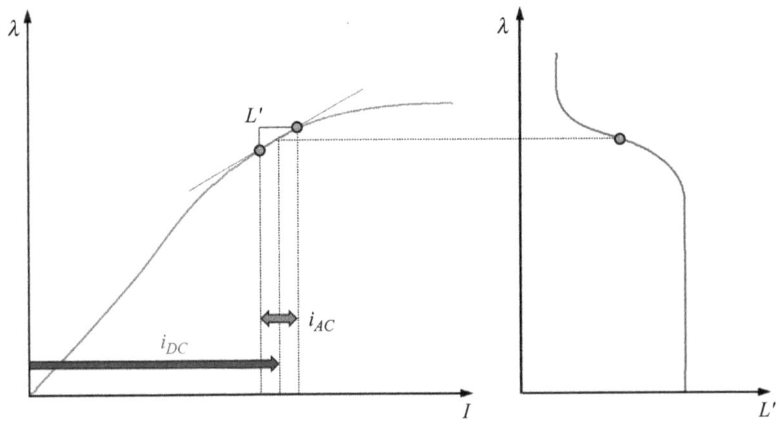

Figure 6.38 Generalized relationship between flux linkage and current, and flux linkage and differential inductance for a motor with concentric rotor

To obtain an excitation at the knee point of the BH-curve, a DC component is added to the pulsating voltage.

$$
\begin{aligned}
v_{as}^*(\theta, \omega t) &= V_{DC} \cos(\theta) + V_{ac} \cos(\theta) \operatorname{sgn}(\sin(\omega t)) \\
v_{bs}^*(\theta, \omega t) &= V_{DC} \cos(\theta - 2\pi/3) + V_{ac} \cos(\theta - 2\pi/3) \operatorname{sgn}(\sin(\omega t)) \\
v_{cs}^*(\theta, \omega t) &= V_{DC} \cos(\theta + 2\pi/3) + V_{ac} \cos(\theta + 2\pi/3) \operatorname{sgn}(\sin(\omega t))
\end{aligned}
$$

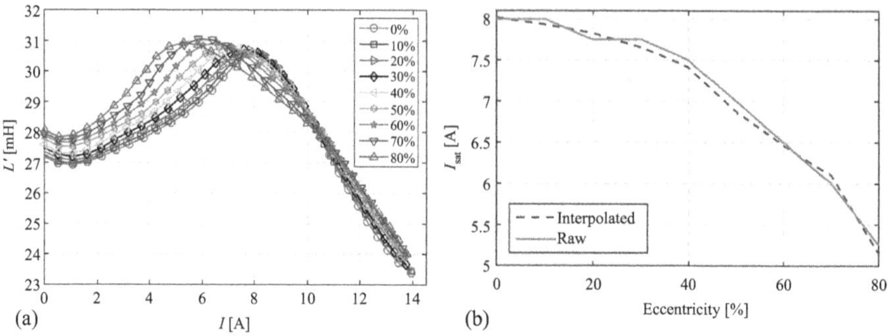

Figure 6.39 Experimental evaluation of the proposed fault indicators, (a) $L'(I)$ and (b) I_{sat}

V_{DC} is the magnitude of the DC component, V_{ac} is the magnitude of the AC component, and θ is the angle of the pulsating voltage excitation. The authors show that the method is insensitive to the relative angle between the injected voltage and the mechanical rotor position. This simplifies the execution of the test, and it is not necessary to conduct a measurement for multiple injection angles as in [93]. A simple choice for the injection angle is to set $\theta = 0$. However, multiple measurements at increasing DC voltage level have to be performed to obtain the required $L'(I)$ relation. One difference between the generalized relationship shown in Figure 6.38, and the actual measurements of the differential inductance as shown in Figure 6.39(a) is that the differential inductance is not flat for excitation currents below the knee point. Instead, the differential inductance has a peak before it decreases due to an increasing level of saturation. The current where the differential inductance peaks is labeled I_{sat}. This current is used as the indicator for the degree of eccentricity. Experimental results for an induction machine with a varying degree of static and dynamic eccentricities are presented for validation. The effective eccentricity is varied from 0% to 80% in 10% increments. For comparison, the measurements are evaluated by using the traditional spectral methods based on the monitoring of frequencies specified in (6.35) and (6.36). It proves to be difficult to make an accurate assessment when the spectral methods are used, especially, since the different fault-indicating frequencies in (6.35) do not show a large degree of correlation. In contrast to the spectral methods, the fault indicator based on the dynamic inductance (I_{sat}) shows a clear trend with respect to the level of rotor eccentricity as shown in Figure 6.39. Compared to the previously suggested method [93], the sensitivity is increased by five to ten times, and even lower levels of eccentricity can be identified with a higher reliability.

A similar strategy for detecting eccentricity based on the measurement of the differential inductance is applied to PMSMs [94]. In this application, the differential d-axis inductance serves as a fault indicator. There is no need to provide a DC voltage excitation (as in [95]) due to the PM flux Φ_m. The injection of a small AC voltage into the d-axis is sufficient for excitation. Unlike in the induction machine case, the orientation of the d-axis with respect to the eccentricity has a noticeable influence on

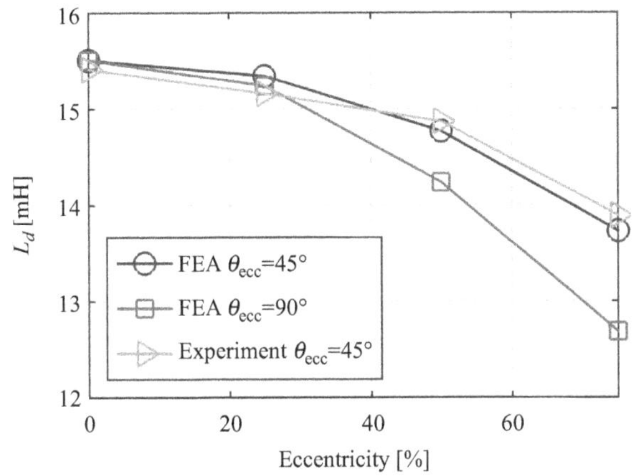

Figure 6.40 Experimental ($\theta_{ecc} = 45°$) and FEA results ($\theta_{ecc} = 45°, 90°$) for L_d estimation for 0%, 25%, 50%, and 75% dynamic eccentricity

the sensitivity of the method. The authors point out that the optimal implementation of the algorithm has to be adjusted for different motor designs. Experimental results show the impact of the angle between d-axis and the orientation of the eccentricity (see Figure 6.40) on the ability of the method to identify the eccentricity condition.

The benefits of the two previous methods described in [93,95] apply to this approach as well.

6.4.2 Broken rotor bars and end rings

Rotor asymmetries such as broken rotor bars and end rings are another group of faults commonly found in induction machines that can be detected by means of spectral analysis. Most diagnostic approaches are based on the stator current, and the required sampling frequency as well as the data length can be supported by modern drive units. Therefore, implementation on an inverter-based platform is an obvious application for these methods. Some of the earlier works from the 1980s and early 1990s [88,96,97] show that broken rotor bars can be identified by monitoring specific harmonics in the stator current:

$$f_{ra} = \left[k \left(\frac{1-s}{p} \right) \pm s \right] f_s \tag{6.40}$$

where $k = 1, 2, \ldots ; p$ is the number of pole pairs; s is the rotor slip; and f_s the fundamental operating frequency. Another commonly used approach for finding broken bars analyzes frequencies that are in close proximity to the fundamental operating frequency [98,99]

$$f_{bb} = (1 \pm 2ks) f_s \tag{6.41}$$

with $k = 1, 2, \ldots$, s is the rotor slip, and f_s is the fundamental operating frequency. An appropriate frequency resolution is required in order to separate these frequencies from the fundamental operating frequency.

In [89], a monitoring strategy based on (6.40) is introduced, and some alternative high-resolution spectral analysis tools are discussed as an alternative approach to the classical FFT.

In [100], a detection method for broken bars based on (6.41) is proposed. The application of a pattern-recognition technique based on the Bayes minimum error classifier is devised in order to automate the process and to avoid ambiguity when the frequency analysis is done. No more than two pieces of information are required: (1) the number of rotor slots used for the calculation of the speed to determine the rotor slip and (2) the number of pole pairs. The authors point out that the method can be easily modified to detect other faults such as eccentricities or phase unbalances.

A method that is based on the envelope analysis of the time domain current waveforms is presented in [101]. Instead of performing a frequency analysis, this method exploits the modulation of the current envelope by the broken bar at twice the slip frequency and is indirectly based on (6.41). The decision-making is based on a Bayes classifier. The method is also applied to detect shorted turns which cause an amplitude modulation of the envelope at the fundamental operating frequency. Another approach that is not based on spectral analysis but rather on time series data mining is presented in [102].

Drif *et al.* [103] suggest analyzing the instantaneous reactive power to identify broken rotor bars instead of using only the measured current. Due to the multiplication of the current and voltage in the calculation of the reactive power,

$$q(t) = v_\alpha i_\beta - v_\beta i_\alpha$$

the fundamental component in current and voltage appears as a DC component in the power spectrum, and the fault-indicating frequency (6.41) is moved to $2ksf_s$. The authors state that one advantage is the removal of the supply frequency in the spectrum. However, this argument seems weak since the separation between the dominating component and the fault-indicating component is still $2ksf_s$ as it is in the analysis of the stator current. Furthermore, the measurement of the line voltage is required. Usually, this is directly available in neither line-fed nor inverter-fed machines.

Similar to the detection of stator turn faults described in Section 6.3.1, Briz *et al.* [47] show that the high-frequency injection-based method can also be applied to rotor faults. The fault-indicating component for this failure mode is twice the rotor frequency $2\omega_r$ in the negative sequence carrier-signal reference frame, which is at $-\omega_c + 2\omega_r$ in the stationary reference frame. In Figure 6.41, a measurement result for a healthy rotor and for an induction machine with one broken bar is displayed over a wide load range.

The fault-indicating component is barely influenced by the load condition, and the signature from the faulty machine can clearly be separated from the healthy case.

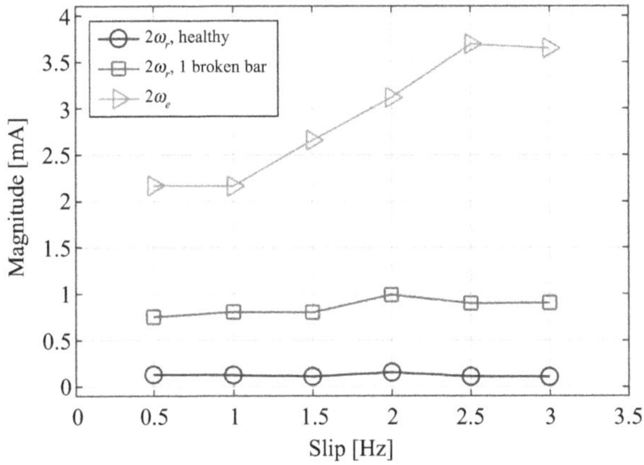

Figure 6.41 Measurement of the negative sequence carrier-signal current at $2\omega_e$ and $2\omega_r$ using a motor with semi-open rotor slots operated at rated flux for a healthy rotor and for a rotor with a broken rotor bar

The authors point out that the diagnostic results can be significantly impacted by the design of the machine. For machines with closed rotor slots, the separation between the healthy and faulty case is much less pronounced, which reduces the sensitivity and reliability of the method. Another drawback of the method compared to the stator turn fault diagnostic is that it needs a relatively precise knowledge of the rotor speed (n_r in rpm) since it directly yields the frequency ($\omega_r = (2\pi p/60)n_r$ with p being the number of pole pairs). A similar high-frequency injection approach is taken in [104]. The authors show that the high frequency resistance of the rotor windings is a good indicator of broken rotor bars, and that the high frequency resistance can be identified via an HF-voltage injection. This method has some advantages compared to the previously suggested HF injection approaches because no rotor speed information is required and the method works under all load conditions. Also, it is not affected by the rotor topology and can be applied to machines with closed rotor slots.

Another paper [105] proposes a method for testing inverter-fed motors during intervals when they are not operating. The approach is similar to the method for the identification of rotor eccentricities introduced in [93] and described in Section 6.4.1. The authors show that the equivalent motor impedance at standstill is modified when broken rotor bars occur. The basic pattern of a healthy motor and a motor with a broken bar is shown in Figure 6.42.

The impedance can be identified by applying the test voltages described by (6.38) and by then using (6.39) to calculate the equivalent impedance. Compared to the eccentricity case, there are a few important differences. When a rotor bar breaks, both, the rotor leakage inductance and the rotor resistance are affected. Therefore, either the equivalent resistance R_{eq}, or the equivalent impedance X_{eq}, or a combination

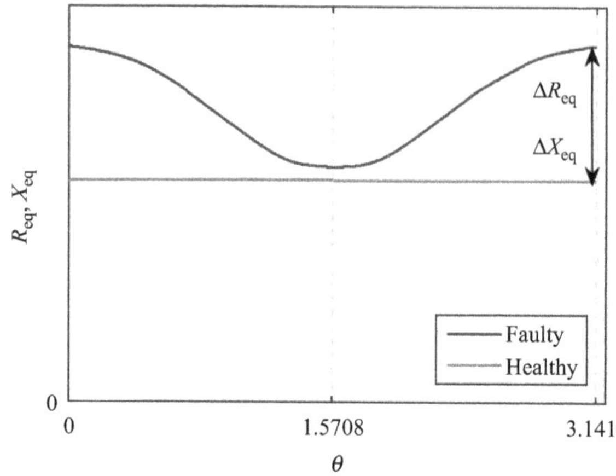

Figure 6.42 Pattern of the fault indicators R_{eq} and X_{eq} with respect to the injection angle θ for a motor with a healthy and a faulty rotor

of the two, can be used as fault indicators. A second difference is the dependence of the impedance on the angle of the injected voltage. When the injection angle is aligned with the location of the broken rotor bar, the fault signature is not sufficiently excited, and the difference between the healthy case and the faulty motor is barely noticeable. The signature is the strongest when the angle between the injected voltage and the location of the broken bar(s) is 90°. This is also illustrated in Figure 6.42. The raw values of R_{eq} and X_{eq} are further processed to get a simple fault indicator. A baseline measurement from the healthy motor (R_{eq0}, X_{eq0}) is subtracted from the measurements collected during the monitoring phase in order to obtain

$$R'_{eq}(\theta) = R_{eq}(\theta) - R_{eq0}(\theta)$$
$$X'_{eq}(\theta) = X_{eq}(\theta) - X_{eq0}(\theta).$$

The suggested fault indicators $\Delta R'_{eq}$ and $\Delta X'_{eq}$ are the maximum of $R'_{eq}(\theta)$ and $X'_{eq}(\theta)$ in percent with respect to the average values of $R_{eq0}(\theta)$ and $X_{eq0}(\theta)$, respectively.

An induction machine with up to three broken bars at the $\theta = 90°$ position is used for validation. The result ($R'_{eq}(\theta)$ and $X'_{eq}(\theta)$) for a pulsating voltage of 50 Hz is shown in Figure 6.43. As expected, the fault shows as an increase in R'_{eq} and X'_{eq} when the injection angles of 0° and 180° are approached. $\Delta R'_{eq}$ and $\Delta X'_{eq}$ for three different excitation frequencies are displayed in Figure 6.44.

When the frequency of the injected voltage rises, the sensitivity of the equivalent resistance increases while the sensitivity of the equivalent inductance decreases. In any of the test cases, the faulty motor is identified by the approach suggested. Since the test procedure can be implemented in the inverter software and applied on a regular basis during intervals when the motor is not operating, this method provides a powerful

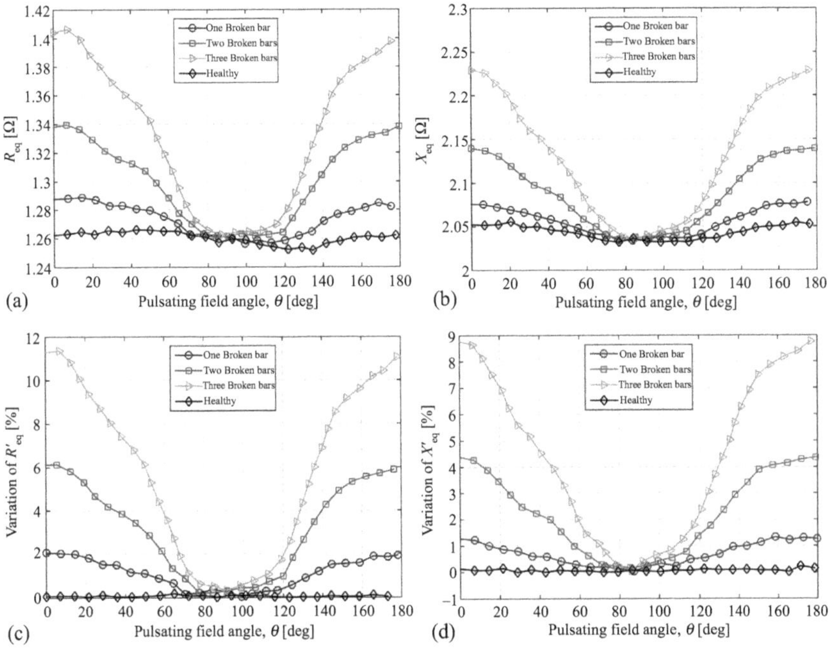

Figure 6.43 Measurement results for motor with zero, one, two, and three broken bars under 50 Hz excitation: (a) R_{eq} versus angle of applied voltage θ, (b) X_{eq} versus angle of applied voltage θ, (c) R'_{eq} versus angle of applied voltage θ in %, and (d) X'_{eq} versus angle of applied voltage θ in %

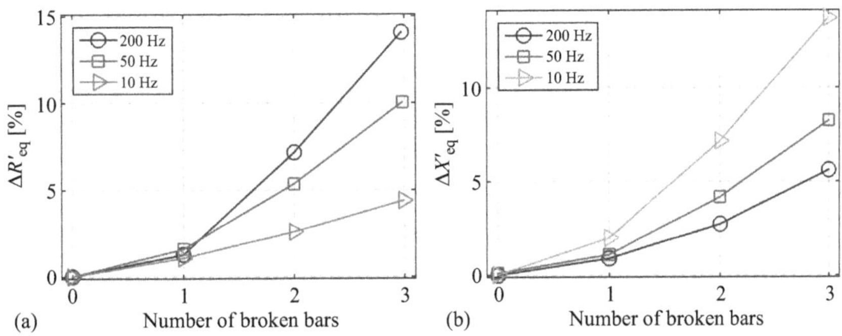

Figure 6.44 Measurement results for zero, one, two, and three broken bars under 10, 50, and 200 Hz excitation: (a) $\Delta R'_{eq}$ versus number of broken bars and (b) $\Delta X'_{eq}$ versus number of broken bars

monitoring tool without adding any additional hardware. Furthermore, the proposed method is not influenced by the operating conditions as the online counterparts are.

6.4.3 Demagnetization

One of the main concerns in permanent magnet motors is the possibility of demagnetization. Demagnetization occurs when a large field opposing the permanent magnet field is induced during operation or when the temperature exceeds a certain threshold. In most applications, the temperatures will never get close to the temperatures required for demagnetization. A demagnetization due to overcurrent is far more likely and usually happens during faults, when the stator current is not well-controlled. The demagnetization can be local or can affect all magnets uniformly. Factors that impact the demagnetization characteristics are the magnet material, the temperature, and the shape of the magnet. Demagnetization of the rotor can affect the motor performance significantly.

An online detection method for demagnetization faults is proposed in [92]. Initially, the authors suggest using harmonic components at the rotating frequency to diagnose rotor faults such as eccentricity and demagnetization:

$$f_{demag} = f_e\left(1 \pm \frac{k}{p}\right) \tag{6.42}$$

where f_e is the fundamental operating frequency; $k = 1, 2, 3, \ldots$; and p is the number of pole pairs. Simulation and experimental results show that this indicator can reliably detect dynamic eccentricity and fails to detect static eccentricity. Demagnetization is identified for some operating conditions only, which limits the usefulness of this method. Another drawback is that there is no reliable way to distinguish between demagnetization and a dynamic eccentricity. To generate a more reliable indicator for demagnetization, the authors suggest using the estimated magnet flux $\hat{\lambda}^e_{d(pm)}$ as a fault indicator instead:

$$\hat{\lambda}^e_{d(pm)} = \frac{v^e_{qs} - r_s i^e_{qs}}{\omega_e} - L^e_{ds} i^e_{ds}$$

where e denotes that the variable is in the synchronous reference frame, v_{qs} is the q-axis stator voltage, i_{qs} is the q-axis stator current, i_{ds} the d-axis stator current, r_s the stator resistance, L_{ds} the d-axis stator inductance, and ω_e the synchronous speed. For the validation of the proposed fault indicator, experiments with a modified machine are conducted: a part of the magnet is removed from the rotor. Figure 6.45 shows the results of the estimated flux for the healthy rotor, as well as for the damaged rotor under several operating conditions. The measurements obtained from the defective machine can clearly be separated from the results obtained from the healthy machine for all operating conditions. The method should be suitable for detecting uniform demagnetization as well, even though only local demagnetization is evaluated in the paper. A major drawback is the requirement for a voltage measurement in order to estimate the magnet flux. It could be sufficient to use the d- and q-axes voltage commands generated by the current controller to avoid the installation of a voltage sensing circuit. However, this issue is not further addressed.

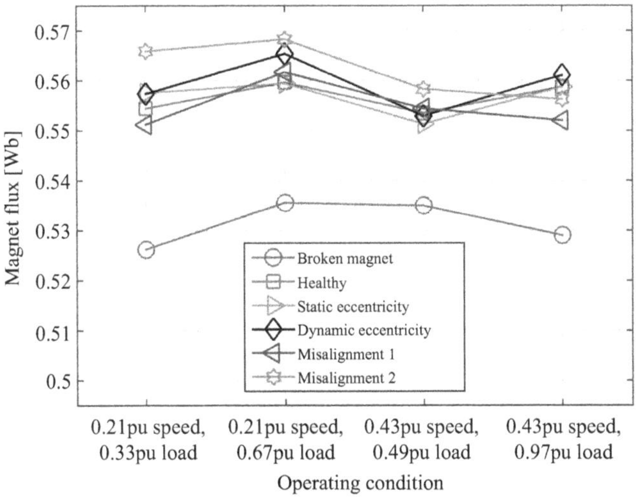

Figure 6.45 Estimated magnet flux for different rotor faults at various operating conditions

The condition monitoring in [92] requires steady-state conditions for the algorithm to work properly. In [106], a method based on the WT is introduced. It is capable of detecting rotor magnet faults even under nonstationary conditions. Instead of decomposing a signal into sine or cosine components, the WT uses wavelets. Both, the continuous WT as well as the discrete WT are successfully used to detect a defective magnet, even when the motor speed is changing. Another approach that can also monitor the magnet condition under nonstationary conditions is introduced in [107]. Instead of using WTs, the method is based on the Hilbert–Huang transform.

The use of the zero sequence voltage component to detect demagnetization in surface-mounted permanent magnet machines with symmetrical concentrated stator windings is proposed in [108]. The authors develop an analytical model for the back-emf of one slot. They assume that one pole of the rotor experienced partial demagnetization. The model is then extended to obtain the back-emf of each of the phases. The main difference between the per-slot back-emf and the back-emf of the full phase is that demagnetization causes harmonics as defined in (6.42) in the per-slot back emf, while the phase back-emf will have a reduced magnitude but does not contain any of the fault harmonics. An obvious drawback of this method is that the measurement of the stator voltages is necessary.

A method that uses the inverter during periods when the machine is not operated is introduced in [109]. Whenever the motor is at standstill and not operating, a test signal can be injected. The authors suggest injecting a pulsating voltage as defined in (6.38). The angle of the excitation with respect to the A-phase of the stationary reference frame is θ, the magnitude of the excitation is V, and the angular frequency of the pulsating voltage is ω. One advantage of a pulsating voltage is that the net torque is zero, and the vibration due to the pulsating torque can be kept at a low level

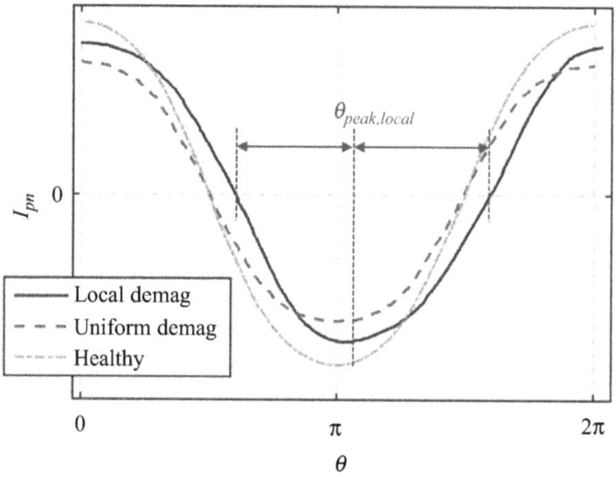

Figure 6.46 Pattern of I_{pn} for healthy magnet and magnets with uniform and local demagnetizations

if V is chosen sufficiently low and the excitation frequency ω is set sufficiently high. The indicator of the magnet condition is based on the current response to the injected voltage with respect to θ. If the flux produced by the pulsating voltage is aligned with the rotor magnet flux (d-axis), the positive and negative peaks of the current response are not symmetrical. The current will be larger at the positive peak than at the negative peak since the equivalent impedance decreases due to an increased level of saturation. When the injected voltage and the current response are in quadrature with the magnetic axis of the rotor, the current response will have perfectly symmetrical positive and negative peaks. The authors suggest using the sum of the negative and the positive peaks of the current response as an indicator of the magnet condition:

$$I_{pn}(\theta) = I_p(\theta) + I_n(\theta)$$

where I_p is the positive peak of the current response, and I_n the negative peak of the current response for the voltage angle θ. Figure 6.46 shows I_{pn} for three different cases: no demagnetization, uniform demagnetization, and local demagnetization.

A uniform demagnetization results in a uniform reduction of the saturation level. If the same test voltage is applied, the impedance seen from the motor terminals is larger than the impedance for the healthy magnet, and the magnitude of the injected current will be smaller for all voltage angles θ. This results in a uniform decrease in I_{pn}. If there is a local demagnetization, this will result not only in a reduction of the magnitude of I_{pn} but also in a phase shift or a distortion of the current response. Some of the practical issues the authors point out are the temperature dependence of the magnets, the motor-to-motor variation of the impedance, and the determination of an appropriate test voltage level. The stator flux Φ_s induced by the test voltage has to be large enough to ensure an acceptable sensitivity of the method, without inducing any movement or a noticeable level of vibration. Figure 6.47 shows a plot of the peak of I_{pn} ($I_{pn,peak}$) with respect to the stator flux Φ_s, which is a measure of the sensitivity for the machine used in this paper.

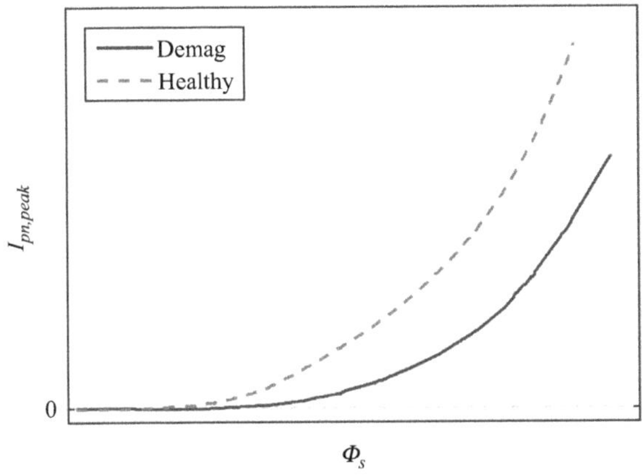

Figure 6.47 Fault indicator, $I_{pn,peak}$, as a function of the stator winding flux Φ_s for a motor with a healthy and a demagnetized PM, respectively

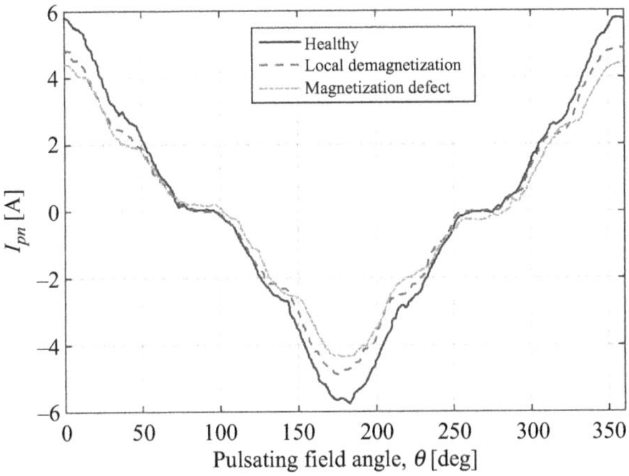

Figure 6.48 I_{pn} measurement using a 60 V, 100 Hz excitation for a healthy magnet, local demagnetization, and uniform demagnetization

The method offers numerous advantages compared to existing procedures. No additional hardware is needed, and the method can be integrated seamlessly without affecting the operation of the motor. Regular checks can be conducted to keep track of the magnet condition unless the application requires the motor to run nonstop around the clock. The method does not require the knowledge of any motor parameters. Also, it is not affected by the operating conditions or other defects such as rotor eccentricity that could cause ambiguity in the measurement results. Both, uniform as well as local demagnetization can be identified. For validation, the authors present some measurement results that verify the capability of the method to identify local as well as uniform demagnetization. For example, Figures 6.48 and 6.49 show the experimental

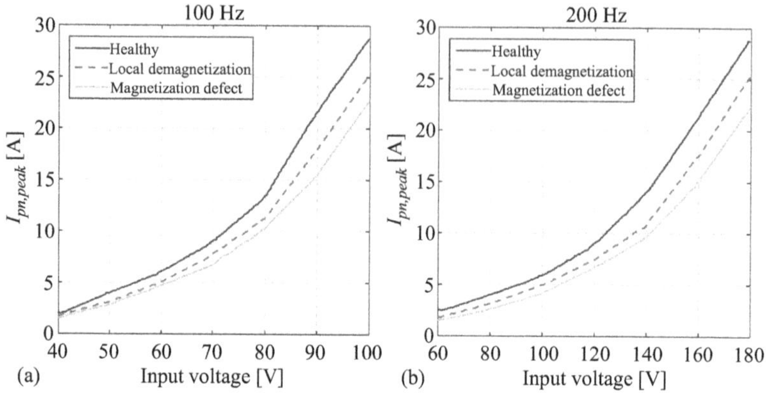

Figure 6.49 Fault indicator $I_{pn,peak}$ as a function of excitation voltage magnitude for a healthy magnet, local demagnetization, and uniform demagnetization: (a) 100 Hz excitation and (b) 200 Hz excitation

results for I_{pn} and $I_{pn,peak}$ of a healthy motor, one with a local demagnetization and one with uniform demagnetization, respectively.

6.5 Bearings, gearbox, and other mechanical problems

This section introduces some applications that not only pertain to the monitoring of the electric machine itself but also include approaches for monitoring components connected to the electric machine, such as a gearbox or a pump. If the health of a connected asset is analyzed, the electric machine is used as a sensor.

The most commonly used approaches for the detection of mechanical defects are based on a spectral analysis of the measurements. The spectral analysis can be performed by using the discrete Fourier transform, the FFT, the STFT, or the WT. For a fast response with the spectral techniques, a considerable computational power is needed, which might be available only on advanced platforms. Yet, as a rule, mechanical defects of this nature usually are progressing slowly, and the time between the detection of an incipient fault and a catastrophic failure is much larger than that for electrical faults, for example. In this case, the computation can be carried out at a slower rate and can be performed on a less powerful platform. Crucial parameters for the spectral analysis are the sampling rate and the duration of the measurement. To achieve a good spectral resolution, a large measurement window is needed. The frequency resolution is inversely proportional to the measurement duration. The monitoring of faults that occur at high frequencies requires a sufficiently high sampling rate. In theory, the sampling rate has to be at least two times higher than the highest frequency to be measured (Nyquist Theorem). For more reliable results a sampling frequency that is at least five times higher than the largest frequency to be monitored is recommended. The maximum sampling frequency requirement depends on the frequency content of the measured quantity and the analog filtering (anti-aliasing filter) that is applied prior to the spectral analysis.

Instead of running the monitoring algorithm on the inverter platform, the inverter can be used as a data-acquisition system only, and the data analysis can be performed on an external platform such as a server. One advantage of performing the computation on an external platform is that the data from a large number of units can be processed and then compared. This would be very helpful for the management of larger fleets. The health of the assets could be ranked and then used to improve the efficiency of the maintenance process.

Traditionally, vibration analysis has been used in combination with the spectral techniques [110], which requires the installation of expensive monitoring equipment. The same concept can be used in combination with the electrical measurements that are already available (e.g., stator currents or terminal voltages). The low cost and simplicity of spectral monitoring using electrical signals makes these approaches very attractive for practical application.

6.5.1 Bearings faults

The failure of the rolling element bearings is one of the main failure mechanisms in small machines (see Section 6.1). Bearing failures can be classified into different types. Some failures occur at specific locations and are restricted to a small area. Those types of defects are labeled as single-point defects. An example of single-point defects is brinelling. Other failures due to flaking, fluting, or corrosion stretch over a larger area and cause a defect that is sometimes referred to as generalized roughness. The early work on the monitoring of rolling element bearings by means of electrical signature analysis established the relationship between the vibration frequencies introduced by bearing defects and the corresponding components in the stator current [111]:

$$f_b = f_e \pm k \cdot f_{vib} \tag{6.43}$$

where f_e is the electrical supply frequency; $k = 1, 2, 3, \ldots$; and f_{vib} is the vibration frequency caused by the defect. There are several components of a bearing that can fail, and each of them produces a set of characteristic frequencies. A defect of the inner race will show at

$$f_{inner} = \frac{N_b}{2} f_m \left(1 + \frac{B}{P} \cos\alpha\right) \tag{6.44}$$

a defect of the outer race is found at

$$f_{outer} = \frac{N_b}{2} f_m \left(1 - \frac{B}{P} \cos\alpha\right) \tag{6.45}$$

and a ball defect will result in a mechanical fault frequency of

$$f_{ball} = \frac{B}{P} f_m \left(1 - \left[\frac{B}{P} \cos\alpha\right]^2\right) \tag{6.46}$$

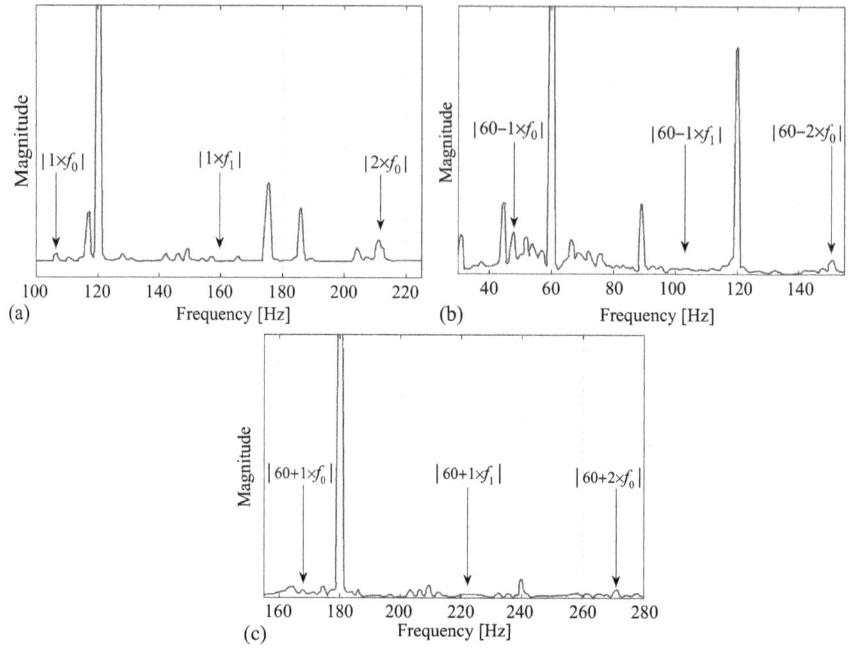

Figure 6.50 Vibration- and current-based diagnostic results for a loaded four-pole induction machine with a hole in the outer race of the shaft-end bearing: (a) vibration spectrum (via acceleration) and frequencies, (b) stator current spectrum with frequencies for $f_b = f_e - k \cdot f_{vib}$*, and (c) stator current spectrum with frequencies for* $f_b = f_e + k \cdot f_{vib}$

where N_b is the number of rolling elements; B is the ball diameter; P is the bearing pitch diameter; f_m is the mechanical rotor frequency; and α is the contact angle of the balls on the race. The required geometrical bearing data can be obtained from data sheets. Unless otherwise stated or implied by the application, the contact angle is usually assumed to be zero.

In [111], the basic concept of incipient bearing fault detection by using spectral current analysis is evaluated. As a reference, a vibration measurement is conducted. Some experimental results for an outer race fault obtained with an induction machine are shown in Figure 6.50. The defect was manufactured by drilling a hole in the outer race. Even though no comparison between the healthy and damaged bearing is made, an outer race defect seems to be present at the frequencies predicted by (6.45) and (6.43). A second type of bearing defect, the brinelling of the inner and outer race, is tested as well. The tests show that an increasing load reduces the sensitivity of the fault indicators due to the damping provided by the load. In addition, the bearing frequencies shift with the load level due to the change in the slip frequency. Higher fault frequencies in the electrical spectrum experience a higher level of attenuation than

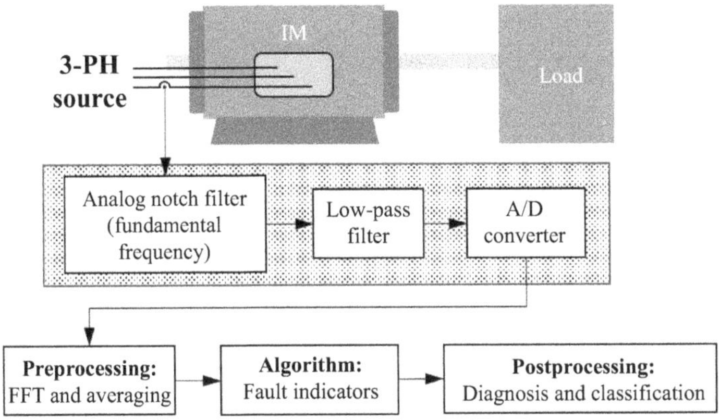

Figure 6.51 Block diagram of stator current-based monitoring scheme

the corresponding fault-indicating frequencies in the vibration spectrum. The attenuation at higher frequencies likely stems from the increase of the motor impedance with increasing excitation frequency.

The work in [111] is complemented by [18], where a possible route for the automation of the monitoring process is suggested. The authors propose using a combination of preconditioning through digital signal processing (FFT, averaging, filtering) and a clustering and classification of the data by a neural network. The classification algorithm flags the data as faulty or healthy. A flowchart of the suggested algorithm is shown in Figure 6.51.

A similar approach for the monitoring of bearing defects is taken in [89]. Additional analog and digital signal-processing steps are suggested, and different spectral analysis techniques are analyzed in order to improve the reliability of the monitoring algorithm. The suggested processing steps are the same as shown in Figure 6.51. After removing the fundamental frequency with an analog notch filter and applying an anti-aliasing filter, the measurement is converted to the frequency domain. To reduce the noise in the spectrum, the authors suggest averaging multiple measurements in the frequency domain. Then the same calculation of fault indicators as suggested in [111] (see (6.43)–(6.46)) is performed. A post-processor decides on the condition of the bearing. Instead of using an FFT, the authors suggest using a high-resolution spectral technique called multiple signal classification (MUSIC) estimator. This technique exhibits a high spectral resolution, while maintaining the stability of the transformation. Measurement results are presented, but no bearing defects are evaluated in this paper.

An approach to finding fault signatures in the motor current spectrum that is specifically targeting applications with lower level platforms is introduced in [90]. Instead of using conventional signal-processing methods such as FFT, MUSIC, or the WT, the authors suggest an approach that is based on the lock-in principle. The proposed technique is capable of accurately detecting the magnitude and phase of a specified frequency component even in a noisy environment with a low signal-to-noise ratio. The measured current I_{meas} is multiplied by the reference signal I_{ref} in order to extract the fault signature. The reference signal can be defined as

$$I_{ref}[n] = I_{ref}\cos(\omega_{ref}n + \varphi_{ref}) \tag{6.47}$$

where n is the sample number; ω_{ref} is the angular frequency of interest (in this application, the frequency related to the bearing defect); and φ_{ref} is the phase offset of the reference signal. All the components in the signal except for the fault component including the fundamental and its harmonics are considered to be noise. The measured current can then be defined as

$$I_{meas}[n] = I_{fault}\cos(\omega_{fault}n + \varphi_{fault}) + \sum I_{noise}\cos(\omega_{noise}n + \varphi_{noise})$$

where n is the sample number; ω_{fault} is the frequency created by the defect (in this application, the frequency is related to the bearing defect); φ_{fault} the phase offset of the fault-producing component; and I_{fault} is the magnitude due to the defect. The product of the two signals yields

$$\begin{aligned} I_{ref}[n]I_{meas}[n] &= I_{ref}I_{fault}\cos(\omega_{ref}n - \omega_{fault}n + \varphi_{ref} - \varphi_{fault}) \\ &\quad + I_{ref}I_{fault}\cos(\omega_{ref}n + \omega_{fault}n + \varphi_{ref} + \varphi_{fault}) \\ &\quad + \sum I_{ref}I_{noise}\{\cos(\omega_{ref}n - \omega_{noise}n + \varphi_{ref} - \varphi_{noise}) \\ &\quad + \cos(\omega_{ref}n + \omega_{noise}n + \varphi_{ref} + \varphi_{noise})\} \end{aligned} \tag{6.48}$$

If the reference frequency is set to the fault-indicating frequency ($\omega_{ref} = \omega_{fault}$) and the signal is low-pass filtered, the following fault indicator I_{FI} is obtained:

$$I_{FI}(\varphi) = K_1\cos(\varphi_{ref} - \varphi_{fault}) + K_2\sum\cos(\varphi_{ref} - \varphi_{noise}) \tag{6.49}$$

The signal-to-noise ratio of the fault signature can be maximized if the phase of the reference signal is set to the same value as the phase of the fault signal. The phase of the fault signal is identified by tracking the correlation of the sinusoidal and cosinusoidal components of the reference signal with the measurement signal. This is accomplished by decomposing (6.47) into sine and cosine and by calculating the arctangent of the correlation ratio. The authors show that the computational requirement is $O(N)$, where N is the number of data points. This computational burden is lower than the one of an FFT for a large number of data points.

Simulation examples evaluate the robustness of the method with respect to parameters such as sampling frequency and the duration of the data sample. A set of experiments is conducted including an outer race bearing fault that is emulated by drilling a hole through the outer raceway. Detection by the proposed method is compared with results from an FFT analysis which are used as a benchmark. Both methods can identify the bearing defect using the measured current. The magnitude of the fault-indicating components obtained by either method match closely for the healthy and the faulty bearing. The results are displayed in Table 6.1.

For the setting of the fault threshold, an adaptive method, which considers the operating condition of the machine, might be necessary since the operating conditions

Table 6.1 Normalized experimental results

Signature frequency		143.3 Hz	167.5 Hz	383.3 Hz	395.4 Hz
Healthy	FFT	−99.55 dB	−94.99 dB	−97.80 dB	−94.84 dB
Healthy	Lock-in	−98.83 dB	−96.17 dB	−98.44 dB	−96.60 dB
Faulty	FFT	−82.72 dB	−87.58 dB	−84.03 dB	−82.95 dB
Faulty	Lock-in	−82.93 dB	−88.02 dB	−84.94 dB	−82.78 dB

have a strong influence on the fault signature. A continuation of this work and an in-depth approach for the determination of the fault threshold are presented in [112,113].

Another popular topic related to inverters and bearings is the investigation of bearing currents. Currents through the bearing lead to a deterioration of its condition and shorten the bearing life [114–117]. An analysis of bearing currents in machines that are operated with hard- and soft-switching inverters is presented in [118]. The authors conclude that there is no significant difference in terms of bearing currents between the two inverter topologies.

6.5.2 Gearbox faults

The data-acquisition system and/or data-processing unit of the inverter platform can also be used for the detection of defects in wind turbine drivetrains. There is a number of defects that can potentially be detected with this technology. This includes mechanical defects of the generator (defective bearings), gearbox defects (damaged gears and bearings), or blade defects. In [119], the investigation is targeted toward defective bearings in the gearbox. The vibration frequencies in rolling element bearings are known from previous research (6.44)–(6.46) and can be adopted for this application. The same applies to the frequencies in the stator current corresponding to the vibration frequencies that can be calculated by using (6.43). A processing-step labeled electrical multiphase imbalance separation technique (eMIST) (MUSIC) is suggested to improve the sensitivity of the method. This method is supposed to remove all the balanced components that are present in a balanced system without any defects and to amplify unbalanced components in a three-phase system. The removal of the balanced components is achieved by the following processing step:

$$V_{eMIST} = V_{ah}^2 + V_{bh}^2 + V_{cm}^2 \tag{6.50}$$

$$I_{eMIST} = I_{ah}^2 + I_{bh}^2 + I_{cm}^2 \tag{6.51}$$

The signals for the A-phase V_{ah} and B-phase V_{bh} are taken from a baseline measurement. The C-phase measurement V_{cm} is taken during the monitoring process under similar operating conditions as the baseline measurement. If the system develops a fault, this fault will be present only in the measurement taken during the monitoring phase (V_{cm}). The fault component introduces an unbalance in the three-phase system. Even though this method can improve the signal-to-noise ratio and amplify the fault signature, there are several practical challenges that can diminish the benefit or even

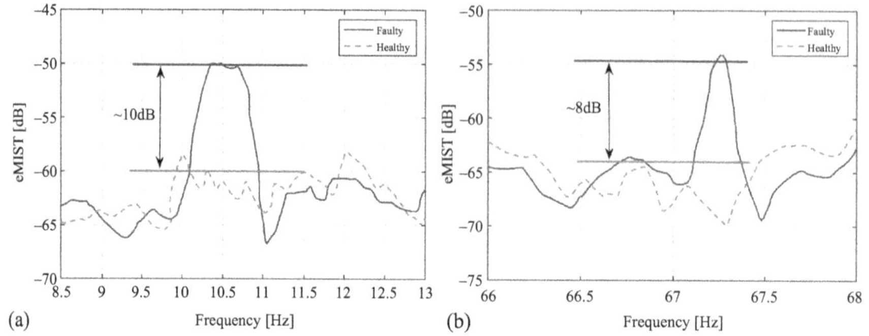

Figure 6.52 eMIST spectrum of the stator current of a 1.5-MW wind turbine: (a) wind turbine with inner race damage on the high-speed intermediate shaft and (b) wind turbine with outer race damage on the planet bearing

compromise the fault detection. One of the main challenges is the proper alignment of the C-phase signal with the other two phases from the baseline measurement. Another source of error can come from small differences in the load level between the baseline and the monitoring phase. Further problems can occur due to the presence of other frequency components that can cause an overlap during the modulation process that is inherent to the eMIST processing steps [(6.50) and (6.51)]. Multiple experimental results are presented for an inner race single point defect on one of the gearbox bearings at the high-speed end. First, a lab validation is presented and then a similar fault is detected in a field application. The turbine with the faulty bearing can be identified through the proposed method. A result for a 1.5 MW wind turbine with an inner-race bearing fault of the high speed intermediate shaft is shown in Figure 6.52(a). The result for a similar wind turbine with an outer-race damage on the planet bearing is displayed in Figure 6.52(b). The baseline data set for the healthy machine was obtained from a turbine without any defects.

The detection of mechanical imbalances such as a blade imbalance via spectral analysis of the stator current is presented in [17]. The technique is applied to wind turbines with direct drives and synchronous generators. A blade defect produces a harmonic frequency in the rotor speed signal at the fundamental rotating frequency. This rotor speed can be monitored by using a stator current measurement since

$$f_1(t) = p \cdot f_r(t) \tag{6.52}$$

where f_1 is the fundamental operating frequency of the stator current; p is the number of pole pairs of the synchronous generator; and f_r is the shaft rotating frequency, i.e., the rotor speed. The problem is that the speed of the turbine and consequently the frequency changes constantly, so that the classical spectral analysis methods like FFT or PSD are not effective for monitoring the fault signature. Additional signal-processing steps are required to extract a reliable fault signature. If the sampling frequency is varied proportionally to the rotor speed, the harmonics in the spectrum

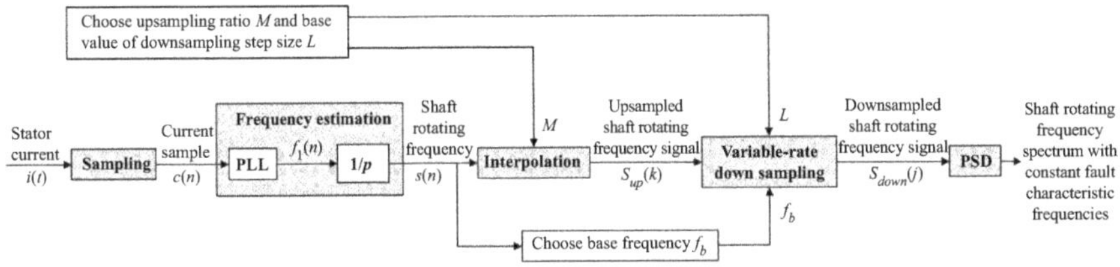

Figure 6.53 Schematic diagram of the stator current-based diagnostic approach for the detection of wind turbine defects

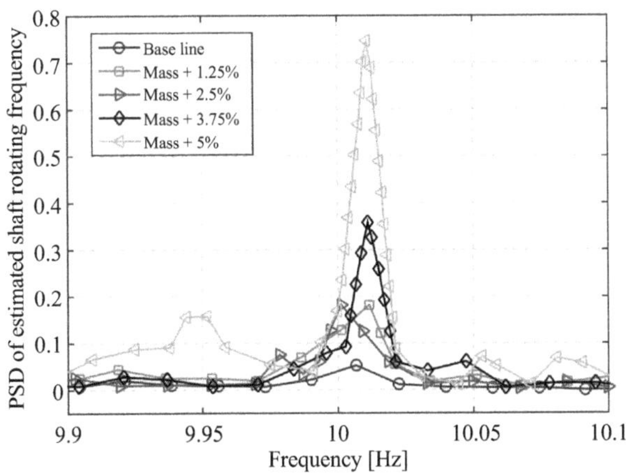

Figure 6.54 Comparison of the 1P-invariant PSD of the estimated shaft-rotating frequency for various blade imbalance scenarios against the baseline case in a frequency range around 1P

will be constant with respect to the shaft speed. To achieve this effect, the signal is sampled at a high rate first, and the rotating speed is determined by using (6.52). Then a base speed is chosen and the signal is resampled so that the ratio between the actual-rotating speed and the sampling frequency is constant. After properly adjusting the sampling of the current measured, the spectral analysis via PSD can be performed and the imbalance condition can be monitored by observing the first-order harmonic, i.e., the base frequency. The approach including the adjustment of the sampling is labeled as invariant PSD, and this process is often referred to as synchronous sampling. The processing steps are shown in Figure 6.53.

Simulation and experimental results are presented for validation. The experimental results for a blade imbalance experiment, in which the mass of one blade is varied by 1.25%, 2.5%, 3.75%, and 5%, respectively, are shown in Figure 6.54.

In this case, 10 Hz was chosen as the base frequency, and the increased amount of imbalance is clearly reflected in an increasing fault signature. In later publications

[120,121], the implementation of the synchronous sampling approach is modified and refined. The computational efficiency has been largely improved, which makes it more feasible for the real-time application on the inverter platform. Furthermore, the application space has expanded to mechanical defects such as rotor eccentricity and bearing faults.

References

[1] H. W. Penrose, "Test methods for determining the impact of motor condition on motor efficiency and reliability," ALL-TEST Pro, LLC, Old Saybrook, CT, Tech. Rep., 2002.

[2] K. Younsi, P. Neti, M. Shah, *et al.*, "On-line capacitance and dissipation factor monitoring of ac stator insulation," *IEEE Transactions on Dielectrics and Electrical Insulation*, vol. 17, no. 5, pp. 1441–1452, 2010.

[3] S. Cheng, Y. Du, J. A. Restrepo, P. Zhang, and T. G. Habetler, "A nonintrusive thermal monitoring method for induction motors fed by closed-loop inverter drives," *IEEE Transactions on Power Electronics*, vol. 27, no. 9, pp. 4122–4131, 2012. [Online]. Available: http://ieeexplore.ieee.org/document/6153070/.

[4] S. Grubic, J. Restrepo, J. M. Aller, B. Lu, and T. G. Habetler, "A new concept for online surge testing for the detection of winding insulation deterioration in low-voltage induction machines," *IEEE Transactions on Industry Applications*, vol. 47, no. 5, pp. 2051–2058, 2011.

[5] S. Nandi, H. A. Toliyat, and X. Li, "Condition monitoring and fault diagnosis of electrical motors – a review," *IEEE Transactions on Energy Conversion*, vol. 20, no. 4, pp. 719–729, 2005.

[6] P. Zhang, Y. Du, T. G. Habetler, and B. Lu, "A survey of condition monitoring and protection methods for medium-voltage induction motors," *IEEE Transactions on Industry Applications*, vol. 47, no. 1, pp. 34–46, 2011.

[7] M. E. H. Benbouzid and G. B. Kliman, "What stator current processing-based technique to use for induction motor rotor faults diagnosis?" *IEEE Transactions on Energy Conversion*, vol. 18, no. 2, pp. 238–244, 2003.

[8] M. E. H. Benbouzid, "A review of induction motors signature analysis as a medium for faults detection," *IEEE Transactions on Industrial Electronics*, vol. 47, no. 5, pp. 984–993, 2000.

[9] F. Filippetti, G. Franceschini, C. Tassoni, and P. Vas, "Recent developments of induction motor drives fault diagnosis using AI techniques," *IEEE Transactions on Industrial Electronics*, vol. 47, no. 5, pp. 994–1004, 2000. [Online]. Available: http://ieeexplore.ieee.org/document/873207/.

[10] G. Singh and S. A. S. A. Kazzaz, "Induction machine drive condition monitoring and diagnostic researcher – a survey," *Electric Power Systems Research*, vol. 64, no. 2, pp. 145–158, 2003. [Online]. Available: http://www.sciencedirect.com/science/article/pii/S0378779602001724.

[11] A. Siddique, G. S. Yadava, and B. Singh, "A review of stator fault monitoring techniques of induction motors," *IEEE Transactions on Energy Conversion*, vol. 20, no. 1, pp. 106–114, 2005.

[12] P. Tavner, "Review of condition monitoring of rotating electrical machines," *IET Electric Power Applications*, vol. 2, no. 4, pp. 215–247, 2008.

[13] S. Grubic, J. M. Aller, B. Lu, and T. G. Habetler, "A survey on testing and monitoring methods for stator insulation systems of low-voltage induction machines focusing on turn insulation problems," *IEEE Transactions on Industrial Electronics*, vol. 55, no. 12, pp. 4127–4136, 2008. [Online]. Available: http://ieeexplore.ieee.org/document/4624549/.

[14] A. Gandhi, T. Corrigan, and L. Parsa, "Recent advances in modeling and online detection of stator interturn faults in electrical motors," *IEEE Transactions on Industrial Electronics*, vol. 58, no. 5, pp. 1564–1575, 2011.

[15] Y. Duan and H. Toliyat, "A review of condition monitoring and fault diagnosis for permanent magnet machines," in *IEEE Power and Energy Society General Meeting, 2012*, 2012. [Online]. Available: http://ieeexplore.ieee.org/document/6345545/.

[16] R. Tallam, T. Habetler, and R. Harley, "Stator winding turn-fault detection for closed-loop induction motor drives," *IEEE Industry Applications Society*, vol. 39, no. 3, pp. 720–724, 2003. [Online]. Available: http://ieeexplore.ieee.org/document/1201539/.

[17] X. Gong and W. Qiao, "Imbalance fault detection of direct-drive wind turbines using generator current signals," *IEEE Transactions on Energy Conversion*, vol. 27, no. 2, pp. 468–476, 2012.

[18] R. R. Schoen, B. K. Lin, T. G. Habetler, J. H. Schlag, and S. Farag, "An unsupervised, on-line system for induction motor fault detection using stator current monitoring," *IEEE Transactions on Industry Applications*, vol. 31, no. 6, pp. 1280–1286, 1995.

[19] X. Z. Gao, S. J. Ovaska, and Y. Dote, "Motor fault detection using Elman neural network with genetic algorithm-aided training," in *Systems, Man, and Cybernetics, 2000 IEEE International Conference on*, vol. 4, 2000, pp. 2386–2392.

[20] Y. L. Murphey, M. A. Masrur, Z. Chen, and B. Zhang, "Model-based fault diagnosis in electric drives using machine learning," *IEEE/ASME Transactions on Mechatronics*, vol. 11, no. 3, pp. 290–303, 2006.

[21] S. Khomfoi and L. M. Tolbert, "Fault diagnostic system for a multilevel inverter using a neural network," *IEEE Transactions on Power Electronics*, vol. 22, no. 3, pp. 1062–1069, 2007.

[22] M. A. S. K. Khan and M. A. Rahman, "Development and implementation of a novel fault diagnostic and protection technique for IPM motor drives," *IEEE Transactions on Industrial Electronics*, vol. 56, no. 1, pp. 85–92, 2009.

[23] S. M. A. Cruz, H. A. Toliyat, and A. J. M. Cardoso, "DSP implementation of the multiple reference frames theory for the diagnosis of stator faults in a DTC induction motor drive," *IEEE Transactions on Energy Conversion*, vol. 20, no. 2, pp. 329–335, 2005.

[24] A. Boglietti, E. Carpaneto, M. Cossale, and S. Vaschetto, "Stator-winding thermal models for short-time thermal transients: definition and validation," *IEEE Transactions on Industrial Electronics*, vol. 63, no. 5, pp. 2713–2721, 2016.

[25] Z. Gao, T. G. Habetler, R. G. Harley, and R. S. Colby, "A sensorless adaptive stator winding temperature estimator for mains-fed induction machines with continuous-operation periodic duty cycles," *IEEE Transactions on Industry Applications*, vol. 44, no. 5, pp. 1533–1542, 2008.

[26] B. Lu, T. G. Habetler, and R. G. Harley, "A nonintrusive and in-service motor-efficiency estimation method using air-gap torque with considerations of condition monitoring," *IEEE Transactions on Industry Applications*, vol. 44, no. 6, pp. 1666–1674, 2008.

[27] B. Lu, T. G. Habetler, and R. G. Harley, "A survey of efficiency-estimation methods for in-service induction motors," *IEEE Transactions on Industry Applications*, vol. 42, no. 4, pp. 924–933, 2006.

[28] J. M. Aller, A. Bueno, and T. Paga, "Power system analysis using space-vector transformation," *IEEE Transactions on Power Systems*, vol. 17, no. 4, pp. 957–965, 2002.

[29] K. D. Hurst and T. G. Habetler, "Sensorless speed measurement using current harmonic spectral estimation in induction machine drives," *IEEE Transactions on Power Electronics*, vol. 11, no. 1, pp. 66–73, 1996.

[30] J. F. Moreno, F. P. Hidalgo, and M. D. Martinez, "Realisation of tests to determine the parameters of the thermal model of an induction machine," *IEE Proceedings – Electric Power Applications*, vol. 148, no. 5, pp. 393–397, 2001.

[31] S.-B. Lee and T. Habetler, "An online stator winding resistance estimation technique for temperature monitoring of line-connected induction machines," *IEEE Transactions on Industry Applications*, vol. 39, no. 3, pp. 685–694, 2003. [Online]. Available: http://ieeexplore.ieee.org/document/1201535/.

[32] P. Zhang, Y. Du, B. Lu, and T. G. Habetler, "A dc signal injection-based thermal protection scheme for soft-starter-connected induction motors," *IEEE Transactions on Industry Applications*, vol. 45, no. 4, pp. 1351–1358, 2009. [Online]. Available: http://ieeexplore.ieee.org/document/4957039/.

[33] P. Zhang, B. Lu, and T. G. Habetler, "An active stator temperature estimation technique for thermal protection of inverter-fed induction motors with considerations of impaired cooling detection," *IEEE Transactions on Industry Applications*, vol. 46, no. 5, pp. 1873–1881, 2010. [Online]. Available: http://ieeexplore.ieee.org/document/5508416/.

[34] P. R. Matic, M. A. Gecic, D. M. Lekic, and D. P. Marcetic, "Thermal protection of vector-controlled IM drive based on dc current injection," *IEEE Transactions on Industrial Electronics*, vol. 62, no. 4, pp. 2082–2089, 2015.

[35] L. He, S. Cheng, Y. Du, R. G. Harley, and T. G. Habetler, "Stator temperature estimation of direct-torque-controlled induction machines via active flux or torque injection," *IEEE Transactions on Power Electronics*, vol. 30, no. 2, pp. 888–899, 2015. [Online]. Available: http://ieeexplore.ieee.org/document/6767114/.

[36] N. Z. Popov, S. N. Vukosavic, and E. Levi, "Motor temperature monitoring based on impedance estimation at PWM frequencies," *IEEE Transactions on Energy Conversion*, vol. 29, no. 1, pp. 215–223, 2014.

[37] N. Popov and S. N. Vukosavic, "Estimator of the rotor temperature of induction machines based on terminal voltages and currents," *IEEE Transactions on Energy Conversion*, vol. 32, no. 1, pp. 155–163, 2016.

[38] G. C. Stone, E. A. Boulter, I. Culbert, and H. Dhirani, *Electrical Insulation for Rotating Machines: Design, Evaluation, Aging, Testing, and Repair.*, M. E. El-Hawary, Ed. Piscataway, NJ: Wiley-IEEE Press, 2004.

[39] R. M. Tallam, S. B. Lee, G. C. Stone, *et al.*, "A survey of methods for detection of stator-related faults in induction machines," *IEEE Transactions on Industry Applications*, vol. 43, no. 4, pp. 920–933, 2007.

[40] S. Williamson and K. Mirzoian, "Analysis of cage induction motors with stator winding faults," *IEEE Transactions on Power Apparatus and Systems*, vol. PAS-104, no. 7, pp. 1838–1842, 1985.

[41] J. L. Kohler, J. Sottile, and F. C. Trutt, "Alternatives for assessing the electrical integrity of induction motors," *IEEE Transactions on Industry Applications*, vol. 28, no. 5, pp. 1109–1117, 1992.

[42] G. B. Kliman, W. J. Premerlani, R. A. Koegl, and D. Hoeweler, "A new approach to on-line turn fault detection in ac motors," in *Industry Applications Conference, 1996. Thirty-First IAS Annual Meeting, IAS '96., Conference Record of the 1996 IEEE*, vol. 1, 1996, pp. 687–693.

[43] F. Trutt, J. Sottile, and J. Kohler, "Online condition monitoring of induction motors," *IEEE Transactions on Industry Applications*, 2002. [Online]. Available: http://ieeexplore.ieee.org/document/1058119/.

[44] S. B. Lee, R. Tallam, and T. Habetler, "A robust, on-line turn-fault detection technique for induction machines based on monitoring the sequence component impedance matrix," *IEEE Transactions on Power Electronics*, 2003. [Online]. Available: http://ieeexplore.ieee.org/document/1198066/.

[45] R. Tallam, T. Habetler, and R. Harley, "Self-commissioning training algorithms for neural networks with applications to electric machine fault diagnostics," *IEEE Transactions on Power Electronics*, vol. 17, no. 6, pp. 1089–1095, 2002. [Online]. Available: http://ieeexplore.ieee.org/document/1159001/.

[46] F. Briz, M. W. Degner, A. Zamarron, and J. M. Guerrero, "Online stator winding fault diagnosis in inverter-fed ac machines using high-frequency signal injection," *IEEE Transactions on Industry Applications*, vol. 39, no. 4, pp. 1109–1117, 2003.

[47] F. Briz, M. W. Degner, A. B. Diez, and J. M. Guerrero, "Online diagnostics in inverter-fed induction machines using high-frequency signal injection," *IEEE Transactions on Industry Applications*, vol. 40, no. 4, pp. 1153–1161, Jul 2004.

[48] F. Briz, M. W. Degner, J. M. Guerrero, and P. Garcia, "Stator windings fault diagnostics of induction machines operated from inverters and soft-starters using high-frequency negative-sequence currents," *IEEE Transactions on Industry Applications*, vol. 45, no. 5, pp. 1637–1646, 2009.

[49] S. M. A. Cruz and A. J. M. Cardoso, "Diagnosis of stator inter-turn short circuits in DTC induction motor drives," *IEEE Transactions on Industry Applications*, vol. 40, no. 5, pp. 1349–1360, 2004.

[50] A. Stavrou, H. G. Sedding, and J. Penman, "Current monitoring for detecting inter-turn short circuits in induction motors," *IEEE Transactions on Energy Conversion*, vol. 16, no. 1, pp. 32–37, 2001.

[51] G. Joksimovic and J. Penman, "The detection of interturn short circuits in the stator windings of operating motors," in *Industrial Electronics Society, 1998. IECON '98. Proceedings of the 24th Annual Conference of the IEEE*, vol. 4, 1998, pp. 1974–1979.

[52] S. Nandi and H. A. Toliyat, "Novel frequency-domain-based technique to detect stator interturn faults in induction machines using stator-induced voltages after switch-off," *IEEE Transactions on Industry Applications*, vol. 38, no. 1, pp. 101–109, 2002.

[53] B. Du, S. Wu, S. Han, and S. Cui, "Interturn fault diagnosis strategy for interior permanent-magnet synchronous motor of electric vehicles based on digital signal processor," *IEEE Transactions on Industrial Electronics*, vol. 63, no. 3, pp. 1694–1706, 2016.

[54] I. F. El-Arabawy, M. I. Masoud, and A. E. K. Mokhtari, "Stator inter-turn faults detection and localization using stator currents and Concordia patterns – neural network applications," in *2007 Compatibility in Power Electronics*, 2007, pp. 1–7.

[55] S. Grubic, J. Restrepo, and T. G. Habetler, "Online surge testing applied to an induction machine with emulated insulation breakdown," *IEEE Transactions on Industry Applications*, vol. 49, no. 3, pp. 1358–1366, 2013.

[56] F. Perisse, P. Werynski, and D. Roger, "A new method for ac machine turn insulation diagnostic based on high frequency resonances," *IEEE Transactions on Dielectrics and Electrical Insulation*, vol. 14, no. 5, 2007. [Online]. Available: http://ieeexplore.ieee.org/document/4339494/.

[57] F. Perisse, D. Mercier, E. Lefevre, and D. Roger, "Robust diagnostics of stator insulation based on high frequency resonances measurements," *IEEE Transactions on Dielectrics and Electrical Insulation*, vol. 16, no. 5, pp. 1496–1502, 2009. [Online]. Available: http://ieeexplore.ieee.org/document/5293965/.

[58] P. Nussbaumer, M. A. Vogelsberger, and T. M. Wolbank, "Induction machine insulation health state monitoring based on online switching transient exploitation," *IEEE Transactions on Industrial Electronics*, vol. 62, no. 3, pp. 1835–1845, 2015. [Online]. Available: http://ieeexplore.ieee.org/document/6915865/.

[59] C. Zoeller, M. A. Vogelsberger, T. M. Wolbank, and H. Ertl, "Impact of sic semiconductors switching transition speed on insulation health state monitoring of traction machines," *IET Power Electronics*, vol. 9, no. 15, pp. 2769–2775, 2016. [Online]. Available: http://ieeexplore.ieee.org/document/7765302/.

[60] P. Nussbaumer, M. A. Vogelsberger, and T. M. Wolbank, "Exploitation of induction machine's high-frequency behavior for online insulation monitoring," in *9th IEEE International Symposium on Diagnostics for Electric Machines, Power Electronics and Drives (SDEMPED)*, 2013. [Online]. Available: http://ieeexplore.ieee.org/document/6645773/.

[61] P. Nussbaumer, T. M. Wolbank, and M. A. Vogelsberger, "Separation of disturbing influences on induction machine's high-frequency behavior to ensure accurate insulation condition monitoring," in *Twenty-Eighth Annual IEEE Applied Power Electronics Conference and Exposition (APEC)*, 2013. [Online]. Available: http://ieeexplore.ieee.org/document/6520445/.

[62] *IEEE Std 1434-2000 – IEEE Guide for the Measurement of Partial Discharges in AC Electric Machinery*, IEEE Std. [Online]. Available: http://ieeexplore.ieee.org/document/6973042/.

[63] *IEEE 286-2000 – IEEE Recommended Practice for Measurement of Power Factor Tip-Up of Electric Machinery Stator Coil Insulation*, IEEE Std. [Online]. Available: http://ieeexplore.ieee.org/document/913583/.

[64] N. Hayakawa and H. Okubo, "Partial discharge characteristics of inverter-fed motor coil samples under ac and surge voltage conditions," *IEEE Electrical Insulation Magazine*, 2005. [Online]. Available: http://ieeexplore.ieee.org/document/1389265/.

[65] H. Okubo, N. Hayakawa, and G. C. Montanari, "Technical development on partial discharge measurement and electrical insulation techniques for low voltage motors driven by voltage inverters," *IEEE Transactions on Dielectrics and Electrical Insulation*, 2007. [Online]. Available: http://ieeexplore.ieee.org/document/4401236/.

[66] S. B. Lee, K. Younsi, and G. Kliman, "An online technique for monitoring the insulation condition of ac machine stator windings," *IEEE Transactions on Energy Conversion*, vol. 20, no. 4, pp. 737–745, 2005. [Online]. Available: http://ieeexplore.ieee.org/document/1546065/.

[67] S. B. Lee, J. Yang, K. Younsi, and R. Bharadwaj, "An online groundwall and phase-to-phase insulation quality assessment technique for ac-machine stator windings," *IEEE Transactions on Industry Applications*, 2006. [Online]. Available: http://ieeexplore.ieee.org/document/1658323/.

[68] P. Zhang, K. Younsi, and P. Neti, "A novel online stator ground-wall insulation monitoring scheme for inverter-fed ac motors," *IEEE Transactions on Industry Applications*, vol. 51, no. 3, pp. 2201–2207, 2014. [Online]. Available: http://ieeexplore.ieee.org/document/6998065/.

[69] J. Yang, S. B. Lee, J. Yoo, S. Lee, Y. Oh, and C. Choi, "A stator winding insulation condition monitoring technique for inverter-fed machines," *IEEE Transactions on Power Electronics*, vol. 22, no. 5, pp. 2026–2033, 2007. [Online]. Available: http://ieeexplore.ieee.org/document/4300912/.

[70] J. Yang, J. Cho, S. B. Lee, J.-Y. Yoo, and H. D. Kim, "An advanced stator winding insulation quality assessment technique for inverter-fed machines," *IEEE Transactions on Industry Applications*, vol. 44, no. 2, 2008. [Online]. Available: http://ieeexplore.ieee.org/document/4475314/.

[71] *IEEE 43-2013 – IEEE Recommended Practice for Testing Insulation Resistance of Electric Machinery*, IEEE Std. [Online]. Available: http://ieeexplore.ieee.org/document/6754111/.

[72] J. Aguayo, A. Claudio, L. Vela, S. Lesecq, and A. Barraud, "Stator winding fault detection for an induction motor drive using actuator as sensor principle," in *IEEE 34th Annual Power Electronics Specialist Conference, 2003. PESC '03*. IEEE, 2003. [Online]. Available: http://ieeexplore.ieee.org/document/1216783/.

[73] Q. T. An, L. Sun, and L. Z. Sun, "Current residual vector-based open-switch fault diagnosis of inverters in PMSM drive systems," *IEEE Transactions on Power Electronics*, vol. 30, no. 5, pp. 2814–2827, 2015.

[74] C. Choi and W. Lee, "Design and evaluation of voltage measurement-based sectoral diagnosis method for inverter open switch faults of permanent magnet synchronous motor drives," *IET Electric Power Applications*, vol. 6, no. 8, pp. 526–532, 2012.

[75] R. Peuget, S. Courtine, and J. P. Rognon, "Fault detection and isolation on a PWM inverter by knowledge-based model," *IEEE Transactions on Industry Applications*, vol. 34, no. 6, pp. 1318–1326, 1998.

[76] A. M. S. Mendes, M. B. Abadi, and S. M. A. Cruz, "Fault diagnostic algorithm for three-level neutral point clamped ac motor drives, based on the average current Park's vector," *IET Power Electronics*, vol. 7, no. 5, pp. 1127–1137, 2014.

[77] D. Kastha and B. K. Bose, "Investigation of fault modes of voltage-fed inverter system for induction motor drive," *IEEE Transactions on Industry Applications*, vol. 30, no. 4, pp. 1028–1038, 1994.

[78] M. A. Masrur, Z. Chen, and Y. Murphey, "Intelligent diagnosis of open and short circuit faults in electric drive inverters for real-time applications," *IET Power Electronics*, vol. 3, no. 2, pp. 279–291, 2010.

[79] K. H. Kim, D. U. Choi, B. G. Gu, and I. S. Jung, "Fault model and performance evaluation of an inverter-fed permanent magnet synchronous motor under winding shorted turn and inverter switch open," *IET Electric Power Applications*, vol. 4, no. 4, pp. 214–225, 2010.

[80] D. R. Espinoza-Trejo, D. U. Campos-Delgado, E. Barcenas, and F. J. Martinez-Lopez, "Robust fault diagnosis scheme for open-circuit faults in voltage source inverters feeding induction motors by using non-linear proportional-integral observers," *IET Power Electronics*, vol. 5, no. 7, pp. 1204–1216, 2012.

[81] D. R. Espinoza-Trejo, D. U. Campos-Delgado, G. Bossio, E. Barcenas, J. E. Hernandez-Diez, and L. F. Lugo-Cordero, "Fault diagnosis scheme for open-circuit faults in field-oriented control induction motor drives," *IET Power Electronics*, vol. 6, no. 5, pp. 869–877, 2013.

[82] J. O. Estima and A. J. M. Cardoso, "A new algorithm for real-time multiple open-circuit fault diagnosis in voltage-fed PWM motor drives by the reference current errors," *IEEE Transactions on Industrial Electronics*, vol. 60, no. 8, pp. 3496–3505, 2013.

[83] N. M. A. Freire, J. O. Estima, and A. J. M. Cardoso, "A voltage-based approach without extra hardware for open-circuit fault diagnosis in closed-loop PWM ac regenerative drives," *IEEE Transactions on Industrial Electronics*, vol. 61, no. 9, pp. 4960–4970, 2014.

[84] J. H. Choi, S. Kim, D. S. Yoo, and K. H. Kim, "A diagnostic method of simultaneous open-switch faults in inverter-fed linear induction motor drive for reliability enhancement," *IEEE Transactions on Industrial Electronics*, vol. 62, no. 7, pp. 4065–4077, 2015.

[85] I. Jlassi, J. O. Estima, S. K. E. Khil, N. M. Bellaaj, and A. J. M. Cardoso, "A robust observer-based method for IGBTs and current sensors fault diagnosis in voltage-source inverters of PMSM drives," *IEEE Transactions on Industry Applications*, vol. 53, no. 3, pp. 2894–2905, 2016.

[86] K. Lee, J. Hong, K. W. Lee, S. B. Lee, and E. J. Wiedenbrug, "A stator-core quality-assessment technique for inverter-fed induction machines," *IEEE Transactions on Industry Applications*, vol. 46, no. 1, pp. 213–221, 2010.

[87] S. B. Lee, J. Yang, J. Hong, *et al.*, "A new strategy for condition monitoring of adjustable speed induction machine drive systems," *IEEE Transactions on Power Electronics*, vol. 26, no. 2, pp. 389–398, 2011.

[88] G. Kliman and J. Stein, "Induction motor fault detection via passive current monitoring," in *International Conference in Electrical Machines*, Cambridge, MA, 1990, pp. 13–17.

[89] M. E. H. Benbouzid, M. Vieira, and C. Theys, "Induction motors' faults detection and localization using stator current advanced signal processing techniques," *IEEE Transactions on Power Electronics*, vol. 14, no. 1, pp. 14–22, 1999.

[90] B. Akin, U. Orguner, H. A. Toliyat, and M. Rayner, "Phase-sensitive detection of motor fault signatures in the presence of noise," *IEEE Transactions on Industrial Electronics*, vol. 55, no. 6, pp. 2539–2550, 2008.

[91] S. M. A. Cruz, A. J. M. Cardoso, and H. A. Toliyat, "Diagnosis of stator, rotor and airgap eccentricity faults in three-phase induction motors based on the multiple reference frames theory," in *38th IAS Annual Meeting on Conference Record of the Industry Applications Conference, 2003*, vol. 2, 2003, pp. 1340–1346.

[92] W. le Roux, R. G. Harley, and T. G. Habetler, "Detecting rotor faults in low power permanent magnet synchronous machines," *IEEE Transactions on Power Electronics*, vol. 22, no. 1, pp. 322–328, 2007. [Online]. Available: http://ieeexplore.ieee.org/document/4052417/.

[93] D. Hyun, J. Hong, S. B. Lee, *et al.*, "Automated monitoring of airgap eccentricity for inverter-fed induction motors under standstill conditions," *IEEE Transactions on Industry Applications*, vol. 47, no. 3, pp. 1257–1266, 2011.

[94] J. Hong, S. B. Lee, C. Kral, and A. Haumer, "Detection of airgap eccentricity for permanent magnet synchronous motors based on the d-axis inductance," *IEEE Transactions on Power Electronics*, vol. 27, no. 5, pp. 2605–2612, 2012.

[95] J. Hong, D. Hyun, S. B. Lee, and C. Kral, "Offline monitoring of airgap eccentricity for inverter-fed induction motors based on the differential inductance," *IEEE Transactions on Industry Applications*, vol. 49, no. 6, pp. 2533–2542, 2013.

[96] I. Kerszenbaum and C. F. Landy, "The existence of large inter-bar currents in three phase squirrel cage motors with rotor-bar and/or end-ring faults," *IEEE Transactions on Power Apparatus and Systems*, vol. PAS-103, no. 7, pp. 1854–1862, 1984.

[97] G. B. Kliman, R. A. Koegl, J. Stein, R. D. Endicott, and M. W. Madden, "Noninvasive detection of broken rotor bars in operating induction motors," *IEEE Transactions on Energy Conversion*, vol. 3, no. 4, pp. 873–879, 1988.

[98] N. M. Elkasabgy, A. R. Eastham, and G. E. Dawson, "Detection of broken bars in the cage rotor on an induction machine," *IEEE Transactions on Industry Applications*, vol. 28, no. 1, pp. 165–171, 1992.

[99] R. F. Walliser and C. F. Landy, "Determination of interbar current effects in the detection of broken rotor bars in squirrel cage induction motors," *IEEE Transactions on Energy Conversion*, vol. 9, no. 1, pp. 152–158, 1994.

[100] M. Haji and H. A. Toliyat, "Pattern recognition – a technique for induction machines rotor broken bar detection," *IEEE Transactions on Energy Conversion*, vol. 16, no. 4, pp. 312–317, 2001.

[101] A. M. da Silva, R. J. Povinelli, and N. A. O. Demerdash, "Induction machine broken bar and stator short-circuit fault diagnostics based on three-phase stator current envelopes," *IEEE Transactions on Industrial Electronics*, vol. 55, no. 3, pp. 1310–1318, 2008.

[102] R. J. Povinelli, J. F. Bangura, N. A. O. Demerdash, and R. H. Brown, "Diagnostics of bar and end-ring connector breakage faults in polyphase induction motors through a novel dual track of time-series data mining and time-stepping coupled FE-state space modeling," *IEEE Transactions on Energy Conversion*, vol. 17, no. 1, pp. 39–46, 2002.

[103] M. Drif and A. J. M. Cardoso, "The use of the instantaneous-reactive-power signature analysis for rotor-cage-fault diagnostics in three-phase induction motors," *IEEE Transactions on Industrial Electronics*, vol. 56, no. 11, pp. 4606–4614, 2009.

[104] S. K. Kim and J. K. Seok, "High-frequency signal injection-based rotor bar fault detection of inverter-fed induction motors with closed rotor slots," *IEEE Transactions on Industry Applications*, vol. 47, no. 4, pp. 1624–1631, 2011.

[105] B. Kim, K. Lee, J. Yang, S. B. Lee, E. J. Wiedenbrug, and M. R. Shah, "Automated detection of rotor faults for inverter-fed induction machines under standstill conditions," *IEEE Transactions on Industry Applications*, vol. 47, no. 1, pp. 55–64, 2011.

[106] J.-R. R. Ruiz, J. A. Rosero, A. G. Espinosa, and L. Romeral, "Detection of demagnetization faults in permanent-magnet synchronous motors under nonstationary conditions," *IEEE Transactions on Magnetics*, vol. 45, no. 7, pp. 2961–2969, 2009. [Online]. Available: http://ieeexplore.ieee.org/document/5089909/.

[107] A. G. Espinosa, J. A. Rosero, J. Cusidó, L. Romeral, and J. A. Ortega, "Fault detection by means of Hilbert–Huang transform of the stator current in a PMSM with demagnetization," *IEEE Transactions on Energy Conversion*, vol. 25, no. 2, pp. 312–318, 2010. [Online]. Available: http://ieeexplore.ieee.org/abstract/document/5404255/.

[108] J.-C. Urresty, J.-R. Riba, M. Delgado, and L. Romeral, "Detection of demagnetization faults in surface-mounted permanent magnet synchronous motors by means of the zero-sequence voltage component," *IEEE Transactions on Energy Conversion*, vol. 27, no. 1, pp. 42–51, 2012. [Online]. Available: http://ieeexplore.ieee.org/document/6104129/.

[109] J. Hong, D. Hyun, S. B. Lee, J. Y. Yoo, and K. W. Lee, "Automated monitoring of magnet quality for permanent-magnet synchronous motors at standstill," *IEEE Transactions on Industry Applications*, vol. 46, no. 4, pp. 1397–1405, 2010.

[110] S. A. McInerny and Y. Dai, "Basic vibration signal processing for bearing fault detection," *IEEE Transactions on Education*, vol. 46, no. 1, pp. 149–156, 2003.

[111] R. R. Schoen, T. G. Habetler, F. Kamran, and R. G. Bartfield, "Motor bearing damage detection using stator current monitoring," *IEEE Transactions on Industry Applications*, vol. 31, no. 6, pp. 1274–1279, 1995.

[112] S. Choi, B. Akin, M. M. Rahimian, and H. A. Toliyat, "Implementation of a fault-diagnosis algorithm for induction machines based on advanced digital-signal-processing techniques," *IEEE Transactions on Industrial Electronics*, vol. 58, no. 3, pp. 937–948, 2011.

[113] S. Choi, B. Akin, M. M. Rahimian, and H. A. Toliyat, "Performance-oriented electric motors diagnostics in modern energy conversion systems," *IEEE Transactions on Industrial Electronics*, vol. 59, no. 2, pp. 1266–1277, 2012.

[114] J. Erdman, R. Kerkman, D. Schlegel, and G. Skibinski, "Effect of PWM inverters on ac motor bearing currents and shaft voltages," *IEEE Transactions on Industry Applications*, vol. 32, no. 2, pp. 250–259, 1996. [Online]. Available: http://ieeexplore.ieee.org/document/491472/.

[115] Y. Jiang, W. Wu, Y. He, H. S.-H. Chung, and F. Blaabjerg, "New passive filter design method for overvoltage suppression and bearing currents mitigation in a long cable based PWM inverter-fed motor drive system," *IEEE Transactions on Power Electronics*, vol. 32, no. 10, pp. 7882–7893, 2017. [Online]. Available: http://ieeexplore.ieee.org/document/7782436/.

[116] R. K. Singleton, E. G. Strangas, and S. Aviyente, "The use of bearing currents and vibrations in lifetime estimation of bearings," *IEEE Transactions on Industrial Informatics*, vol. 13, no. 3, pp. 1301–1309, 2017. [Online]. Available: http://ieeexplore.ieee.org/document/7795228/.

[117] M. Asefi and J. Nazarzadeh, "Survey on high-frequency models of PWM electric drives for shaft voltage and bearing current analysis," *IET Electrical Systems in Transportation*, vol. 7, no. 3, pp. 179–189, 2017. [Online]. Available: http://ieeexplore.ieee.org/document/8019959/.

[118] S. Chen and T. A. Lipo, "Bearing currents and shaft voltages of an induction motor under hard- and soft-switching inverter excitation," *IEEE Transactions on Industry Applications*, vol. 34, no. 5, pp. 1042–1048, 1998.

[119] P. Zhang and P. Neti, "Detection of gearbox bearing defects using electrical signature analysis for doubly fed wind generators," *IEEE Transactions on Industry Applications*, vol. 51, no. 3, pp. 2195–2200, 2015. [Online]. Available: http://ieeexplore.ieee.org/document/6998076/.

[120] X. Gong and W. Qiao, "Current-based mechanical fault detection for direct-drive wind turbines via synchronous sampling and impulse detection," *IEEE Transactions on Industrial Electronics*, vol. 62, no. 3, pp. 1693–1702, 2015. [Online]. Available: http://ieeexplore.ieee.org/document/6926835/.

[121] X. Jin, W. Qiao, Y. Peng, F. Cheng, and L. Qu, "Quantitative evaluation of wind turbine faults under variable operational conditions," *IEEE Transactions on Industry Applications*, vol. 52, no. 3, pp. 2061–2069, 2016. [Online]. Available: http://ieeexplore.ieee.org/document/7386632/.

Chapter 7
Battery storage

Andres Salazar Llinas[1], Antonio Ginart[2], and Javad Mohammadpour Velni[3]

7.1 Introduction

7.1.1 Batteries principle of operation

In the last decade, the high penetration of renewable energy sources and the accelerated demand for electric vehicles (EVs) has resulted in the batteries to play a critical role in the worldwide energy future.

The chemical energy contained in batteries active materials is converted into electrical energy through an electrochemical oxidation–reduction (redox) reaction. This type of reaction involves the transfer of electrons from one material to another through an electrical circuit.

The operating principle of a single cell (the basic building block of a battery) is visualized in Figure 7.1. Here, two electrodes (anode and cathode) are submerged in an electrolyte component, which provides the medium for the transfer of charge. During the electrochemical reaction, the anode is oxidized while it is giving up electrons to the external circuit; meanwhile, the cathode is reduced while accepting electrons from the external circuit. Depending on the type of electrolyte and cathode, the process can be reversible or not. Non-rechargeable batteries are known as primary batteries, whereas rechargeable batteries are called secondary batteries.

The general advantages of primary batteries are good shelf life, high energy density at low to moderate discharge rates, and low maintenance [1]. Secondary batteries are characterized (in addition to their ability to be recharged) by high power density, high discharge rate, flat discharge curves, and good low-temperature performance [1]. Secondary batteries can also be classified by the type of chemistry they use. Among them, most common are as follows:

- Lead-acid batteries: This was the first battery type that could be recharged by passing a reverse current through it. The lead-acid battery uses lead dioxide

[1]Sonnen Inc., USA
[2]Smart Wires Inc., USA
[3]School of Electrical and Computer Engineering, University of Georgia, USA

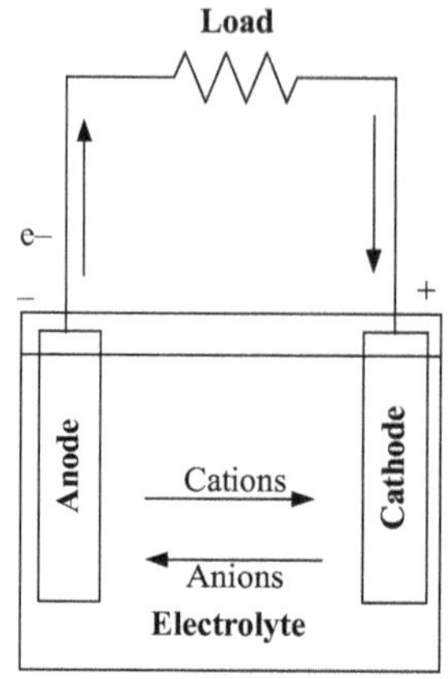

Figure 7.1 *Basic operation of a single battery cell during discharge*

(PbO_2) as the active material of the positive electrode and metallic lead (Pb) in a high-surface-area porous structure as the negative active material. The electrolyte is a sulfuric acid solution:

$$Pb + PbO_2 + 2H_2SO_4 \xrightarrow{\text{discharge}} 2PbSO_4 + 2H_2O$$

$$Pb + PbO_2 + 2H_2SO_4 \xleftarrow{\text{charge}} 2PbSO_4 + 2H_2O$$

Lead-acid batteries can be found in energy storage systems (ESSs), emergency power systems, and electric and hybrid vehicles. They are also used in combustion engine vehicles' lighting and engine ignition circuits.

- Nickel–cadmium batteries: This is the most popular type of alkaline secondary batteries. The charge/discharge mechanism in the alkaline electrolyte involves only the transport of oxygen or hydroxyl ions from one electrode to the other; hence, the concentration of the electrolyte does not change during charge and discharge. Nickel–cadmium (NiCd) batteries are found in different design forms and sizes, such as sealed pocket plate and vented pocket plate. The reactions of charge and discharge can be illustrated by the following simplified equations:

$$2NiOOH + 2H_2O + Cd \xrightarrow{\text{discharge}} 2Ni(OH)_2 + Cd(OH)_2$$

$$2NiOOH + 2H_2O + Cd \xleftarrow{\text{charge}} 2Ni(OH)_2 + Cd(OH)_2$$

The main application of NiCd is with cordless electronic tools. It is also used in applications, such as aircraft engine starting, as well as communications and electronics equipment.

- Lithium-ion batteries: These are the most popular type of secondary batteries nowadays. Lithium-ion (Li-ion) batteries are comprised of cells that employ lithium intercalation compounds as the positive and negative materials. The positive electrode material is typically a metal oxide with a layered structure, such as lithium cobalt oxide ($LiCoO_2$), or a material with a tunneled structure, such as lithium manganese oxide ($LiMn_2O_4$), on a current collector of aluminum foil. The negative electrode material is typically a graphitic carbon, also a layered material, on a copper current collector. In the charge/discharge process, lithium ions are inserted or extracted from interstitial space between atomic layers within the active materials.

Applications for lithium-ion batteries range from commercial electronics, such as cell phones and laptops, to EVs and ESSs.

This chapter focuses on lithium-ion batteries. It also covers relevant aspects of this technology, focusing on the impedance measurement as a diagnostic tool mechanism for detecting aging on this type of batteries. Furthermore, we divide such mechanisms into online and offline methods based on their means of operation during the service time of the battery.

7.1.1.1 Important parameters of batteries

- Capacity. This measures the amount of charge that the battery can deliver at the rated voltage. It is measured in ampere-hours (A h). The energy capacity relates the capacity to the battery voltage, and it is measured in watt-hours (W h).
- C-rate. This is a measure of the rate of the battery's discharge relative to its capacity. It is defined as the multiple of the current over the discharge current that the battery can sustain over 1 h. The C-rate of 1 for a 10 A h battery corresponds to a discharge current of 10 A over 1 h.
- SoC. This measures the percentage of the battery capacity available for charging or discharging.
- Power. This describes the capability of the battery for charge or discharge at a specific current level. There are two parameters to describe power: (1) power density: power per volume and (2) specific power: power per weight. Depending on the application, one of these might be more important than the other. For example, for a cell phone, the power density is more important, whereas for an EV, specific power has more value.
- Cycle lifetime aging. It is defined as the number of charging and discharging cycles after which the battery capacity drops below 80% of the nominal value.
- Calendar life aging. Calendar aging comprises all aging processes that lead to a degradation of a battery cell independent of charge–discharge cycling.

7.1.2 Li-ion batteries

Lithium-ion cells are composed of four components: a positive electrode, a negative electrode, an electrolyte, and a separator in between them (Figure 7.2). The positive electrode is made of a metal oxide material (such as lithium cobalt oxide $LiCoO_2$), a material with a tunneled structure (such as lithium manganese oxide $LiMn_2O_4$), or a material with an olivine structure (such as lithium iron phosphate $LiFePO_4$). The

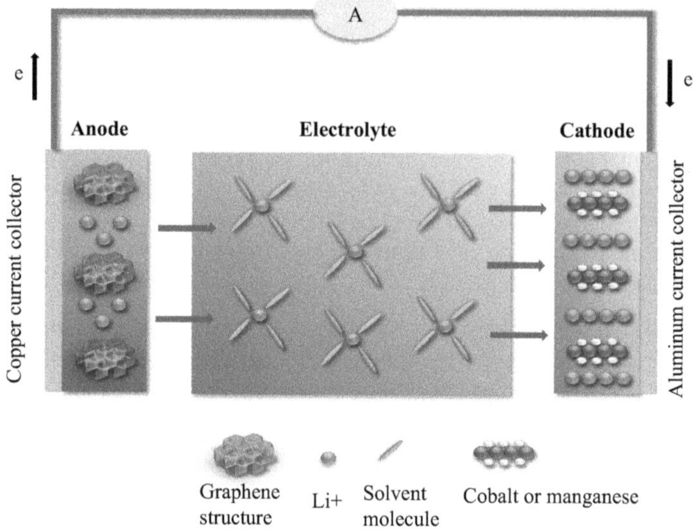

Figure 7.2 Discharging process of a lithium-ion cell. Adapted, with permission, from Reference [2]

negative electrode is built up of carbon (graphite), and the electrolyte varies depending on the type of battery technology. The separator prevents physical contact between the anode and the cathode, while facilitating ion transport in the cell [2,3]. Throughout the charge process of the battery, the lithium positive electrode pulls out some of its lithium ions, which move through the electrolyte to reach the negative electrode and remain there. When the cell is discharging, the lithium ions move back across the electrolyte to the positive electrode, producing the energy required by the load connected to the battery. In both cases, electrons flow in the opposite direction from the ions around the outer circuits. Electrons do not flow through the electrolyte, which behaves as an insulating barrier for them. The flows of ions (through the electrolyte) and electrons (around the external circuit, in the opposite direction) are interconnected processes, and if either one of them stops, the other also stops. If ions stop moving through the electrolyte because the battery completely discharges, then the electrons cannot move through the outer circuit either, so the power is lost. Furthermore, at no-load conditions, the electron flow stops and consequently the flow of ions also stops, forcing the battery to stop discharging [4].

Evolution of cathodes technologies: It took around 30 years to reach the basis of what we know today as commercial rechargeable high-energy density Li-ion batteries. The research of different compounds and materials to improve the efficiency and power density of the cell started in the early 1970s, when Matsushita introduced a lithium–carbon monofluoride (Li–CFx) primary cell. In 1975, Sanyo was the first company to commercialize lithium–manganese dioxide cells (Li–Mn_2). It was not a rechargeable cell but was able to produce a voltage of 3.5 V. The basic usage of these cells was for cameras and memory backup applications. In the following years, great efforts were made to convert lithium cells into rechargeable cells with high energy density. Different materials were studied, without achieving any competitive advantage, except by the polyacene (PAS) battery, for which the main use was in the production of coin cells for memory backups [2]. Between 1973 and 1976, Whittingham at Exxon developed a novel material for the cathode based on a titanium disulfide cathode and started the principle of lithium-ion rechargeable batteries, which he called intercalation compounds. Exxon commercialized this first rechargeable battery with a lithium–aluminum anode, but its low-voltage profile (only around 2 V) made it insufficient for many applications [3]. In 1980, Goodenough developed the Li–cobalt oxide as a cathode material at Oxford. This new compound offered a higher energy density than lithium-ion cells [2,5]. The applications and market were not foreseen at that time, and it was only in 1991 when Sony, using the developments made by Yoshino [4], started the commercialization of the first rechargeable $LiCoO_2$ battery. The cell was based on the Goodenough's discovery but with the addition of a carbonized material in the anode. Another cathode structure was developed in 1983 at Oxford within the Goodenough research group. Later, Thackeray developed the structure called spinel and used a compound based on manganese dioxide (MnO_2) [6]. This new compound would offer price advantages over the cobalt compound; however, its lower capacity and stability issues gave priority to $LiCoO_2$, enabling the latter to take over the market. In 1997, Goodenough, this time with the University of Texas at Austin, developed a new material for the cathode: Li-ion iron phosphate ($LiFePO_4$) (Figure 7.2). This material gave the cell better stability, flattening the discharge characteristic profile and bringing the following advantages [7]:

- High output performance with standard discharge for 2–5C and continuous discharge high current capacity of up to 10C and the instantaneous discharge pulse up to 20C.
- Good performance is observed at high temperatures from 65 to 95 °C keeping the battery in good safe condition.
- It shows excellent life cycles as after 6,000 cycles also it shows discharge capacity to be above 80% [10].
- It gets quickly charged with much less time compared to other batteries.
- It is an environment-friendly battery which does not produce any waste.

Although the advances in new materials for anode and cathode are being made in different laboratories around the world, commercially speaking, $LiCoO_2$ and $LiFePO_4$ are the leading technologies for EV and grid ESSs.

7.1.3 High power applications for Li-ion batteries

7.1.3.1 Electrical vehicles

In the last decade, the demand for lithium-ion batteries for EVs has increased almost 100%. This is mainly due to the progress in technology that has allowed for the improvement in energy density, making it possible to travel greater distances without

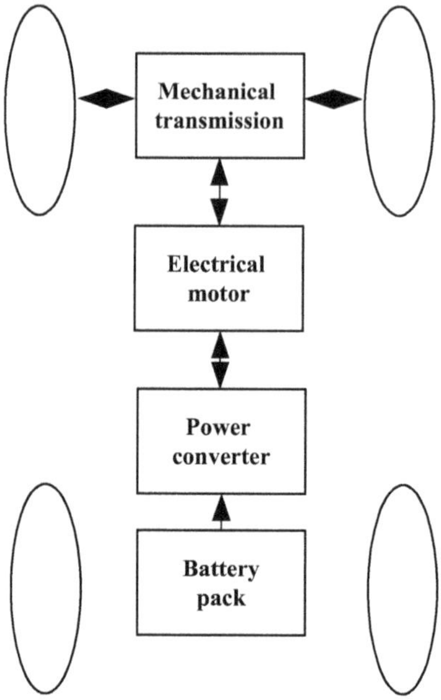

Figure 7.3 Block diagram of an EV power conversion

compromising costs and/or vehicle size. A basic block diagram of an EV application with lithium-ion batteries is shown in Figure 7.3. Here, the battery pack constitutes the main source of power for the vehicle. The electric motor replaces the combustion motor, and a power converter (DC/DC converter) is responsible for providing adequate power levels to the motor, based on different mechanical conditions such as acceleration, torque, speed, etc.

7.1.3.2 Residential energy storage

Residential AC nano-grids are playing an important role in the expandability of distributed renewable energy generation. These types of systems are generally composed of a battery ESS (BESS), photovoltaic (PV) systems, and interconnection and protection components (see Figure 7.4). In recent years, BESSs have been fabricated using lithium-ion (Li-ion) batteries [8]. Li-ion batteries can act in two ways: first, as a buffer between the existing PV system installed in homes, accumulating surplus energy and minimizing the grid energy consumption and second, as a backup system, where in case of power grid outages, the energy consumed by the home is provided by the batteries. In both cases, batteries interact with a sinusoidal power generation source through a power converter. In the residential US market, the power converters in charge of this interaction are single-phase DC/AC converters.

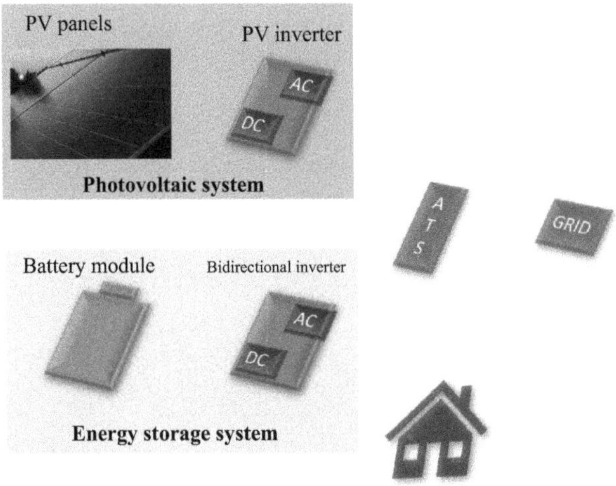

Figure 7.4 Schematic of a typical AC nano-grid for residential applications

7.2 Electrical model of Li-ion batteries

Figure 5a illustrates a suitable AC impedance model for a battery, with specific characteristics parameters for Li-ion. The components of this circuit are as follows: (1) parasitic inductor L_e; (2) ohmic resistance R_o; (3) charge transfer resistance R_{ct}; (4) double-layer capacitor C_{dl} as a result of activation polarization; and (5) Warburg impedance Z_w as a result of concentration polarization [9]. The battery AC impedance Z_{ac} can then be expressed as follows:

$$Z_{ac}(\omega) = jL_e\omega + R_o + (R_{ct} + Z_w)\|\frac{1}{jC_{dl}\omega}. \tag{7.1}$$

A Nyquist plot for the battery impedance is shown in Figure 7.5b. Normally, this plot is obtained by an electrochemical impedance spectroscopy (EIS) analysis, measuring the phase displacement between current and voltage when an oscillatory current, at different frequencies, is injected into the battery. The parameters in (7.1) can be estimated by different techniques as described in [10,11]. In the middle frequencies, the effect of the Warburg impedance and the inductance component are not that evident; hence, for injected currents with frequencies in that range, the AC impedance can be modeled as

$$Z_{ac}(\omega) = R_o + \left(R_{ct}\|\frac{1}{jC_{dl}\omega}\right). \tag{7.2}$$

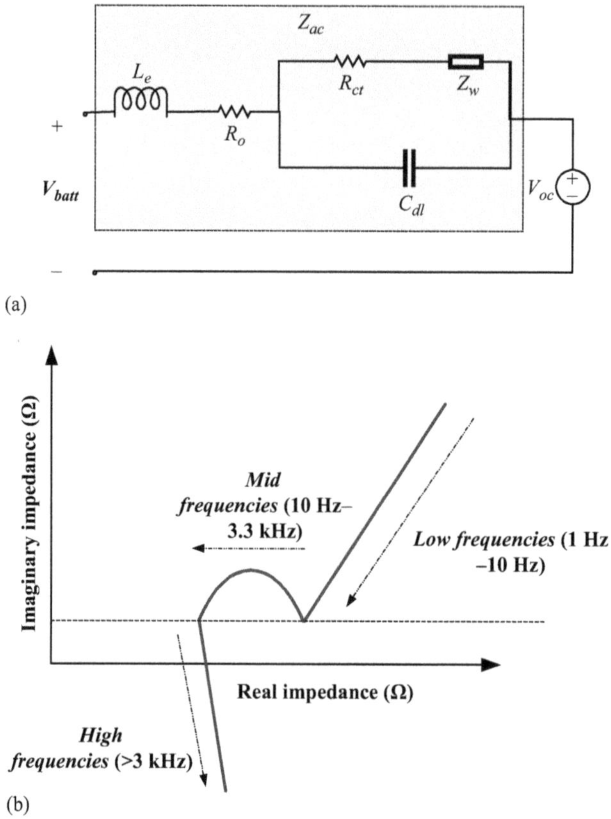

Figure 7.5 Li-ion battery model: (a) Li-ion cell electrical circuit representation and (b) Nyquist plot representation for the AC-battery impedance

7.3 Aging of Li-ion batteries

We distinguish between two types of aging happening in a battery: cycle aging, occurring mainly due to regular battery operation, and calendar aging, occurring even if the battery is not under operation. Some of the consequences of aging in Lithium-ion batteries are described below.

Corrosion in the copper collector may lead to contact loss and hence to an increase in the contact resistance. At the same time, the anode will start changing its morphological structure due to cycle aging. Due to high charging/discharging currents in combination with low temperatures, the Li+ ions cannot intercalate fast enough into the anode material, and lithium deposits are formed on the graphite surface. The growth of these deposits and the formation of dendrites can destroy the separator, leading to thermal runaway in the cell. Over the cycle life of the battery, formation of a solid–electrolyte interphase (SEI) occurs typically at the negative electrode during recharging. Initially, SEI formation protects the electrode against solvent decomposition at large negative voltage, but over time it leads to a gradual decrease in capacity as the SEI layer thickens. Above a certain threshold, films become resistive, which is considered an aging effect. On the other hand, decomposition of the SEI can result in lithium corrosion and thus in irreversible energy loss. Especially at high temperatures (≥60 °C), the SEI layer begins to decompose, causing the anode's passivation to fail. Similarly to the anode, the current collector suffers from corrosion; furthermore, changes in the cathode morphology and decomposition of the binder are some of the effects resulting from cycle aging of the cell [12].

The quantification of the current battery life is described by the state of health (SOH). This factor can be studied from two different points of view: first, by comparing the actual capacity of the battery to the capacity of a new battery and expressing this in a percentage. Normally, when the capacity decreases to 80% of the manufacturer's rated capacity, the battery is considered to be at the end of its operational life. Second point of view is achieved by means of measuring the internal resistance. An increase of the internal resistance resulting from aging effects in the cell leads to a reduction of the battery's power capabilities.

7.3.1 Method for detection of aging in batteries using impedance measurement

It has been shown in previous studies that the real (or ohmic) part of the battery increases with the increase of battery life cycles [10,13,14]. Figure 7.6 shows a typical Nyquist plot of a Li-ion battery at the beginning of and end of its life. This illustrates that it is possible to examine the battery lifetime by measuring the battery impedance.

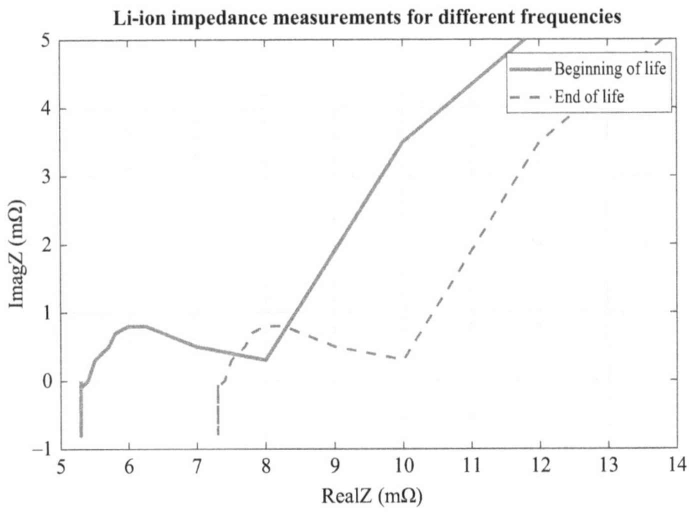

Figure 7.6 AC-battery impedance before and after aging

There exist different ways to measure the battery impedance in order to determine the SOH. The majority of those methods are based on the measure of voltage and current in the battery and the calculation of the battery impedance as

$$Z(\omega) = \frac{V(\omega)}{I(\omega)}. \tag{7.3}$$

We can distinguish between two methodologies for measuring the battery impedance. Offline methods are based on a parameter estimation of the battery packs or cells while they are not operational. The measurement occurs after the battery is neither charging nor discharging, and the parameters are fitted using the battery electrical model. On the other hand, online methods are used while the battery is operating, meaning that non-interruption of the charge/discharge process is needed. These methods are more suitable for implementation as part of the battery management system algorithms. The aging is then detected by comparing the baseline impedance measurement at the beginning of the battery life against the obtained impedance calculation, as shown in (7.4), where R_{eol} is the resistance at the end of life, R_{bol} is the resistance at the beginning of life, and R is the measured resistance.

$$SOH_R = \frac{R_{eol} - R}{R_{eol} - R_{bol}}. \tag{7.4}$$

7.3.1.1 Offline measurement methods

Electrochemical impedance spectroscopy

One of the most common methods for measuring the internal impedance of Li-ion batteries is the EIS. EIS is a nonintrusive and reliable method that can be applied for characterizing and modeling the dynamic behavior of Li-ion batteries [10]. The technique consists of applying a sinusoidal current or voltage of a certain amplitude and frequency. Then, the phase shift and amplitude of the output current or voltage is measured. This process is repeated for different frequencies, and finally the impedance spectrum of the battery is obtained. At very high frequencies, the impedance spectra show an inductive behavior, while at low frequencies, the impedance has a capacitive behavior that corresponds to the diffusion process. The small semicircle in the spectra represents the presence of the SEI layer; the larger semicircle is associated with the charge transfer process and the double-layer capacitance. An electrical model of the impedance is obtained after the measured Nyquist diagram is fitted by a complex nonlinear least square algorithm.

DC current pulses

This is a more straightforward technique which consists in applying a current pulse ΔI and measuring the resulting change in the output voltage ΔV. The impedance of the battery is obtained by applying the Ohm's law as

$$R = \frac{\Delta V}{\Delta I} = \frac{V_1 - V_0}{I_1 - I_0}. \tag{7.5}$$

Two distinctions are made at the moment of measuring the impedance. First, the resistance is determined during the current ripple (current injection method); second, the resistance is obtained after the current pulse is off (current interruption method).

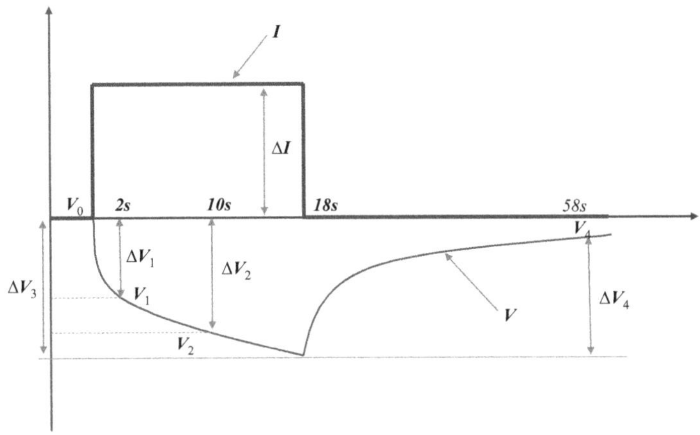

Figure 7.7 DC current pulse waveform representation. Adapted, with permission, from Reference [15]

Usually, the values of the impedance obtained with the current injection method are higher than those obtained with the current interruption method. This difference can be mitigated by allowing a relaxation time after cutting off the current, which would allow the equilibrium to be established.

Figure 7.7 shows the battery voltage response after the current pulse is introduced. This method is used by [15] to develop a model to detect the increase of the battery resistance after the aging effects.

7.3.1.2 Online measurement methods

Online methods are characterized by exploiting the characteristics of the systems in which the batteries are used. These methods are mainly applicable and suitable for EV applications or ESS applications in which online prognosis of the batteries is used for determining levels of system maintenance.

Quasi-electrochemical impedance spectroscopy

Quasi-EIS (QEIS) is a novel alternative to the EIS. In QEIS, high-frequency oscillations occurring in the battery current, and voltage during acceleration and regenerative braking in EVs, are exploited to perform the impedance measurement across the frequency spectrum. Contrary to EIS, QEIS does not perform a periodical current or voltage perturbation. The method instead uses the real current and voltage values of the module during an in-use load cycling scenario as input data in order to calculate an impedance spectrum, or so-called QEIS, which could be fitted to an equivalent electrical circuit model [16]:

$$Z(f) = \frac{V(f)}{I(f)} = \frac{(2/T)V(f) * I(f)}{(2/T)I(f) * I(f)} = \frac{\psi_{IV}(f)}{\psi_I(f)}, \tag{7.6}$$

where ψ denotes the power spectral density.

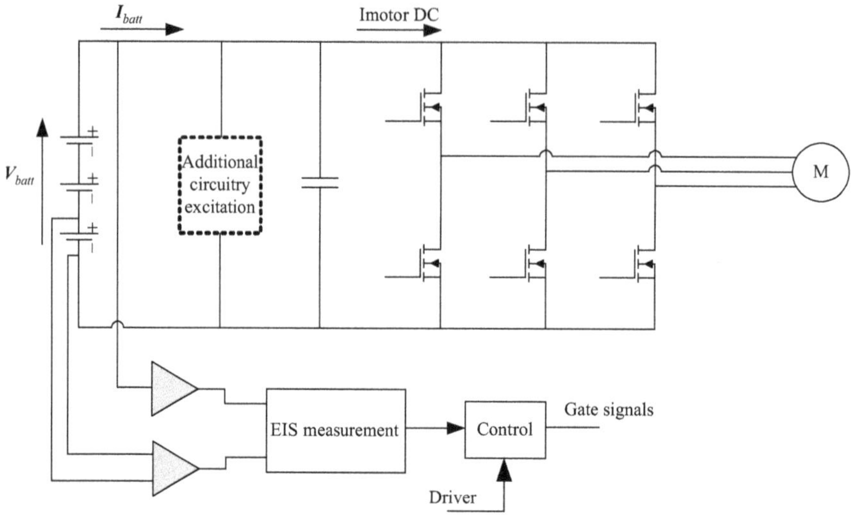

Figure 7.8 Battery impedance using motor control excitation. Adapted, with permission, from Reference [17]

Motor control excitation

This method is also suitable for EV applications. It uses an excitation current generated by a motor controller. To mitigate the presence of noise in the measurement signals, a statistical correlation approach is used. The excitation in the current signal comes from variations in either the main traction current to the driver or the controller response. The authors in [17] demonstrate the capability of this method by obtaining the impedance measurement in the presence of noise in the measurement signals and then applying a QEIS for calculating the impedance spectrum (Figure 7.8).

Duty cycle perturbation in DC/DC converter

This battery impedance measurement is performed during system operation. In this method, instead of injecting the AC signal through an external generator, the duty-cycle value of the DC/DC power converter, which is used to interface the battery with the load, is sinusoidally perturbed at a given frequency around its steady-state DC value (the duty-cycle value needed to achieve a desired output voltage). This duty-cycle perturbation results in sinusoidal variations of the battery voltage and the battery current around their corresponding steady-state DC values as

$$d(t) = D_{dc} + D_{ac} \sin(2\pi f_p t). \tag{7.7}$$

The sinusoidal ripples of the battery voltage and the battery current are then measured and used to determine the AC impedance of the battery at the perturbation frequency. The proposed method can be either continuously or periodically performed without interrupting the normal operation of the battery system and the power converter. Figures 7.9 and 7.10 show, respectively, the circuit and waveforms presented in [18,19] to demonstrate the efficiency of the method.

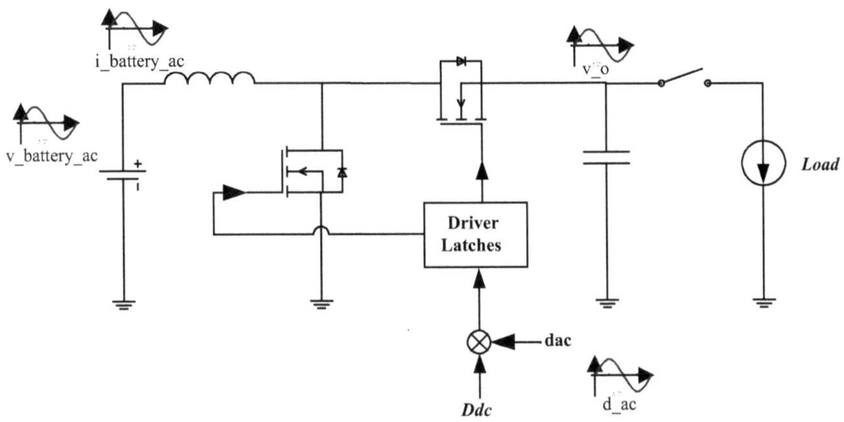

Figure 7.9 Duty cycle perturbation in DC converters. Adapted, with permission, from Reference [18]

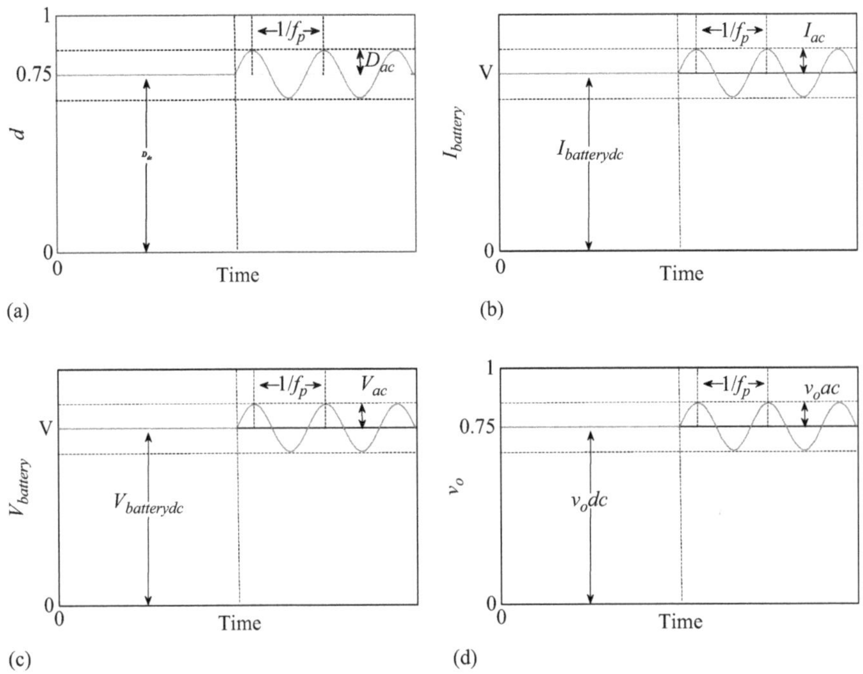

Figure 7.10 Battery impedance measurement using duty-cycle perturbation: (a) duty cycle, (b) battery current, (c) battery voltage, and (d) output voltage. Adapted, with permission, from Reference [18]

Non-perturbing method for residential ESS

The most recent online method for estimating a Li-ion battery AC impedance at twice the fundamental grid frequency has been proposed by the authors of this chapter. The proposed method presented in [20] does not require perturbing the power electronics control of the inverter used in AC nano-grid applications interacting with the AC power grid through single-phase inverters. This method differentiates from those found in the literature, because those methods focus only on EVs and other DC applications in which no interaction with AC sources is required. Furthermore, the above methods involve the use of a perturbation in the power electronics duty cycle and hence require modifying the closed-loop control of the power converters. A single-phase power system experiences a pulsating power transfer between the batteries and the AC source, which can be computed based on the energy conservation principle as

$$P_{DC} = I_{DC}V_{DC} = \eta I_{RMS}V_{RMS} = P_{AC}. \tag{7.8}$$

Assuming that the efficiency η and the power factor are equal to one, the instantaneous values for the current and voltage are obtained from (7.9), where $\omega_0 = 2\pi f_0$ and f_0 is the fundamental frequency of the power grid [21]:

$$V_{DC}I_{DC} = \sqrt{2}V_{RMS}\sin(\omega_0 t)\sqrt{2}I_{RMS}\sin(\omega_0 t) = 2V_{RMS}I_{RMS}\sin^2(\omega t). \tag{7.9}$$

As a result, the battery DC current should have a quadratic sine form as

$$I_{DC}(t) = \frac{2V_{RMS}I_{RMS}\sin^2(\omega_0 t)}{V_{DC}}. \tag{7.10}$$

A plot of $I_{DC}(t)$ is shown in Figure 7.11b. Assuming no losses in the power conversion process, the battery DC current can be rewritten as

$$I_{DC}(t) = 2I_{DC}\sin^2(\omega_0 t). \tag{7.11}$$

This leads to the conclusion that the current waveform flowing through the battery in a DC/AC energy conversion process is alternating at twice the frequency of the fundamental grid frequency, and hence

$$I_{DC}(t) = I_{DC}(1 + \cos(2\omega_0 t)) = I_{DC} + \hat{I}_{batt}(t), \tag{7.12}$$

where $\hat{I}_{batt}(t)$ is the AC component of the battery current. Figure 7.11c depicts the actual measurement of the battery current and voltage during the DC/AC power conversion that demonstrates the quadratic sine wave form of the current and also its influence over the DC voltage, where a ripple oscillation with the same frequency of the current is present. It is clear then that both measurements have a DC component and AC components.

The following equation is employed to represent the transfer function of the AC battery impedance derived from (7.2):

$$Z_{ac}(s) = \frac{s^2R_aC_{dl} + s(R_oR_aC_{dl}) + R_o + R_a}{sR_aC_{dl} + 1}. \tag{7.13}$$

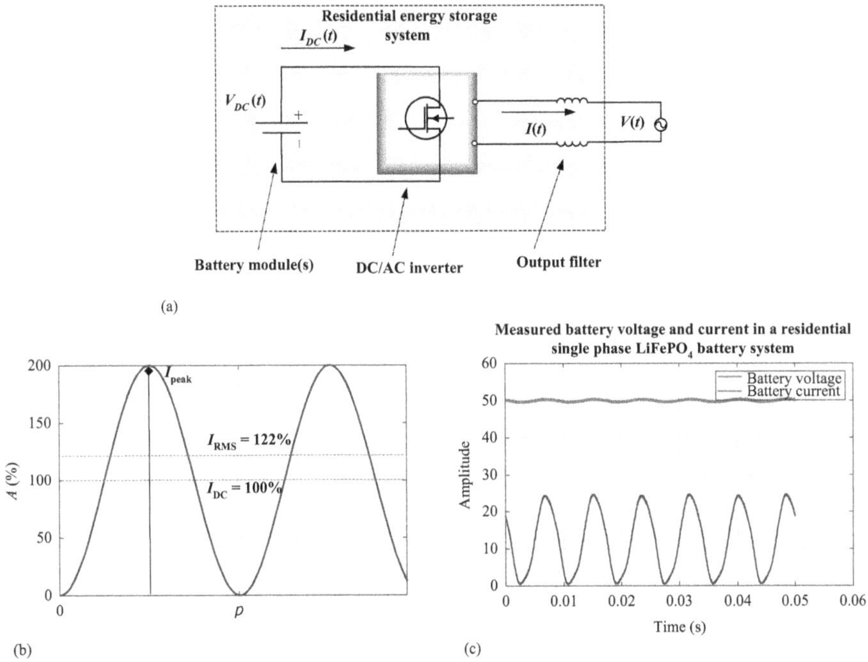

Figure 7.11 Battery current and voltage waveforms during power conversion process (DC/AC–AC/DC) using a single phase inverter: (a) power conversion in a battery energy storage system, (b) battery current shape during the power conversion process, and (c) battery current and voltage during the power conversion

During the DC/AC power conversion process, the battery's over potential voltage (V_{op}) at twice the grid frequency can be obtained by combining (7.13) and (7.12) as

$$V_{op} \equiv |Z_{ac}(2\omega_o)|\, I_{DC} \cos(2\omega_o t + \angle Z_{ac}(2\omega_o)), \tag{7.14}$$

where $\omega_o = 2\pi f_o$ and f_o is the frequency of the power grid.

The AC part of (7.14) represents the voltage ripple caused by the oscillatory behavior of the current at $2f_o$. It is possible to approximate the AC component of the voltage in the battery terminals by V_{op}, i.e., $\hat{V}_{batt} \equiv \hat{V}_{op}$. Then, the AC battery impedance at the frequency $2f_o$ can be expressed as

$$|Z_{ac}(2\omega_o)| = \frac{\mathscr{F}\{\hat{V}_{batt}(t)\}}{\mathscr{F}\{I_{DC} \cos(2\omega_o t + \angle Z(2\omega_o))\}}, \tag{7.15}$$

where $\mathscr{F}$ denotes the Fourier transform. *This shows that by only measuring the current flowing into the battery and the voltage in the battery terminals, an estimation of the internal AC impedance of the cells, for a specific grid frequency, can be achieved.* The proposed online method for battery impedance estimation consists of sampling the voltage $V_{DC}(t)$ and current $I_{DC}(t)$ at the battery terminals and, then, obtaining the high frequency components by removing the DC components using a moving average filter. The following difference equation shows the recursive representation of this filter:

$$y(k) = ax(k) + y(k-1) + bx(k-N), \tag{7.16}$$

where $a = 1/N$, $b = -1/N$ and N, is the number of samples. For the initialization of this filter, i.e., $k = 1$, it is required to store N elements in order to calculate $y(0) = M$, where M is determined by averaging the stored N samples as

$$M = \frac{1}{N}\sum_{k=1}^{N} x(k).$$

Then, the filtered signal is passed through a second order Goertzel filter, which will allow the extraction of the signal magnitude at the desired frequency [22]. The transfer function of this filter is given by

$$H(z) = \frac{v(z)}{y(z)} = \frac{1 - w_N^{-(2f_0N/F_s)}z^{-1}}{1 - (2\cos(4\pi f_0/F_s))z^{-1} + z^{-2}}, \tag{7.17}$$

where F_s is the sampling frequency. With N samples, the filter output will give the signal Fourier transform at the frequency $2f_0$ as

$$\hat{V}(2f_0) = \cos\left(\frac{4\pi f_0}{F_s}\right) v(N-1) - v(N-2) + j\sin\left(\frac{4\pi f_0}{F_s}\right) v(N-1). \tag{7.18}$$

Finally the impedance of the battery at twice the grid frequency is obtained by dividing the real part of the processed battery voltage $\hat{V}_{batt}$ and battery current $\hat{I}_{batt}$ as

$$Z_{ac}(2f_0) = \frac{\Re(\hat{V}_{batt}(2fo))}{\Re(\hat{I}_{batt}(2fo))}. \tag{7.19}$$

References

[1] D. Linden and T. Reddy, *Handbook of Batteries*. New York, NY: McGraw-Hill Handbooks, 2004.

[2] O. Yamamoto and M. Wakihara, *Lithium Ion Batteries Fundamentals and Performance*. Tokyo: Wiley, 1998.

[3] G. Pistoia, *Lithium-Ion Batteries: Advances and Applications*. 1 Ed., Amsterdam: Elsevier, 2014.

[4] A. Yoshino, K. Sanechika, and T. Nakajima, "Secondary battery," Patent 4 668 595, May, 1987.

[5] K. Mizushima, P. Jones, P. Wiseman, and J. Goodenough, "Li_xCoO_2 ($0 < x < -1$): a new cathode material for batteries of high energy density," *Materials Research Bulletin*, vol. 15, no. 6, pp. 783–789, 1980.

[6] M. M. Thackeray, W. I. David, P. G. Bruce, and J. B. Goodenough, "Lithium insertion into manganese spinels," *Materials Research Bulletin*, vol. 18, no. 4, pp. 461–472, 1983.
[7] A. K. Padhi, "Phospho-olivines as positive-electrode materials for rechargeable lithium batteries," *Journal of The Electrochemical Society*, vol. 144, no. 4, p. 1188, 1997.
[8] C. Restrepo, A. Salazar, H. Schweizer, and A. Ginart, "Residential battery storage: is the timing right?" *IEEE Electrification Magazine*, vol. 3, no. 3, pp. 14–21, 2015.
[9] S. Y. Cho, I. O. Lee, J. I. Baek, and G. W. Moon, "Battery impedance analysis considering DC component in sinusoidal ripple-current charging," *IEEE Transactions on Industrial Electronics*, vol. 63, no. 3, pp. 1561–1573, 2016.
[10] D. I. Stroe, M. Swierczynski, A. I. Stan, V. Knap, R. Teodorescu, and S. J. Andreasen, "Diagnosis of lithium-ion batteries state-of-health based on electrochemical impedance spectroscopy technique," *Energy Conversion Congress and Exposition (ECCE), 2014 IEEE*, pp. 4576–4582, 2014.
[11] X. Tang, X. Mao, J. Lin, and B. Koch, "Li-ion battery parameter estimation for state of charge," *American Control Conference (ACC), 2011*, pp. 941–946, 2011.
[12] C. Schlasza, P. Ostertag, D. Chrenko, R. Kriesten, and D. Bouquain, "Review on the aging mechanisms in Li-ion batteries for electric vehicles based on the FMEA method," *2014 IEEE Transportation Electrification Conference and Expo (ITEC)*, pp. 1–6, 2014.
[13] M. Swierczynski, D. I. Stroe, A. I. Stan, R. Teodorescu, and D. U. Sauer, "Selection and performance-degradation modeling of $LiMO_2/Li_4Ti_5O_{12}$ and $LiFePO_4/C$ battery cells as suitable energy storage systems for grid integration with wind power plants: an example for the primary frequency regulation service," *IEEE Transactions on Sustainable Energy*, vol. 5, no. 1, pp. 90–101, 2014.
[14] M. Micea and L. Ungurean, "Online state-of-health assessment for battery management systems," *IEEE Transactions on Instrumentation and Measurement*, vol. 60, no. 6, pp. 1997–2006, 2011.
[15] D. I. Stroe, M. Swierczynski, S. K. Kær, and R. Teodorescu, "Degradation behavior of lithium-ion batteries during calendar ageing – the case of the internal resistance increase," *IEEE Transactions on Industry Applications*, vol. 54, no. 1, 2018, pp. 517–525.
[16] R. Mingant, J. Bernard, and V. Sauvant-Moynot, "Novel state-of-health diagnostic method for Li-ion battery in service," *Applied Energy*, vol. 183, pp. 390–398, 2016.
[17] D. A. Howey, P. D. Mitcheson, S. Member, V. Yufit, G. J. Offer, and N. P. Brandon, "Online measurement of battery impedance using motor controller excitation," *IEEE Transaction on Vehicular Technology*, vol. 63, no. 6, pp. 2557–2566, 2014.

[18] W. Huang and J. A. Qahouq, "An online battery impedance measurement method using DC–DC power converter control," *IEEE Transactions on Industrial Electronics*, vol. 61, no. 11, pp. 5987–5995, 2014.

[19] J. A. Qahouq, "Online battery impedance spectrum measurement method," *Conference Proceedings – IEEE Applied Power Electronics Conference and Exposition – APEC*, pp. 3611–3615, 2016.

[20] A. Salazar, C. Restrepo, Y. Gao, J. M. Velni, and A. Ginart, "An online $LiFePO_4$ battery impedance estimation method for grid-tied residential energy storage systems," *2017 IEEE Energy Conversion Congress and Exposition (ECCE)*, pp. 980–986, 2017.

[21] A. Ginart, A. Salazar, and R. Liou, "Transformerless bidirectional inverter for residential battery storage systems," *IEEE Green Technologies Conference*, 2016, pp. 18–23.

[22] M. A. Jaber and D. Massicotte, "Fast method to detect specific frequencies in monitored signal," *Final Program and Abstract Book – 4th International Symposium on Communications, Control, and Signal Processing, ISCCSP 2010*, 2010.

Chapter 8
Prognostics: a battery case study

Bin Zhang[1]

8.1 Introduction

Modern engineering systems, such as aircraft, industrial plants, power stations, and electrical vehicles, are designed to perform desired functions with the requirement of reliability, productivity, safety, and availability. However, the system will deteriorate and the performance will degrade after it is put into service. For system maintenance, there are two main traditional philosophies, preventive and corrective. Corrective maintenance is conducted when a system breaks down so that the failed component can be restored to an operational condition within the tolerances or limits established for in-service operations. Preventive maintenance, on the other hand, is to perform planned maintenance in a time-based schedule. When preventive maintenance is conducted, components are repaired or replaced before they fail, which often leads to increase of maintenance cost. In fact, one-third of the maintenance cost is caused by preventive maintenance planning. Therefore, these traditional methods result in unexpected downtime, high-cost of labor, and reduction of system availability. Moreover, it lacks accurate estimation of failure rate, and some unexpected failure can still lead to catastrophic events causing equipment damage or even loss of lives.

To overcome this disadvantage, condition-based maintenance (CBM, also known as predictive maintenance) is developed. CBM is defined as the techniques that automatically monitor the condition of in-service system in order to determine the time when maintenance needs be performed based on the real-time measurements. The fundamental capability of CBM consists of sensing, data acquisition and processing, fault diagnosis and prognosis (FDP), and decision-making. The goal of CBM is to minimize disruption of normal system operations, while allowing for optimal maintenance to lower the cost on operation and maintenance (O&M). The main value of CBM is to allow optimal scheduling of preventive maintenance and to prevent unexpected system failures. The key is "the right information at the right time." By knowing the health condition of the system, maintenance can be better planned, and logistics and human resources can be optimized according to the available resources and production plan. With this strategy, unexpected and unplanned downtime can be

[1]College of Engineering and Computing, University of South Carolina, USA

transformed to shorter and fewer expected and planned downtime, which will lead to increasing system availability and throughput.

FDP, as fundamental enabling techniques of CBM, is not a new concept [1–4]. FDP involves fault detection, isolation, identification, and remaining useful life (RUL) prediction. FDP utilizes the real-time data and/or system model to estimate the current health state, predict the evolution of the faults, and estimate the RUL of a failing component. Diagnosis aims to monitor the health state of the system and detect and isolate impending faults in real-time. Usually a fault cannot be measured directly. In Bayes' theory, the fault state is obtained by applying Bayesian estimation with a fault diagnostic model and a real-time measurement [5–11]. Prognosis refers to the generation of long-term predictions that describe the evolution of a fault and the estimation of the RUL of a failing component or subsystem. Compared with many successful cases of diagnosis [6–8,12–17], prognosis is more challenging [2,18–23]. Major contributors to this difficulty include nonlinear nature of fault growth, absence of measurement, hybrid nature of fault modes, and various uncertainties.

In the past decades, many diagnosis and prognosis algorithms were developed. In general, they can be divided into reliability approaches, model-based approaches, and data-driven approaches. Reliability approaches assess the health state based on the historical data from the same component or system to obtain some critical information, such as mean time between failures, to schedule maintenance [24]. Some examples are statistical analysis based on Weibull analysis and Poisson laws [25,26]. Model-based approaches rely on a mathematical representation of system, which describes the system under different operating conditions same as the system. The difference of the system measurement between the model and the system, also known as residuals, is used for diagnosis and prognosis [27]. In these methods, a well-designed model often needs to be developed, which in most cases is not available and limits it application. To address this problem, data-driven approaches were developed. Existing data-driven methods generally include two main steps, i.e., fault feature extraction and fault diagnosis. Commonly used methods include artificial neural network (ANN), support vector machine, and so on [28–30]. Recently, deep learning based approaches were also developed with applications in many different systems [31–33].

While the FDP algorithms become more and more powerful, the development of modern technologies rapidly increases the complexity of systems [34]. Meanwhile, the utilization of embedded systems in those complex systems design, such as vehicles, aircraft, and robots, becomes more and more important. For instance, there are more than 70 distributed microcontrollers (also known as electronic control units – ECUs) in one high-end car [35]. The ECUs in modern cars perform variety of functions such as stability control, cruise control, oil and coolant monitoring, energy-efficient propulsion, turbo on-and-off, navigation, and even autonomous driving. As to aircraft, there are more ECUs that carry more complicated and critical functions to monitor and troubleshoot components for timely and optimal maintenance. FDP is needed to monitor, estimate, and predict the health condition of safety critical components, such as engine, battery, and transmission [36–40].

Traditional ways to design FDP algorithms adopt periodic sampling (also called "Riemann sampling (RS)") where samples are taken in a periodic manner, and the diagnostic and prognostic algorithms are executed at the same rate. A nice feature of traditional RS-based FDP (RS-FDP) with this fixed time interval sampling is the easiness in analysis and design. However, it may be undesirable in many situations, from the computation-efficiency point of view. On the one hand, since the sampling period is determined according to the worst case operating scenario, the FDP algorithm might be executed even if there is little new information actually present in the measurements. This will result in significant over-provisioning of the real-time system hardware. On the other hand, when the fault grows very fast, it is expected to assign more resources to the FDP algorithm so that it can take more frequent actions to provide accurate fault information. For prognosis, RS-FDP usually has a large prediction horizon, from the time that a fault is detected at very early stage to a future time instant that the fault grows to the failure threshold. This long-term prediction not only requires a lot of computation resources but also causes accumulation of uncertainties.

With the development of distributed prognostics and health management, more and more diagnostic and prognostic algorithms are deployed on local processors and embedded systems to mitigate the requirements on communication, power, and computation and improve the system reliability. Since the local processors and embedded systems have very limited computation capabilities, traditional FDP algorithms with heavy computation demands are very difficult, if not impossible, to be deployed. This becomes the bottleneck that prevents the deployment of FDP algorithms in complex systems, and cost-efficient FDP solutions must be developed. With this vision, a Lebesgue sampling-based FDP (LS-FDP) framework is proposed in which an efficient FDP philosophy is developed to reduce the computational cost compared with the traditional RS-FDP algorithms. In this new approach of FDP, the novelty comes from the concept of "Lebesgue sampling (LS)" (or "event-based sampling"). The LS-FDP considers the prediction horizon in the fault dimension axis and is described by the number of Lebesgue states. This provides a straightforward means to conduct prognosis that requires little computation resources [32]. Contrast to conventional periodic sampling-based approaches, the computation in LS-FDP will be triggered only when an event takes place, and the prognosis will be executed based on the LS-based model whose states are predefined according to the quantization level. With the feature of "execution only when necessary" in LS, the computation efforts in LS-FDP can be significantly reduced by eliminating unnecessary computation when fault growth is slow.

The LS-FDP considers the prediction horizon in the fault dimension axis and is described by the number of Lebesgue states. This provides a straightforward means to conduct prognosis that requires little computation resources [41].

8.2 Lebesgue sampling-based fault diagnosis and prognosis

In this section, we will present the overview of the proposed LS-FDP framework, in which the diagnostic and prognostic algorithm is carried out only when new measurements cause state transition. Different from traditional algorithms that are carried out in a fixed time interval, the transition of state cause an event, which justify the

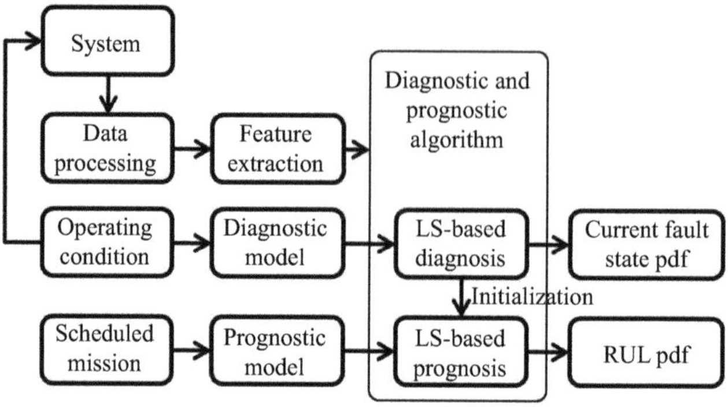

Figure 8.1 The implementation framework of LS-FDP

change of fault state and the execution of the FDP algorithm. Figure 8.1 illustrates the proposed LS-FDP framework, which integrates external inputs, Lebesgue samples of feature and fault dimension, models for diagnosis and prognosis, and diagnostic and prognostic algorithms.

Since our focus is to introduce LS into FDP, data collection, data processing, and feature extraction will not be discussed here. With a feature being extracted from data to indicate the fault growth, a fault dynamic model that describes the fault behavior needs to be established. The performance and efficiency of FDP relies greatly on the fault dynamic model, which will be discussed in the following sections.

8.2.1 Fault mechanism modeling

Suppose the following continuous-time differential equation describes the actual fault growth dynamics:

$$\dot{a} = F(a, u) \tag{8.1}$$

where a is the fault dimension, u is system input including items (such as external environmental factors and operating modes) that have impacts on fault growth, and $F(\cdot)$ is a nonlinear function that describes the fault growth under the current fault dimension with input u. The feature or condition indicator, denoted by y, is extracted from raw measurements and serves as the real-time measurement for FDP algorithm. Note that the mapping between y and a can be described by a nonlinear function $y = h(a)$. In most cases, a is not measurable and $y = a$ is employed such that we can use y to indicate fault a directly. To simplify the description, we take $y = a$ in the following discussion.

To use this model in LS-FDP, the fault measurements need to be quantified. In LS, samples are taken when the difference between the current state and the last sampled state exceeds the predefined Lebesgue state length. Then the LS-based model of the fault dynamics in discrete-time can be described as follows:

$$\hat{a}(t_{k+1}) = \hat{a}(t_k) + f_t(D, \dot{\hat{a}}(t_k)) \tag{8.2}$$

where $\hat{a}(t_k)$ is the Lebesgue state, t_k is the kth sampling instant, D is the Lebesgue length, and $f_t(\cdot)$ is a nonlinear function.

Traditional prognostic algorithm consists of two steps of prognosis. The first step is the generation of a long-term predication for the fault state probability distribution function (pdf) estimation. This is obtained by recursive execution of the fault growth model. The second step is the estimation of RUL, which is essentially related to the probability of failure at future time instants. The RUL pdf is obtained by defining a failure threshold established from historical data or empirical knowledge and comparing this threshold with the long-term prediction of fault state at all the future time instants. Compared to diagnosis, prognosis requires much more computational resources mainly because of long-term predication, especially when the prediction horizon is large, which is not a rare case in FDP applications. To reduce computation time and resources, a new model is developed in the LS-FDP as follows:

$$t_{k+1} = t_k + g_t(D, \dot{\hat{a}}(t_k)) \tag{8.3}$$

Note that $\dot{\hat{a}}(t_k) = f(\hat{a}(t_k), u(t_k))$ and $g_t(D, \dot{\hat{a}}(t_k))$ is a nonlinear function. Rather than conducting a long-term prediction on the time axis, this model calculates the RUL on each Lebesgue state directly so that the prediction horizon is the number of Lebesgue states on the fault dimension axis. Since the number of Lebesgue states on the fault dimension axis is small, the prediction horizon for LS-based prognosis is small and will significantly reduce the computation.

8.2.2 Lebesgue sampling

Figure 8.2 shows an example of battery capacity degradation in 1,000 charging–discharging cycles, which will be used to illustrate the concept of LS.

It is clear that the capacity degradation in $R_1 = [1, 780]$ cycle is slower than that in $R2 = [780, 1,000]$ cycle. Using RS-FDP with fix time interval in Figure 8.2(a), the FDP is executed at each cycle no matter if it is necessary, which results in many unnecessary calculations in range R_1. Since the fix time interval is selected according to the worst case scenario to guarantee tracking accuracy for fault growth in range R_2, there are many unnecessary calculations in range R_1.

Ideally, we expect to reduce the number of FDP execution in range R_1 where the capacity degradation is slow so that more resources can be assigned to other tasks. In the range of R_2 where the capacity degradation becomes fast, we increase the number of FDP execution by assigning more resources to FDP tasks. This setting is desirable in FPGA-based embedded systems where resources are dynamically reconfigured and assigned to different tasks in real-time [42]. With this configuration, a balance between computation and performance can be achieved. This strategy involves time-varying task periods that is not easy for RS, but natural for LS. By defining Lebesgue states on the vertical axis of battery capacity in Figure 8.2, fewer transitions between states are made when the degradation is slow, while more transitions are made when the degradation is fast. In Figure 8.2(b), only four Lebesgue states are visited in the first 780 cycles in R_1 and seven states in the remaining 220 cycles in R_2, which means that the FDP only needs to be executed four times in R_1 and seven times during R_2.

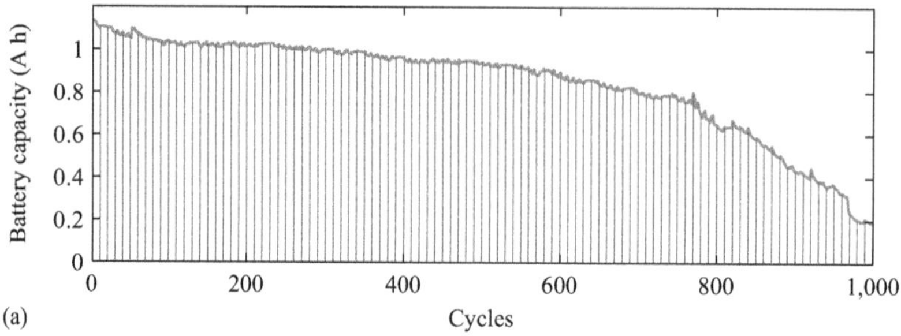

(a)

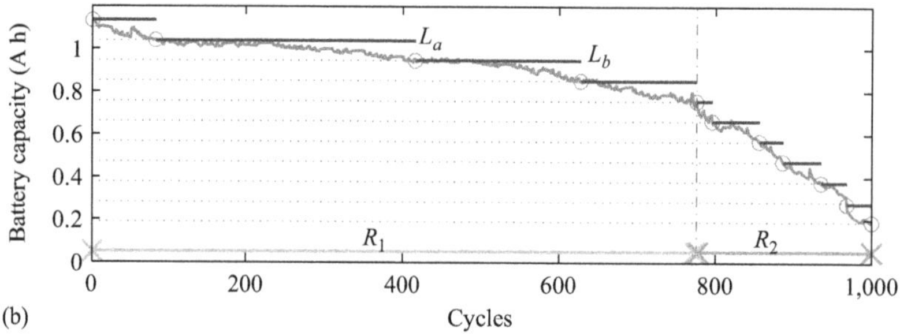

(b)

Figure 8.2 Illustration of LS: (a) RS with fixed time interval and (b) LS with fixed Lebesgue state length

It indicates that FDP needs less resources in R_1 and more resources in R_2, which optimizes the usage of limited resources.

Note that an event is defined by capacity changes from one Lebesgue state to another. For instance, in Figure 8.2(b), if the measurement stays between L_a and L_b, it will not cause an event. As soon as the measurement becomes below L_b, an event is triggered. In our proposed LS-FDP, we only utilize the measurements from existing sensors and the event-checking purely runs in software. Therefore, there is no induced event-monitoring hardware cost.

8.2.3 Lebesgue sampling-based diagnosis

In the LS-FDP, the range of the fault state $a(t)$ is partitioned into Lebesgue states $\{F_1, F_2, \dots, F_f\}$, and the diagnostic model is discretized with these Lebesgue sates. As mentioned above, an event happens when the state $a(t)$ changes from one Lebesgue state to another one [43,44], which triggers the execution of the diagnostic algorithm. The time instant when an event happens is called the "event stamp." The sequence of the event stamps is denoted as $\{t_1, t_2, t_3, \dots, \}$, which defines a time series that can be used as the input of run-time diagnostic algorithms such as a Kalman filter, extended Kalman filter (EKF), or particle filter algorithms [6,7,18,45–47]. Same as diagnostic algorithm in the RS, the diagnostic algorithm in the LS calculates the current fault state distribution at these event stamps and from which the probability of fault detection can be calculated based on a fault-detection threshold. The flowchart shown in Figure 8.3 illustrates the implementation procedure of the LS-based diagnosis.

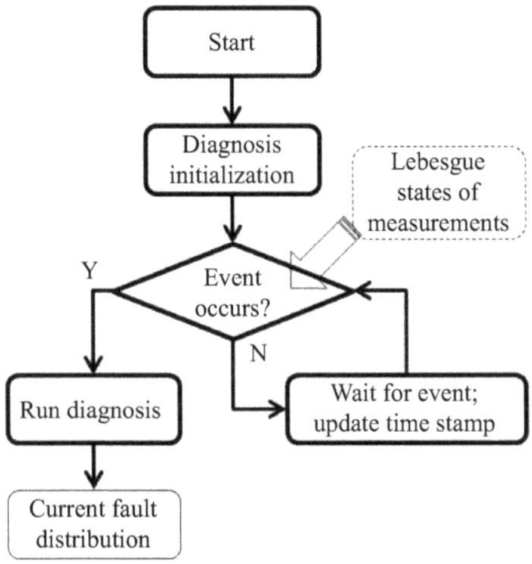

Figure 8.3 Flowchart of Lebesgue sampling-based diagnosis

8.2.4 Lebesgue sampling-based prognosis

When a fault is detected at t_d, the fault state is estimated by a distribution. From this fault state distribution, a time distribution can be initialized and is used as the initial condition for prognosis [47]. By RS-based prognosis, the prediction is conducted from the current time instant t_c to future time instants till t_f, which is defined as the system failure time when the fault state reaches a failure threshold F_f. The prognostic horizon (PH) $[t_c, t_f]$ is usually large, especially at the early stage of the fault or when the fault growth is slow. The prediction calculates the fault state at each fixed time interval, which is demanding on the computational resources. Moreover, since no measurements are available in prognosis, prognostic uncertainty will grow rapidly with the increase of prediction horizon.

With LS, a new prognostic philosophy is proposed. Suppose that the fault is detected at Lebesgue state F_d, then we consider the discretized prognostic model with Lebesgue states $\{F_d, F_{d+1}, \ldots, F_f\}$. The prognostic algorithm is implemented, together with the LS-based prognostic model, to calculate the distributions of operation time when the fault state reaches different Lebesgue states $\{F_d, F_{d+1}, \ldots, F_f\}$. Meanwhile, it will provide an RUL estimation on Lebesgue state F_f. As the state F_d changes to F_{d+1}, the prognostic algorithm will be updated with the PH becoming the total number of Lebesgue states in $\{F_{d+1}, F_{d+2}, \ldots, F_f\}$. This procedure will repeat until the failure threshold F_f is reached.

Note that the PH can be controlled by adjusting Lebesgue state length. Increasing the Lebesgue state length will decrease the number of events, which will reduce the required computational resources. However, there is a trade-off between the number of Lebesgue state (or computation resources) and accuracy of prognosis. Therefore, the selection of Lebesgue state length, or the number of Lebesgue state needs to be determined as problem dependent and make sure the performance of FDP meet the

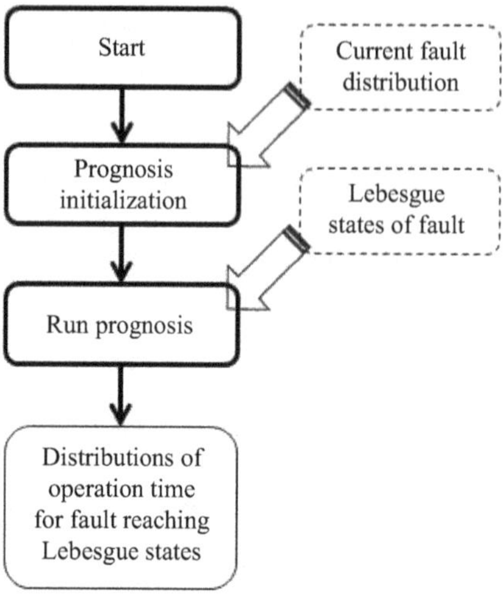

Figure 8.4 Flowchart of Lebesgue sampling-based prognosis

accuracy requirements. The flowchart in Figure 8.4 illustrates the implementation procedure of the LS-based prognosis.

8.3 Applications to batteries

Battery is a safety critical component that provides power to system functions including command, control, communications, computers, and intelligence [41,48]. It is important to improve the performance of the battery-management system (BMS) to make battery O&M safe, reliable, and cost-efficient. BMS is designed to evaluate the instantaneous state-of-charge (SOC) in the battery and at the same time to monitor the slowly varying state-of-health (SOH).

SOC is an indicator that represents the available charge stored in the battery compared to the full capacity charge of the battery, given by a percentage of the current charge to the entire charge [49]. SOH describes the physical condition of a battery. Unlike SOC, there is no widely accepted definition of SOH. A general definition of SOH is that it quantifies the battery's ability to store energy and deliver specified performance compared to a fresh battery [50]. The total capacity of the battery is evaluated in terms of amp-hours (A h), which basically is a dimension of electric charge.

It is essential to accurately estimate the SOC and SOH in order to maximize and optimize the operation of the system. Knowing the SOC also helps prevent overcharge and over discharge of batteries, which is vital for safe operation and long cycle life of lithium-ion batteries. However, both SOC and SOH are not directly observable, which requires estimation and prediction algorithms, such as diagnosis and prognosis, to provide real-time battery health states [9,18,51].

The easiest way to estimate SOH and SOC is based on current integration [52]. Different estimation methods are also developed, including reduced state-space electrochemical models, ANNs, and impedance spectroscopy [53–55]. These methods usually require a large computational cost and very accurate measurements, which make them unsuitable for online application since the estimation of SOC requires much faster and more accurate algorithms[49,56], whereas SOH estimation is not so tough.

EKF is a recursive algorithm that estimates the true state of a nonlinear system based on noisy measurements. It assumes that the nonlinear system dynamic model is linearized using Taylor expansion at a local point, and the noise is subject to Gaussian distribution. EKF has been used for state estimation and RUL prediction of electronic components and battery management [57–59] due to its low-computational requirement.

Traditional EKF algorithms are developed based on RS framework with the advantages of easiness in system analysis and algorithm design. However, it is unfavorable from the computation-efficiency point of view in some cases, especially when the battery SOH degradation is slow. The algorithms are executed as long as a new measurement is available no matter how much new information is included in the measurement. Ideally, it is expected to reduce the executions of algorithm when battery SOH degradation is slow and increase the number of algorithms execution when battery SOH degradation becomes fast. To achieve this goal, LS method is developed [60–62]. In this section, the design of EKF in LS framework will be discussed with an example of diagnosis of battery SOH degradation.

In LS-EKF, the general form of SOH diagnosis model is given as [62]

$$x_k = f(x_{k-1}, u_{k-1}) + \omega_k \tag{8.4}$$

where x_k is the states to be estimated, $f(\cdot)$ is the nonlinear function of states, u_k is the input at time k, w_k is a zero mean Gaussian noises with covariance matrix Q_k. Since most faults cannot be measured directly, it relies on some measurable variables. The observation model that describes the relationship between state x_k and measurements z_k is given by

$$z_k = h(x_k) + v_k \tag{8.5}$$

where z_k is the measurement, $h(\cdot)$ is the measurement function of state, which can be linear or nonlinear, v_k is a zero-mean Gaussian noises with covariance matrix R_k.

Since $f(\cdot)$ and $h(\cdot)$ are nonlinear functions, they cannot be used to calculate the covariance directly. Instead, their Jacobians are calculated as follows:

$$F_k = \frac{\partial f}{\partial x} |_{\hat{x}_{k-1|k-1}} \tag{8.6a}$$

$$H_k = \frac{\partial h}{\partial x} |_{\hat{x}_{k-1|k-1}} \tag{8.6b}$$

Note that the Jacobian needs to be calculated with the predicted state at each instant.

The EKF algorithm includes two steps: the first step (prediction) is to propagate the state vector x into the next time step by using the state transition model; the second step (update) is to correct the prediction from the first step by using the measurement z. The prediction step can be described as

$$\hat{x}_{k|k-1} = f(\hat{x}_{k-1|k-1}, u_k) \tag{8.7a}$$

$$P_{k|k-1} = F_k P_{k-1|k-1} F_k^T + Q_k \tag{8.7b}$$

where P_k and Q_k are the covariance matrices of the predicted state and the process noises.

The equations for the update state are expressed as

$$\tilde{y}_k = z_k - h(\hat{x}_{k|k-1}) \tag{8.8a}$$

$$S_k = H_k P_{k|k-1} H_k^T + R_k \tag{8.8b}$$

$$K_k = P_{k|k-1} H_k^T S_k^{-1} \tag{8.8c}$$

$$P_{k|k} = (I - K_k H_k) P_{k|k-1} \tag{8.8d}$$

$$\hat{x}_{k|k} = \hat{x}_{k|k-1} + K_k \tilde{y}_k \tag{8.8e}$$

where $\tilde{y}_k$ is the measurement residual, S_k is the residual covariance, K is the near-optimal Kalman gain, R_k is the covariance matrix of the observation noises, $P_{k|k}$ is the updated covariance estimate, $\hat{x}_{k|k}$ is the updated state estimate, and I is the identity matrix.

LS is introduced in diagnosis and prognosis by [60,61]. Different from RS method, LS method divides the state axis by a number of predefined states (also called Lebesgue states). The FDP is triggered only when the feature value changes from one Lebesgue state to another, or an event happens. The prediction horizon is described by the number of the Lebesgue states instead of the periodic sampling intervals in RS framework. This LS-based FDP is need-based, and its execution is justified by an event. These features enable it to significantly reduce the computation demands and the uncertainty accumulation by eliminating unnecessary computation and decreasing the prediction horizon.

When LS-based prognosis is activated, prognostic algorithm is executed based on the LS model (LSM) to estimate distributions of operating time for the fault state reaching each Lebesgue state. The LSM is described as

$$t_{k+1} = t_k + g_t(D(t_k), \hat{x}(t_k), \dot{\hat{x}}(t_k)) + \omega_t(t_k) \tag{8.9}$$

where t_k is the operation time for fault state reaching the kth Lebesgue state, D is the Lebesgue state length, $\omega_t(t_k)$ represents the uncertainties, and $g_t(\cdot)$ is a nonlinear function that describes the time distribution of the fault state arriving at each Lebesgue

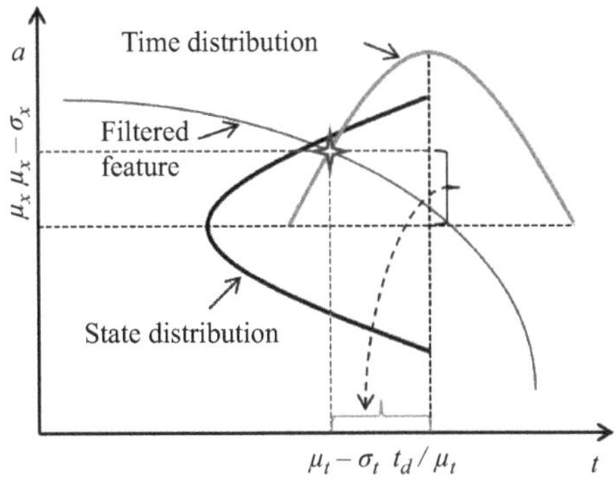

Figure 8.5 Conversion from state distribution to time distribution

state. Note that the output of diagnosis at the time instant of the current event t_d is a fault state distribution defined on the state axis, which cannot be used in LS-based prognosis, which is based on (8.9) and needs an initial distribution of operation time.

To obtain the time distribution on the current Lebesgue state, the time instant of the current event t_d is set to be the mean of the time distribution μ_t, the sigma point $(\mu_x - \sigma_x)$ of the state distribution has a cross point with the filtered feature curve, which is marked in Figure 8.5. The coordinates of the cross point $(\mu_t - \sigma_t, \mu_x - \sigma_x)$ represent that sigma point $(\mu_x - \sigma_x)$ reaches the threshold at time instant $(\mu_t - \sigma_t)$. The time interval between t_d and the marked point in Figure 8.5 is set to be the variance of the time distribution. By this method, the variance of state distribution σ_x is approximately converted to that of the time distribution σ_t. Here, μ_x, σ_x, μ_t, and σ_t are the mean and variance of the state estimation, the mean and variance of the time distribution, respectively. The details are shown in Figure 8.5.

8.4 Performance metrics for prognosis

8.4.1 Prognostic horizon

PH is an accuracy metrics [63], which is given by the first time when the prediction of RUL fall into a predefined zone centered at the ground true RUL. As shown in Figure 8.6, the accuracy zone is the shaded area with width of 2α and is centered at the solid line, which is the ground truth RUL. PH is determined by the difference of the end of life (EoL) and time T, where the prediction for the first time reaches the accuracy zone. For a prognostic algorithm, it is desirable to have a large PH. The PH is defined as

$$PH = EoL - T \tag{8.10}$$

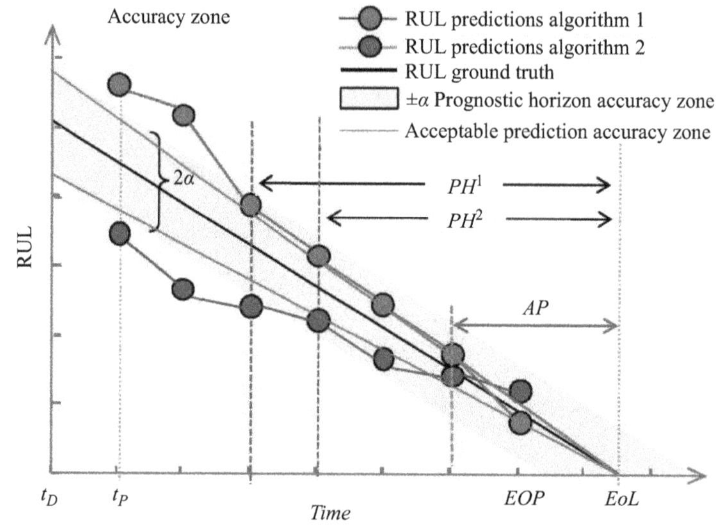

Figure 8.6 Prognostic horizon and acceptable prediction [63]

8.4.2 Acceptable predictions

Acceptable predictions (AP) is similar to the PH. Different from PH in which the accuracy zone is defined by two parallel lines, AP uses a cone accuracy zone as shown in Figure 8.6. It is mathematically defined as [64]

$$\delta_i = \begin{cases} 1 & \text{if}\,(1-\alpha)r_i^* \leq r_i \leq (1+\alpha)r_i^* \\ 0 & \text{otherwise} \end{cases} \tag{8.11}$$

$$AP = mean(\delta_i \mid t_H \leq t_i \leq t_{EoP}) \tag{8.12}$$

where t_{EoP} is the end point of prognosis.

8.4.3 α-λ Metrics

Same as PH, α–λ measures whether the predictions of RUL fall into an accuracy zone, defined by the shaded cone zone given in Figure 8.7 [63]. In this figure, the shaded cone zone is defined by $\alpha \times 100\%$ of the ground truth RUL at all the time instants. For a given time instant t_λ, with λ varies from 0 to 1, this metrics determines whether the prediction at this time instant (a fraction of the entire prediction horizon defined by λ) fall with the accuracy zone (an accuracy defined by α and ground truth RUL). Mathematically, it is defined as follows:

$$[1-\alpha]\cdot r^*(t_\lambda) \leq r(t_\lambda) \leq [1+\alpha]\cdot r^*(t_\lambda) \tag{8.13}$$

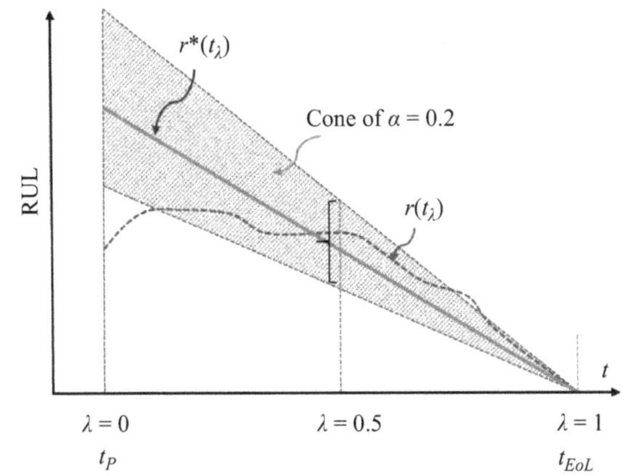

Figure 8.7 α–λ Accuracy with the accuracy cone shrinking with the time on RUL vs. time plot [63]

where α defines the shaded accuracy zone around the ground truth RUL, λ defines the time span $t_\lambda = t_P + \lambda(t_{EoL} - T_P)$ with t_P and t_{EoL} being the starting and ending time instants of prediction, $r^*(t_\lambda)$ and $r(t_\lambda)$ are the ground truth RUL and the predicted RUL, respectively.

8.4.4 Relative accuracy

Relative accuracy (RA) quantitatively measure the accuracy by studying the mean value absolution percentage errors for the RUL prediction in the prediction horizon, which is given from a defined time instant where prognostic results are taken into consideration to the time instant when prognostic ends. It is mathematically described as

$$RA_\lambda = 1 - \frac{1}{t_{EoP} - t_H} \frac{| r_*(t_\lambda) - r^l(t_\lambda) |}{r_*(t_\lambda)} \tag{8.14}$$

8.4.5 Convergence

Convergence (CG) evaluates the speed that a given prediction performance improves as more data become available. Suppose the nonnegative performance metrics is M, which can be evaluated at any time instant in the prediction, the CG is defined as [64]

$$CG = 1 - \frac{1}{t_{EoP} - t_H} \left(\frac{\frac{1}{2} \sum_{i=P}^{EoP} (t_{i+1}^2 - t_i^2) M_i}{\sum_{i=P}^{EoP} (t_{i+1} - t_i) M_i} - t_P \right) \tag{8.15}$$

8.4.6 Performance score

In order to compare these different performance metrics to evaluate the overall performance, a performance score is calculated as the weighted sum of AP, RA, and CG [64,65]:

$$score = w_1 \cdot mean(AP) + w_2 \cdot mean(RA) + w_3 \cdot mean(CG) \tag{8.16}$$

where $w_1 = 0.6, w_2 = 0.3, w_3 = 0.1$ are weighting factors and $mean(\cdot)$ calculates the mean value. It is desirable for a prognostic algorithm to have a larger performance score.

8.4.7 Experimental results

The experiment investigates the SOH of four identical Li-ion batteries with 1.1 A h rated capacity. The degradation of the capacities is obtained from charge–discharge tests on an Arbin BT2000 battery test system under room temperature at a discharge current of 1.1 A [66]. The battery capacity degradation speed is the decrease of the capacity with respect to charging–discharging cycles. The charge–discharge cycle is cutoff at predetermined cutoff voltages of 2.7 V for discharging and 4.2 V for charging, respectively. The failure threshold for the SOH is set to be 0.35 A h, and the battery capacity reaches this threshold at the 840th, 810th, 954th, and 981st cycle, respectively. From this series of experimental data, an empirical model will be established and will be used in the diagnosis and prognosis, as discussed below.

8.4.7.1 EKF in Riemann sampling framework

In traditional RS framework, samples are taken equivalently along the time axis. The diagnosis and prognosis model is identically developed and is written as

$$C(t+1) = C(t) - \gamma \cdot (p_1 + p_2 \cdot t + p_3 \cdot t^2)^{p_4} + \omega_C(t) \tag{8.17}$$

where C is the battery capacity, t is the time index given by cycle number, $p = [0.01, 45, 0.2, 1.25]$ are parameters, γ is a hyper model parameter with mean of $1.6e^{-8}$ and variance of $5e^{-12}$, and ω_C is a model noise.

Figure 8.8 shows the diagnostic results at the 400th cycle. The mean of capacity estimation is 0.9418, and the 95% confidence interval (CI) is [0.939, 0.9446]. The upper sub-figure is the comparison of the capacity from Arbin system against the capacity estimation from EKF. The bottom sub-figure shows the comparison of initial baseline pdf (baseline capacity distribution) compared with the real-time estimated pdf (real time capacity estimation) at the 400th cycle. Note that the diagnostic algorithm is executed 400 times in the past 400 cycles, i.e., every time when a new measurement becomes available.

With an estimation of the current battery capacity as the initial condition, the prognosis is executed to conduct the long-term prediction and estimation of RUL. Figure 8.9 shows the expected value, upper and lower bounds of 95% CI of the battery capacity pdf at each future cycle.

The time-to-failure (TTF) distribution from EKF-based prognosis is a Gaussian distribution with mean value of 833 cycles when the mean predicted fault size reaches

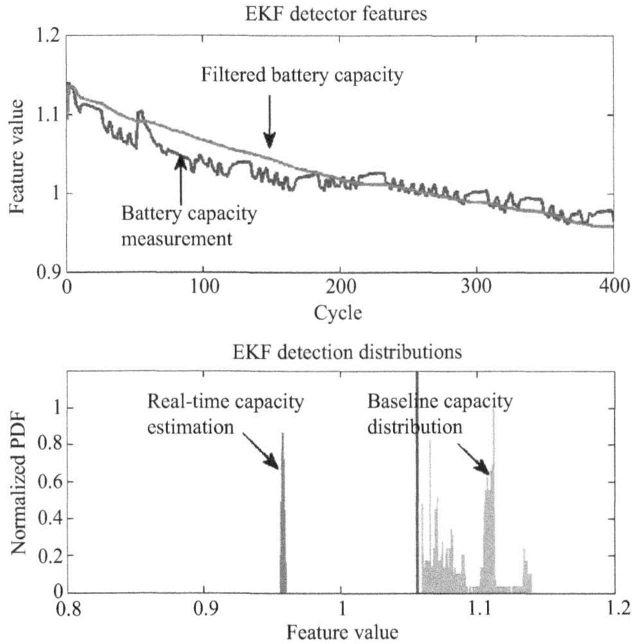

Figure 8.8 RS-EKF diagnosis for battery at 400th cycle

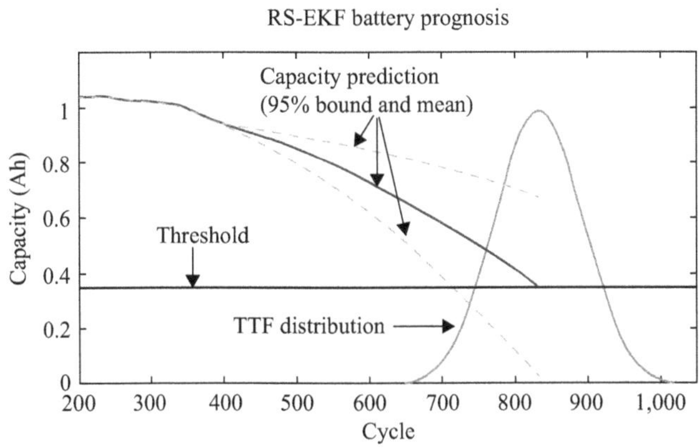

Figure 8.9 RS-EKF prognosis for battery at 400th cycle

the failure threshold. The upper bound and lower bound of the distribution are calculated based on EKF algorithm and can be simply approximated as the time instant when the upper bound and lower bound of the capacity distribution reach the failure threshold; details are shown in Figure 8.5. By this means, the standard deviation of the TTF distribution is approximated to be 59.25 cycles.

In this figure, the predicted RUL is 433 cycles. The distance between the prediction and ground truth is 23 cycles. The 95% CI of the RUL pdf is [714.5 951.5], which indicates that the uncertainty accumulated along the prediction horizon is very large.

8.4.7.2 EKF in Lebesgue sampling framework

In LS-FDP, the feature value range is divided into a series of Lebesgue states. If a new measurement causes a transition of Lebesgue state, i.e., an event happens, the diagnostic algorithm is executed.

To implement an LS-FDP for the battery capacity degradation, initially, 40 uniformly distributed Lebesgue states are defined in the battery's full capacity of 1.1 A h. With this setting, the diagnostic algorithm is executed only when the capacity degrades from one Lebesgue state to another. During the diagnosis process, the length of the Lebesgue states is optimally adjusted according to the fault growth speed. If the fault grows faster, the next Lebesgue length will be decreased; otherwise, the next Lebesgue length will be increased. The adjustment details are illustrated in [67]. The number of Lebesgue states changes during this optimization process.

The model for diagnosis is given by

$$C(t_{k+1}) = C(t_k) - p_d \cdot C(t_k) \cdot D \cdot sgn(C(t_k) - C(t_{k-1})) + \omega_C(t_k) \tag{8.18}$$

where t_k is the time instant of the kth event, p_d is a constant, D is the Lebesgue length, which is given by the difference between two successive Lebesgue states, *sgn* is a sign function, and $\omega_C(t_k)$ is the noise term.

Figure 8.10 shows the diagnostic results at the 400th cycle based on LS-EKF. The mean of capacity estimation is 0.9407 and the 95% CI is [0.925, 0.9608]. Same as the example in RS-EKF, the upper sub-figure shows the comparison of capacity from Coulomb counting (blue) against the estimated mean value from diagnosis (magenta). The lower sub-figure shows the comparison of initial baseline pdf compared with the real-time estimated pdf at the 400th cycle. Note that the diagnostic algorithm is only executed 70 times in the past 400 cycles.

The output of fault diagnosis is the fault state distributions at the current time instant, which cannot be used for LS-based prognosis directly and has to be transformed into the operation time distribution. To implement prognosis in LS-EKF framework, the operation time distribution is achieved as discussed in Figure 8.5, which is used as the initial condition for prognosis.

Prognostic part in LS-EKF is conducted on fault dimension axis to predict the time-to-Lebesgue-state directly. The model for prognosis is given as

$$t_{k+1} = t_k + p_p \cdot C(t_k) \cdot D \cdot \exp\left(-\dot{C}(t_k)\right) + \omega_t(t_k) \tag{8.19}$$

where p_p is a constant and ω_t is the model noise.

Figure 8.11 shows the prognostic results at the 400th cycle. To make the figure clear, only the time distribution pdf at a few selected Lebesgue state are plotted. The prediction horizon is 59 Lebesgue states, which is very small compared to the

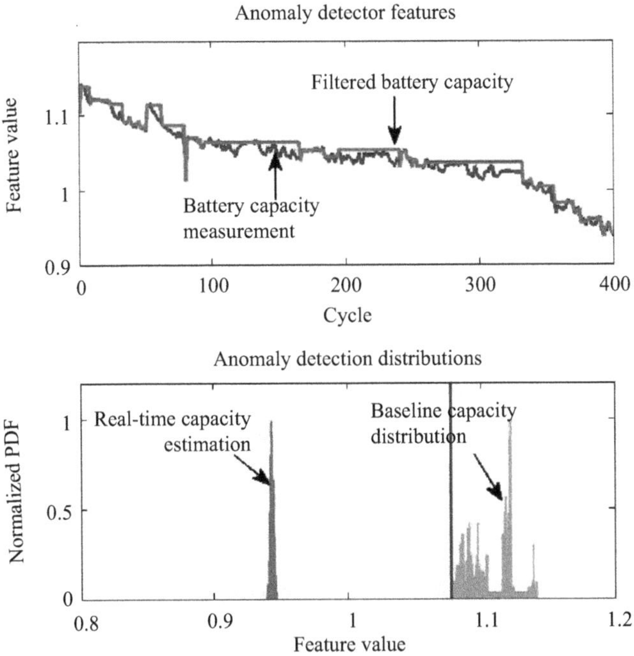

Figure 8.10 LS-EKF diagnosis for battery at 400th cycle

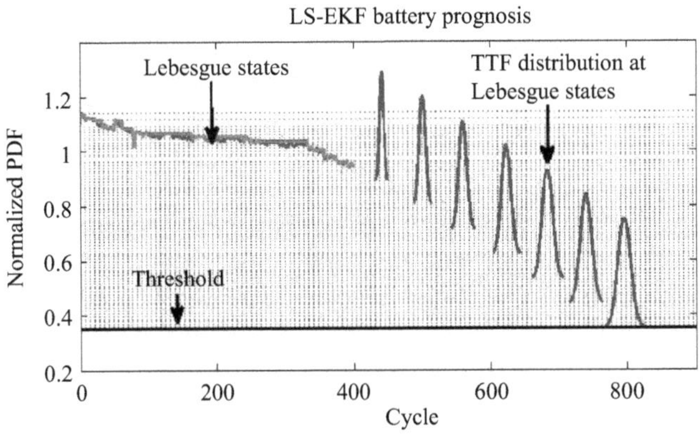

Figure 8.11 LS-EKF prognosis for battery at 400th cycle

433 cycles in RS-EKF. The predicted TTF for this battery is 795.7 and the RUL is 395.7 cycles. The 95% CI of the TTF distribution is [786.7 804.7]. The uncertainty is much smaller than that of RS-based prognosis due to the small prediction horizon. Compared with the ground truth TTF of 810, the difference between ground truth and the prediction is 14.3 cycles.

Table 8.1 Comparison of traditional RS-EKF and LS-EKF for battery

Diagnosis results	**RS-EKF**	**LS-EKF**
Capacity expectation	0.9418	0.9406
Capacity 95% CI	[0.939 0.9446]	[0.9205 0.9607]
Execution numbers	400 (100%)	70 (17.5%)
Prognosis results	RS-EKF	LS-EKF
True TTF	810	810
Estimate TTF	833	795.7
95% CI of TTF	[714.5 951.5]	[786.7 804.7]
prediction horizon	433	59
Computation time	0.55146 (100%)	0.007064 (12.8%)

8.4.7.3 Comparison of RS-EKF and LS-EKF

Compared to RS-EKF prognosis with large horizon (433 cycles), the LS-EKF prognosis shown in Figure 8.11 only has a PH of 59 Lebesgue states. The reduction of computation time is (0.055146−0.007064)/0.055146=87.19% and the computation is about 7.8 times faster.

Diagnostic and prognostic results of RS-EKF and LS-EKF algorithms are compared in Table 8.1. Compared with RS-EKF prognosis with a horizon of 433 cycles at the 400th cycle, the LS-EKF prognosis has a horizon of 59 Lebesgue states. The computation time for every LS-EKF prognosis routine is 0.007064 s, which is only 1.28% of that of the RS-EKF prognosis given as 0.55146 s. Note that the Lebesgue state length in the LS-EKF prognosis is changed according to the fault growth speed to keep a closer monitoring on the SOH. If the fault growth speed becomes faster along the prediction steps, the Lebesgue state length for the following prediction steps will be decreased; otherwise, it will be increased. The computational sources is optimally distributed during the diagnosis and prognosis process by increasing the Lebesgue state length and reducing the unnecessary execution when the fault grows slowly. When the fault grows fast, more computational sources is assigned to the FDP algorithm to monitor the health state of the system.

In order to compare the accuracy of RS-EKF and LS-EKF methods, α–λ matrix is used here, as shown in Figure 8.12 with $\alpha = 0.3$.

Judging from Figure 8.12, the mean of the predicted RUL for RS-EKF is as accurate as that of LS-EKF. However, the variance of predicted RUL of LS-EKF is much smaller as shown in Table 8.1, which is the natural benefit from LS methodology, since the prediction horizon in LS-EKF is much smaller than that of RS-EKF, the uncertainty accumulation during the prediction process is much smaller. Based on these advantages, LS-EKF can provide strong support for decision-making. More importantly, the LS-EKF required less calculation sources compared with RS-EKF, which makes it more feasible for distributed FDP with limited computational sources.

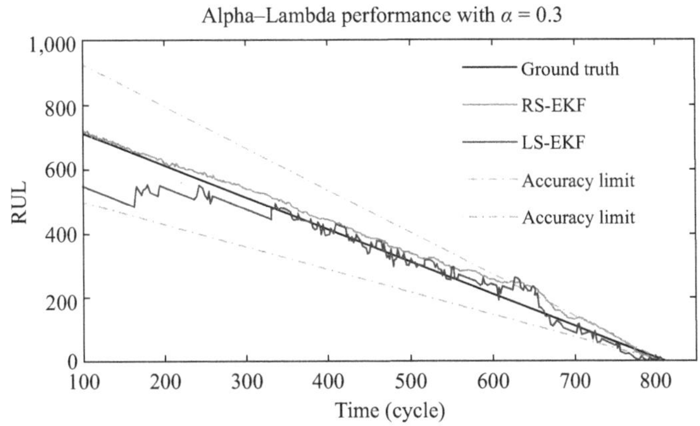

Figure 8.12 Prediction accuracy comparison between RS-EKF and LS-EKF

Table 8.2 Comparison of prognosis results from RS-EKF and LS-EKF by different performance metrics

RS-EKF	**CS2-35**	**CS2-36**	**CS2-37**	**CS2-38**
P.H.	767	726	862	902
A.P.	0.5548	0.6802	0.7676	0.7141
C.G.	0.3907	0.3889	0.4196	0.4144
R.A.	0.9084	0.9371	0.9222	0.9229
Total score	0.6445	0.7281	0.7792	0.7468
LS-EKF	**CS2-35**	**CS2-36**	**CS2-37**	**CS2-38**
P.H.	767	726	862	902
A.P.	0.7599	0.8877	0.7765	0.8457
C.G.	0.6005	0.7185	0.7617	0.7166
R.A.	0.9118	0.9462	0.8954	0.9043
Total score	0.7895	0.8883	0.8107	0.8504

The evaluation of prognosis results from RS-EKF and LS-EKF by other performance metrics are listed in Table 8.2.

8.5 Conclusion

FDP methods were developed in RS framework with great achievement for its simplicity. With the trend of distributed FDP, LSM is introduced with a philosophy of "execution when needed" to reduce the computation and make the long-term online prognosis possible on embedded systems. In this chapter, EKF algorithm is developed in the LS framework. An experiment of lithium-ion battery SOH diagnosis and prognosis with comparison against traditional RS-based approach is presented. It is demonstrated that the proposed approach is able to reduce the requirement on computational sources. This proposed approach combines the advantages of EKF and LS method, which results in low computation and small uncertainty accumulation.

References

[1] I. Tumer and A. Bajwa. A survey of aircraft engine health monitoring systems. In *Proceedings of the 35th AIAA/ASME/SAE/ASEE Joint Propulsion Conference*, pages 620–625, 2004.

[2] G. Vachtsevanos, F. Lewis, M. Roemer, A. Hess, and B. Wu. *Intelligent Fault Diagnosis and Prognosis for Engineering Systems*. John Wiley and Sons, Hoboken, NJ, 2006.

[3] S. Yin, S. Ding, X. Xie, and H. Luo. A review on basic data-driven approaches for industrial process monitoring. *IEEE Transactions on Industrial Electronics*, 61(11):6418–6428, 2014.

[4] S. Yin, X. Zhu, and O. Kaynak. Improved PLS focused on key-performance indicator-related fault diagnosis. *IEEE Transactions on Industrial Electronics*, 62(3):1651–1658, 2015.

[5] P. Boskoski and A. Urevc. Bearing fault detection with application to PHM data challenge. *International Journal of Prognostics and Health Management*, 2(1):1–10, 2011.

[6] B. Zhang, T. Khawaja, R. Patrick, G. Vachtsevanos, M. Orchard, and A. Saxena. A novel blind deconvolution de-noising scheme in failure prognosis. *Transactions of the Institute of Measurement and Control*, 32(1):3–30, 2010.

[7] B. Zhang, C. Sconyers, C. Byington, R. Patrick, M. Orchard, and G. Vachtsevanos. A probabilistic fault detection approach: Application to bearing fault detection. *IEEE Transactions on Industrial Electronics*, 58(5):2011–2018, 2011.

[8] B. Zhang, T. Khawaja, R. Patrick, G. Vachtsevanos, M. Orchard, and A. Saxena. Application of blind deconvolution de-noising in failure prognosis. *IEEE Transactions on Instrumentation and Measurement*, 58(2):303–310, 2009.

[9] A. Sidhu, A. Izadian, and S. Anwar. Adaptive nonlinear model-based fault diagnosis of li-ion batteries. *IEEE Transactions on Industrial Electronics*, 62(2):1002–1011, 2015.

[10] K. Goebel, W. Yan, N. Eklund, et al. *Defect classification of highly noisy NDE data using classifier ensembles*, In *Proceedings of SPIE, Smart Structures and Materials + Nondestructive Evaluation and Health Monitoring*, volume 6167, 2006.

[11] K. Goebel, B. Saha, and A. Saxena. A comparison of three data-driven techniques for prognostics. In *62nd Meeting of the Society For Machinery Failure Prevention Technology (MFPT)*, pages 119–131, 2008.

[12] R. Isermann. Model-based fault detection and diagnosis-status and applications. *Annual Reviews in Control*, 29:71–85, 2005.

[13] M. Zhong, H. Fang, and H. Ye. Fault diagnosis of networked control system. *Annual Reviews in Control*, 31(1):55–68, 2007.

[14] A. Hess and S. Wells. Sliding mode control applied to reconfigurable flight control design. *Journal of Guidance, Control, and Dynamics*, 26(3): 452–462, 2003.

[15] B. Zhang, T. Khawaja, R. Patrick, and G. Vachtsevanos. Blind deconvolution denoising for helicopter vibration signals. *IEEE/ASME Transactions on Mechatronics*, 13(5):558–565, 2008.

[16] J. Pons-Llinares, J. Antonino-Daviu, M. Riera-Guasp, B. L. Sang, T. Kang, and C. Yang. Advanced induction motor rotor fault diagnosis via continuous and discrete time-frequency tools. *IEEE Transactions on Industrial Electronics*, 62(3):1791–1802, 2015.

[17] S. Choi, E. Pazouki, J. Baek, and H.R. Bahrami. Iterative condition monitoring and fault diagnosis scheme for electric motor for harsh industrial application. *IEEE Transactions on Industrial Electronics*, 62(3):1760–1769, 2015.

[18] M.E. Orchard, P. Hevia-Koch, B. Zhang, and L. Tang. Risk measures for particle-filtering-based state-of-charge prognosis in lithium-ion batteries. *IEEE Transactions on Industrial Electronics*, 60(11):5260–5269, 2013.

[19] D. Edwards, M. Orchard, L. Tang, K. Goebel, and G. Vachtsevanos. Impact of input uncertainty on failure prognostic algorithms: Extending the remaining useful life of nonlinear systems. In *Proceedings of the Annual Conference of the Prognostics and Health Management Society*, Portland, OR, 2010.

[20] P. Singleton, E. Strangas, and S. Aviyente. Extended Kalman filtering for remaining-useful-life estimation of bearings. *IEEE Transactions on Industrial Electronics*, 62(2):1781–1790, 2015.

[21] J. Celaya, A. Saxena, and K. Goebel. Uncertainty representation and interpretation in model-based prognostics algorithms based on Kalman filter estimation. In *Proceedings of the Annual Conference of the Prognostics and Health Management Society*, Minneapolis, MN, 2012.

[22] V. Climene-Alarcon, J. Antonino-Daviu, E. Strangas, and M. Riera-Guasp. Rotor-bar breakage mechanism and prognosis in an induction motor. *IEEE Transactions on Industrial Electronics*, 62(3):1814–1825, 2015.

[23] X. Si. An adaptive prognostic approach via nonlinear degradation modeling: application to battery data. *IEEE Transactions on Industrial Electronics*, 62(8):5082–5096, 2015.

[24] M. Pecht and S. Kumar. Data analysis approach for system reliability, diagnostics, and prognostics. In *Proceeding of Pan Pacific Microelectronics Symposium*, Kauai, HI, pages 1–9, 2008.

[25] K. Medjahar and N. Zerhouni. Residual-based failure prognostic in dynamic systems. In *7th IFAC International Symposium on Fault Detection, Supervision and Safety of Technical Processes*, Spain, 2009.

[26] S. Mathew and M. Pecht. Prognostics of systems: approaches and applications. In *Proceedings of the 24th International Congress on Condition Monitoring and Diagnostics Engineering Management*, Stavanger, Norway, 2011.

[27] A. Xu and Q. Zhang. Residual generation for fault diagnosis in linear time-varying systems. *IEEE Transactions on Automatic Control*, 49(5): 767–772, 2004.

[28] C. Chen, B. Zhang, and G. Vachtsevanos. Prediction of machine health condition using neuro-fuzzy and Bayesian algorithms. *IEEE Transactions on Instrumentation and Measurement*, 61(2):297–306, 2012.

[29] C. Chen, B. Zhang, G. Vachtsevanos, and M. Orchard. Machine condition prediction based on adaptive neuro-fuzzy and high-order particle filtering. *IEEE Transactions on Industrial Electronics*, 58(9):4353–4364, 2012.

[30] B. Zhang, G. Georgoulas, M. Orchard, *et al.* Rolling element bearing feature extraction and anomaly detection based on vibration monitoring. In *2008 Mediterranean Conference on Control and Automation*, Ajaccio, France, pages 1792–1797, 2008.

[31] G. Zhao, X. Liu, B. Zhang, Y. Liu, G. Niu, and C. Hu. A novel approach for analog circuit fault diagnosis based on deep belief network. *Measurement*, 121:170–178, 2018.

[32] G. Zhao, G. Zhang, Y. Liu, B. Zhang, and C. Hu. Lithium-ion battery remaining useful life prediction with deep belief network and relevance vector machine. In *2017 IEEE International Conference on Prognostics and Health Management (ICPHM)*, Denver, CO, 2017.

[33] G. Zhao, G. Zhang, B. Zhang, G. Niu, and C. Hu. Bearing health condition prediction using deep belief network. In *Annual Conference of Prognostics and Health Management Society*, Orlando, FL, 2017.

[34] K. Liu, X. Zhang, and J. Shi. Adaptive sensor allocation strategy for process monitoring and diagnosis in a Bayesian network. *IEEE Transactions on Automation Science and Engineering*, 11(2):452–462, 2014.

[35] K. Pattipati, B. Wang, Y. Zhang, M. Howell, and M. Salman. Fault diagnosis and prognosis in a network of embedded systems in automotive vehicles. In *NSF-NIST-USCAR workshop on Cyber-Physical Systems*, 2011.

[36] S.M. Namburu, M.S. Azam, J. Luo, K. Choi, and K.R. Pattipati. Data-driven modeling, fault diagnosis and optimal sensor selection for HVAC chillers. *IEEE Transactions on Automation Science and Engineering*, 4(3):469–473, 2007.

[37] S. Zhang, K.R. Pattipati, Z. Hu, X. Wen, and C. Sankavaram. Dynamic coupled fault diagnosis with propagation and observation delays. *IEEE Transactions on Systems, Man, and Cybernetics: Systems*, 43(6):1424–1439, 2013.

[38] S. Tornil-Sin, C. Ocampo-Martinez, V. Puig, and T. Escobet. Robust fault diagnosis of nonlinear systems using interval constraint satisfaction and analytical redundancy relations. *IEEE Transactions on Systems, Man, and Cybernetics: Systems*, 44(1):18–29, 2014.

[39] A. Scacchioli, G. Rizzoni, M.A. Salman, W. Li, S. Onori, and X. Zhang. Model-based diagnosis of an automotive electric power generation and storage system. *IEEE Transactions on Systems, Man, and Cybernetics: Systems*, 44(1):72–85, 2014.

[40] R.H. Kwong and D.L. Yonge-Mallo. Fault diagnosis in discrete-event systems with incomplete models: Learnability and diagnosability. *IEEE Transactions on Cybernetics*, 45(7):1236–1249, 2015.

[41] B. Zhang, L. Tang, J. DeCastro, M. Roemer, and K. Goebel. Autonomous vehicle battery state-of-charge prognostics enhanced mission planning. *International Journal of Prognostics and Health Management*, 5(2): 1–12, 2014.

[42] S. Wang, J. Zhou, D. Liu, B. Zhang, and Y. Peng. A run-time dynamic reconfigurable computing system for lithium-ion battery RUL prediction. *Energies*, 9(572):1–19, 2016.

[43] R. McCann and A.T. Le. Lebesgue sampling with a Kalman filter in wireless sensors for smart appliance networks. In *Industry Applications Society Annual Meeting, 2008. IAS '08. IEEE*, pages 1–5, 2008.

[44] K.J. Astrom and B.M. Bernhardsson. Comparison of Riemann and Lebesgue sampling for first order stochastic systems. In *Decision and Control, 2002, Proceedings of the 41st IEEE Conference on*, volume 2, pages 2011–2016, 2002.

[45] R. Morales-Menendez, N. de Freitas, I. Campus Monterrey, O. De Freitas, and D. Poole. *Real-Time Monitoring of Complex Industrial Processes with Particle Filters*. In *Proceedings of the Conference on Neural Information Processing Systems*, Vancover, Canada, pages 1433–1440, 2002.

[46] N. de Freitas. Rao-Blackwellised particle filtering for fault diagnosis. In *Proceedings of the IEEE Aerospace Conference*, volume 4, pages 1767–1772, 2002.

[47] W. Yan and B. Zhang. Extended Kalman filter development in Lebesgue sampling framework with an application to li-ion battery diagnosis and prognosis. In *Proceedings of Third European Conference of the Prognostics and Health Management Society (PHME)*, Bilbao, Spain, 2016.

[48] B. Saha, K. Goebel, S. Poll, and J. Christophersen. Prognostics methods for battery health monitoring using a Bayesian framework. *IEEE Transactions on Instrumentation and Measurement*, 58(2):291–296, 2009.

[49] S. Piller, M. Perrin, and A. Jossen. Methods for state-of-charge determination and their applications. *Journal of Power Sources*, 96(1):113–120, 2001.

[50] B. Pattipati, K. Pattipati, J.P. Christopherson, S. Madhavi Namburu, D.V. Prokhorov, and L. Qiao. Automotive battery management systems. *IEEE*, 2008.

[51] B.E. Olivares, M.A. Cerda Munoz, M.E. Orchard, and J.F. Silva. Particle-filtering-based prognosis framework for energy storage devices with a statistical characterization of state-of-health regeneration phenomena. *IEEE Transactions on Instrumentation and Measurement*, 62(2):364–376, 2013.

[52] Y. He, X.T. Liu, C.B. Zhang, and Z.H. Chen. A new model for state-of-charge (SOC) estimation for high-power li-ion batteries. *Applied Energy*, 101: 808–814, 2013.

[53] O. Barbarisi, F. Vasca, and L. Glielmo. State of charge Kalman filter estimator for automotive batteries. *Control Engineering Practice*, 14(3):267–275, 2006.

[54] J.D. Kozlowski. Electrochemical cell prognostics using online impedance measurements and model-based data fusion techniques. In *Aerospace Conference, 2003. Proceedings. 2003 IEEE*, volume 7, pages 3257–3270. IEEE, 2003.

[55] M. Charkhgard and M. Farrokhi. State-of-charge estimation for lithium-ion batteries using neural networks and EKF. *IEEE transactions on industrial electronics*, 57(12):4178–4187, 2010.

[56] W.-Y. Chang. The state of charge estimating methods for battery: A review. *ISRN Applied Mathematics*, 2013.

[57] P. Lall, J. Wei, and K. Goebel. Comparison of Kalman-filter and extended Kalman-filter for prognostics health management of electronics. In *Thermal and Thermomechanical Phenomena in Electronic Systems (ITherm), 13th IEEE Intersociety Conference on*, pages 1281–1291. IEEE, 2012.

[58] A. Vasebi, M. Partovibakhsh, and S. Mohammad Taghi Bathaee. A novel combined battery model for state-of-charge estimation in lead-acid batteries based on extended Kalman filter for hybrid electric vehicle applications. *Journal of Power Sources*, 174(1):30–40, 2007.

[59] B. Saha, K. Goebel, and J. Christophersen. Comparison of prognostic algorithms for estimating remaining useful life of batteries. *Transactions of the Institute of Measurement and Control*, 31:293–308, 2009.

[60] X. Wang and B. Zhang. Real-time Lebesgue-sampled model for continuous-time nonlinear systems. In *Decision and Control (CDC), IEEE 53rd Annual Conference on*, pages 4367–4372, 2014.

[61] B. Zhang and X. Wang. Fault diagnosis and prognosis based on Lebesgue sampling. In *Annual Conference of the Prognostics and Health Management Society 2014*, volume 5, 2014.

[62] W. Yan, B. Zhang, X. Wang, W. Dou, and J. Wang. Lebesgue-sampling-based diagnosis and prognosis for lithium-ion batteries. *IEEE Transactions on Industrial Electronics*, 63(3):1804–1812, 2016.

[63] A. Saxena, J. Celaya, B. Saha, S. Saha, and K. Goebel. Metrics for offline evaluation of prognostic performance. *International Journal of Prognostics and Health Management*, 1(1):20, 2010.

[64] T. Wang. *Trajectory Similarity Based Prediction for Remaining Useful Life Estimation*. University of Cincinnati, 2010.

[65] J. Lam, S. Sankararaman, and B. Stewart. Enhanced trajectory based similarity prediction with uncertainty quantification. In *Annual Conference of the Prognostics and Health Management Society*, 2014.

[66] W. He, N. Williard, M. Osterman, and M. Pecht. Prognostics of lithium-ion batteries based on Dempster–Shafer theory and the Bayesian Monte Carlo method. *Journal of Power Sources*, 196(23):10314–10321, 2011.

[67] W. Yan, W. Dou, D. Liu, Y. Peng, and B. Zhang. Parameters adaption of Lebesgue sampling-based diagnosis and prognosis for li-ion batteries. In *Annual Conference of the Prognostics and Health Management Society*, volume 6, 2015.

Index

www.ingramcontent.com/pod-product-compliance
Lightning Source LLC
LaVergne TN
LVHW051047210926
848486LV00001B/42